PLANT FOODS OF GREECE

ARCHAEOLOGY OF FOOD

PLANT FOODS *of* GREECE

A Culinary Journey to the Neolithic and Bronze Ages

Soultana Maria Valamoti

THE UNIVERSITY OF ALABAMA PRESS
Tuscaloosa

The University of Alabama Press
Tuscaloosa, Alabama 35487–0380
uapress.ua.edu

Typeface: Arno Pro

Cover image: Cereals displayed with replica Neolithic cooking pots,
and poppies, chamomile, and common fumitory; photographs by
Soultana Maria Valamoti and Georgios Vily Kapetanakis
Cover design: Lori Lynch

This project has received funding from the European Research
Council (ERC) under the European Union's Horizon 2020 research
and innovation programme (grant agreement No 682529).
This book reflects the author's view only. The European Research Council is not
responsible for any use that may be made of the information it contains.

Cataloging-in-Publication data is available from the Library of Congress.
ISBN: 978-0-8173-2159-8
E-ISBN: 978-0-8173-9453-0

To my family for their patience and support:
Georgios-Vily, Dimitris, Vasiliki-Sophia

To my two teachers and mentors:
Glynis Jones and Kostas Kotsakis, *sine quibus non*

Contents

Preface

The earth, moist from the rain, embraces the grain that a careful hand has sown into the ground with trust and hope. The power of the seed surrounded by earth, water, and the warmth of the sun turns into a rich harvest, food for people, the fruit of their labor. This book is the story of seeds that sprouted and fed people of a distant past, seeds that healed and protected, seeds that may have been used for good or evil. To the inquisitive mind of a master's student who carried on with a PhD, these seeds posed the challenge to decipher their stories, their "biographies" from the moment they were tossed on the earth by the farmer's hand to the moment they were brought to the mouth as strengthening food. The charred seeds that have survived the decay of millennia, defeating time, have long lost their power to sprout and feed; yet, in their black silence, charred in some oven some thousands of years ago or, during the preparation of a healing poultice or as offerings to ancestors, they are eager to tell their stories, open up small windows into the prehistoric lives of people who lived in the lands these seeds grew and beyond. Some of these seeds traveled a long journey in pockets and bags, traveled from foreign lands to new territories, with the risks and challenges of surviving, being accepted and adopted.

This book explores the cereals, pulses, fruits, and nuts, wild and cultivated, as well as those plants that altered the senses, and healed or killed, over the course of the six millennia of the Neolithic and the Bronze Age in the tip of southeastern Europe that corresponds to the modern Greek state. The narratives woven from these ancient grains, which survived charred in archaeological deposits from Greece, reveal a long-term interaction between plants and the people who shaped and transformed the landscape and cuisine. The food prepared from grains of which only tiny fragments may survive in archaeological deposits is the quintessence of the fascinating interaction between culture and nature. A deep knowledge of the landscape, a tamed wilderness, and the power of cultivating plants shaped ingredients that, through culinary mastery, became dishes and potions that marked daily events as well as special occasions, births, initiation rites, mating, and death.

When I walk in nature looking at plants I think about those first gatherers who experimented with planting, selecting some wild cereal ears to sow in the earth, taking over and intervening in the natural course of life for some plants. I often wonder about the processes that led to the knowledge and selection of plants that were used as food. Was it the color and smell? Or trial and error? From the prehistoric fields to the wilderness surrounding them, what can plants tell us about the rules that defined the movements and actions of people in the course of the year and the succession of the seasons? In our modern, western world, the concrete environments with electric kitchens, metal pans, and microwave ovens, in a globalized market economy where almost any ingredient can be found throughout the year, is there something in the grains and seeds we eat today that links us to that distant prehistoric past? When I cook lentil soup or bulgur, when I drink wine, I think that these food preparations connect us to people so distant, yet close to us precisely due to the perseverance of the grain or fruit and of the basic recipe for its culinary transformation. I do not have to collect brushwood or dry dung for fuel to cook. I do not grow lentils and wheat, nor do I parboil wheat to make bulgur or *trachanas*, this clever dry mixture of cereals and milk, under the unbearable summer heat; this is done large-scale, and I buy the ready-prepared grain or the split grass pea seeds. Yet, when the smell of the cooked grains reaches my nose, I know that it conveys smells experienced millennia ago. Although the tastes, connotations, context, and etiquette of consumption were different from those of the present, the tiny or larger grains and seeds we find in archaeological deposits, charred accidentally or on purpose, are a thread that connects us with our distant ancestors, those who first cultivated and domesticated plants and the landscape, transforming it into fields, orchards, prairies, and pastures, full of a culinary diversity hidden in the colorful plants in bloom or in seed or fruit, depending on the season. *Plant Foods of Greece* is an archaeobotanical account of plant food ingredients and food preparations of prehistoric Greece. At the same time it is a tribute to these prehistoric ancestors who have contributed to the dishes we enjoy today. I wish to share such simple dishes with you in the end of this book as, in my eyes, they constitute the meeting point where the past encounters the present, safeguarding the legacy of ancient grains for the future—ingredients and recipes that have endured the passage of time and constitute a culinary tradition very much alive and tasty.

Acknowledgments

The archaeobotanical and ethnographic data from Greece presented in this book have been collected over the years with funding awarded to me by the following funding bodies: the Institute of Aegean Prehistory, the British School at Athens Centenary Bursaries, the National Geographic Society, the Aristotle University of Thessaloniki Research Committee, and Thrace Flour Mills. They are all gratefully acknowledged. The writing and preparation of this book have been funded by the European Research Council, in the context of the Project PLANTCULT "Identifying the Food Cultures of Ancient Europe," conducted under the European Union's Horizon 2020 Research and Innovation Program, Grant Agreement no. 682529, Consolidator Grant 2016–2021.

This book owes a great deal to the help and support of very many people as well as to the challenges and difficulties others posed. Writing has undergone several phases; what has kept me going is my stubbornness to see it through no matter what and my passion for archaeobotany. Ancient plant remains to me are messengers from past human lives, their interaction with themselves and their environments, bringing their magic to the present. The roots of my endeavor with archaeobotany are to be found in the undergraduate courses of Kostas Kotsakis, who is now professor emeritus at the Aristotle University of Thessaloniki. His inspiring lectures and encouragement showed me the path to environmental archaeology and palaeoeconomy back in the 1980s. It was, however, the chance meeting with the late Gordon Hillman during breakfast in a residence hall at Southampton University during the 1986 World Archaeological Conference (WAC) that initiated my path toward archaeobotany. At the time I was still an undergraduate exploring Europe alone via interrail, culminating at the WAC conference at the end of the journey. Gordon, with his generosity, inspired and initiated my interest in archaeobotany and was always there to help with the literature and unknown specimens during my PhD years. My formation in archaeobotany owes much to Glynis Jones at the Department of Archaeology at Sheffield

University, UK, who supervised my master's work on the plant remains from Mandalo that she generously made available for me; then, five years later, after I had collected enough data, she agreed to supervise my PhD on archaeobotanical material from Neolithic and Early Bronze Age Greece. Without her constant guidance, firm support both on a scientific and personal level as my PhD coincided with the birth of my two children, I would not have succeeded in completing the task. Glynis provided me with the tools that have helped me all along during the last 20 years since the completion of my thesis. It is very unfortunate that the closure of the Department of Archaeology at Sheffield, announced in the summer of 2021, a tragic consequence of seeing higher education as a money-making tool, coincided with the completion of this book. This book would not have been possible without the guidance of Glynis and the stimulating research environment at Sheffield. Mike Charles, my PhD co-supervisor and friend, provided great support during my training in archaeobotany at Sheffield and generously shared his knowledge with me. I am deeply grateful to all of them; they have no responsibility, however, for any mistakes in this book, as many years have intervened since they kept their watchful eyes over my work and the completion of this book.

I started working on this book in 2009, when I completed my Greek book *An Archaeobotanical Investigation of Prehistoric Diet in Greece* (University Studio Press, Thessaloniki). As years passed and work-related problems prevented the completion of the English version of that book, ideas matured and changed over time. My teaching and interaction with undergraduate and postgraduate students contributed to the content of *Plant Foods of Greece*.

My undergraduate students, with their enthusiasm as well as participation in flotation teams, have kept archaeobotany very lively at the Aristotle University of Thessaloniki, in particular my master's and PhD students whom I wish to thank here: Eugenia Gkatzogia, Katerina Karatasaki, Angeliki Karathanou, Georgia Kasapidou, Aimilia Kokonaki, Dimitra Kotsachristou, Stavroula Michou, Katerina Moniaki, Pelagia Paraskevopoulou, Chryssa Petridou, Eleni Telioridou, and Pelagia Theodosaki; Katerina Karatasaki for her excellent master's dissertation on medicinal plants from prehistoric Greece that offered me a further source of information for Chapter 7. I am grateful to my students, their mothers and grandmothers, and wider families for sharing a knowledge that is getting lost in this generation: thanks to Ioanna Mimi, her late grandmother Sophia Papageorgiou and mother Maria Papageorgiou, for their hospitality and generosity, showing me the secrets of the simple and magic cuisine involving plants such as wild pears and grapes preparing fruit *petimezi*, bulgur and *trachanas* at Kosmati. Thanks to Stella Anastasaki, my student, and her grandmother Chryssi Trakaki for teaching me the secrets of *xinochondros* preparation and their warm Cretan hospitality. Thanks also to Maria Elisavet Samoili and her family for sharing their knowledge about the cultivation of grass pea at Corfu and to Niki Chondrou for her accounts of grass pea cultivation in Boeotia. I am grateful to all the people who shared their precious knowledge on plants on the islands of Crete, Lemnos, Corfu, and Boeotia, in particular Babis Kouris of Pontiglio winery on the island of Corfu, the Ethaleia Hotel people Konstantina Karachalia, Apostolos Kotsaftis, Nikos Kotsilitis, and Panagiotis Panagis as well as Evangelia and Sophie Hapsi, Evangelia Kagki, and the Salamousas family on the island of Lemnos, for their generosity and help with establishing contacts on grass pea and Cyprus vetch cultivation on the two islands. This ethnographic work has been a valuable source of inspiration while interpreting archaeobotanical finds presented here.

Similarly influential toward the preparation of this has been the PlantCult project partners, Andreas Heiss, Stefanie Jacomet, and Hans-Peter Stika as well as the whole PlantCult team who has grown over the years (http://plantcult.web.auth.gr/en/project-eng/teams-eng). I thank these people for their trust in joining the PlantCult project, the fruitful discussions we have had, and the publications they have shared over the years. The PlantCult project has been a constant source of inspiration and a challenge at the same time, during the last five years. Special thanks go to Tasos Bekiaris, Maria Bofill, Laurent Bouby, Danai Chondrou, Anastasia Dimoula, Vasilis Fyntikoglou, Ismini Ninou, Clemence Pagnoux, Lambrini Papadopoulou, Sandra Prevost-Dermarkar, Hara Procopiou, Themis Roustanis, Jean-Fred Terral, and Panos Tokmakidis as well as Zoï Tsirtsoni, who through their PlantCult research, have produced exciting new data that I have been able to discuss with them and cite in this book. The interaction and discussions with stone tools experts Toni Palomo and potter Yannis Stagkidis were very illuminating regarding comprehension of aspects of the preparation of ancient tools and cooking installations. Evi Papadopoulou and Sandra Prevost-Dermarkar provided critical comments on the section on cooking facilities. Any mistakes are my own. Special thanks go to Chryssa Petridou for being an excellent collaborator—reliable, hard-working, always willing to help; her assistance with this book has been precious. While making the final corrections in this book, in the context of PlantCult, all archaeobotanical data from Neolithic northern Greece have been pulled together and analyzed with the constant guidance of Angela Kreuz. I am grateful to her for her thoroughness and trust; such an approach could not have been applied to the whole dataset from Greece for various reasons, therefore statistical data analyses have not been included in this book, only qualitative observations based on the presence of plant species and dense, pure concentrations of archaeobotanical finds. This will be a future endeavor.

Many people have helped me along the years in various ways that have shaped the contents of this book. Nora Skouteri-Didaskalou was an inspiring teacher for me at the Aristotle University of Thessaloniki, back in 1987, introducing me to social anthropology and the work of Lévi-Strauss. James Grieg, during the earlier stages of this book, with generosity and care read some chapters. I am grateful to James for his time, critical comments, and improvements. Bettina Arnold also kindly provided feedback for Chapter 10. The following people through their seminal work on the archaeology of food and by providing references at several stages of writing of this book have been a source of inspiration: Yannis Hamilakis, Christine Hastorf, and Martin Jones. Christine Hastorf in particular I wish to thank very much for her encouragement when my optimism regarding the completion of this book was very low. Paul Halstead and Glynis Jones through their teaching and research have shown the importance of ethnographic work and have opened up the way to investigate prehistoric agricultural life in Greece. Delwen Samuel has been a critical judge and precious mentor offering constant support on various aspects of research on ancient plant foods. I am deeply grateful to Felix Bittmann and the Lower Saxony Institute for Historical Coastal Research in Wilhelmshaven, Germany, for sharing precious information on the Argissa malt finds, that is, the unpublished report by Maria Hopf and for showing me her original material from Argissa, discussed in Chapter 7, during the 2010 Conference of the International Work Group for Paleoethnobotany (IWGP) in Wilhelmshaven. Many thanks to my colleague Lena Manakidou for her constant feeding of references related to ancient Greek religion and medicine. Thanks to Erica Rowan for her invitation to Exeter for a conference on Ancient Greek and Roman medicine that gave me the opportunity to think more and learn about healing practices in antiquity. Elias Sverkos, my colleague at the School of History and Archaeology, kindly provided information on traditional treatment of wild pears for consumption. Rainer Bussmann kindly provided information on the use of Cornelian cherries in the Caucasus area.

Many thanks to Li Liu and Jajing Wang for inviting me to the Stanford symposium on alcohol in 2019; this gave me the opportunity to meet with colleagues working on ancient and traditional alcohol production and learn from them, in particular John Arthur and René Friedman, who generously shared their knowledge and references related to ancient and traditional beer production. Especially illuminating toward the interpretation of the Argissa parallel wall structures was the feedback from René Friedman regarding Egyptian examples, and in this context, I wish to thank Ulirch Hartung for allowing me to use his in press manuscript on Buto in Egypt. In addition to the people mentioned here, the following have been very supportive and helpful with discussions and the provision of references and information: Susan Allen, Ferran Antolín, Robert Arnott, Marian Berihuete-Azorín, Amy Bogaard, Andrea Brandolini, Aylen Capparelli, Oliver Craig, Cathy D'Andrea, Füsun Ertuğ, Yannis Fappas, Dragana Filipović, Paul Halstead, Angela Kreuz, Helmut Kroll, Alexandra Livarda, Joseph Maran, Elena Marinova, Maria Ntinou, Emel Oybak Dönmez, Yiannis Papadatos, Anastasia Papathanasiou, Leonor Peña-Chocarro, Tzvetana Popova, Fritz Preinfalk, Jeremy Rutter, Aurelie Salavert, Peter Tomkins, George Willcox, and Michèle Wollstonecroft. As this work relies on twenty years of research, the list of people who have contributed toward the shaping of this book is much longer and I thank them all, even though they many not be mentioned by name.

I am grateful to all excavators who have entrusted me with the study of the archaeobotanical material from their excavations: Manthos Bessios, Areti Chondrogianni-Metoki, Georgios Chourmouziadis, Anastasia Chrysostomou, Panagiotis Chrysostomou, Pascal Darcque, Jack Davis, Nikos Efstratiou, Anastasia Georgiadou, Dimitrios Grammenos, Ntina Kallintzi, Georgia Karametrou-Mentesidi, Kostas Kotsakis, Stavros Kotsos, Chaido Koukouli-Chrysanthaki, Dimitria Malamidou, Evangelia Malapani, Mariza Marthari, Aikaterini Papanthimou, Maria Pappa, Katerina Peristeri, Angeliki Pilali, Sharon Stocker, René Treuil, Zoï Tsirtsoni, and Christina Ziota.

Special thanks are owed to the following people for making images available from their fieldwork or for expeditions especially for this book. Khwaja Shafiq Ahmad Sedeqy kindly provided images of cooking installations from Afghanistan; he had to travel a long way to the villages of his relatives to obtain these photos for the purposes of this book, so I am deeply grateful to him and his family. It is such a great tragedy that his country and people are under mortal threat as these lines are being written. Shahal Abbo, Eva Alram-Stern, Bettina Arnold, Mustafa Bayram, Tasos Bekiaris, John Bennet, Elisabetta Borgna, Areti Chondrogianni-Metoki, Danai Chondrou, Kostis Christakis, Pascal Darcque, Anastasia Georgiadou, Stefanos Gimatzides, Paul Halstead, Glynis Jones, Yiannis Karliambas, Stavros Kotsos, Chaido Koukouli-Chrysanthaki, Dimitria Malamidou, Evangelia Malapani, Joseph Maran, Jerolyn Morrison, Daniel Noll, Evi Papadopoulou, Aikaterini Papanthimou, Maria Pappa, Leonor Peña-Chocarro, Jörg

Rambach, Nicholas Thompson, Giorgos Toufexis, Zoï Tsirtsoni, Evangelia Tsoukala, Rena Veropoulidou, Diana Wardle, Ken Wardle, and Christina Ziota kindly provided images from their work for my book and are gratefully acknowledged. Ilaria Symiakaki from the Italian School of Archaeology was very helpful on image copyright issues. Danai Chondrou prepared some drawings for this book while Panos Litsios generously prepared Figure 7.4. Stavros Dimakopoulos, Vasiliki Kapetanaki, Georgios-Vily Kapetanakis, Ioanna Mimi, Pavlos Lathiras, Chryssa Petridou, Hans-Peter Stika, and Georgina Pratts contributed images for this book in the context of project PlantCult; their help is gratefully acknowledged here. Kostas Prokos, PlantCult collaborator, worked hard on preparing the photos for this book and is gratefully acknowledged. I am especially thankful to Themis Roustanis for preparing the two maps with help from Chryssa Petridou. Athanasia Papamartzivanou was a precious help throughout the preparation of the manuscript, checking the references and style of the manuscript; her thoroughness as well as her encouragement when optimism was low are much acknowledged.

I am grateful to Wendi Schnaufer and the University of Alabama Press Archaeology of Food Series for championing my work. Wendi in particular I wish to thank for her encouragement and efforts regarding this book. Without her perseverance and support this book would not have been completed. I wish to thank my two reviewers, Alexia Smith and an anonymous colleague, for their very helpful comments on an earlier version of this book. Without her perseverance and support this book would not have been completed.

My husband, Georgios-Vily Kapetanakis, deserves many special thanks for being always there and always supportive, the "volunteer" as we called him in PlantCult, for all his help with going places, taking photos and videos, enjoying the process, and thus providing constant encouragement. This book could not have been completed without his quiet strength attitude that acted as a balance to my impulsive attitude with fluctuations of immense enthusiasm and dark moments of utter disappointment. My two children, Dimitris and Vasiliki, have experienced my endeavors with archaeobotany since their conception, accompanying me at water-sieving and sorting expeditions while still in utero. I thank them for being in my life and making it matter. I thank them for their understanding of the long hours I spent at the computer and the times away at work. Last but not least, my two parents, my late father, Dimitrios Valamotis, an agronomist specialized in animal nutrition, and my mother Vasso Valamoti, a knowledgeable housewife; they both sowed the seeds for my love for nature and taught me the secrets of plants and how to transform them into dishes.

PLANT FOODS OF GREECE

Introduction

Stories Told from Ancient Grains, an Archaeobotanical Exploration of Prehistoric Cuisine

Plants have sustained human communities since Paleolithic times, and the domestication of some led to significant social, economic, and culinary changes. Plant foods have formed elements of daily and ritual meals, of healing and social differentiation through food products such as oil, wine, spices, stimulants, and hallucinogenics, often manipulated by elites in different parts of the world. The pages of this book provide a journey to the plant components of the prehistoric cuisine of Greece, an exploration of a culinary past that takes us back in time, spanning nearly 7,000 years of daily and special meals. The remains of plants, trapped in archaeological deposits from Neolithic and Bronze Age Greece, offer snapshots of prehistoric cooking. The book follows the steps of prehistoric farmers, sweating over their land year-round, growing their plant food ingredients in fields and orchards, and harvesting the fruit of the wilderness of forests, scrubland, and clearings. It is a journey through space and time from prehistoric settlements to fields, from storerooms to hearths and ovens, indoors and outdoors, observing people in those chores that ended in cooked food and commensality. From the heat of the summer harvests, to the gathering of autumn fruits and nuts, the silence of the snow-covered fields in winter or the busy humming of springtime and careful weeding of fields, the book examines the crops harvested and the ways they were transformed into meals. Beyond the fields, that part of the landscape that connected prehistoric farmers to their collective hunter-gatherer past offered a wealth of plant food treasures available to harvest in baskets or sacks: wild fruit, mushrooms, roots, greens, and aromatic and medicinal herbs (see, for example, Hadjichambis et al. 2008 for modern parallels). Prehistoric cooks would have had a rich variety of seasonally available ingredients. Harvests from fields and woods would have marked the sequence of seasons, linking the procurement of plant food ingredients to life cycles and special occasions. Plants would have satisfied several dietary needs, culturally defining different regions and communities. Prehistoric plants were used not only as food, drink, and medicine, but also as fodder, fuel, building material, and the like. Making the distinction between food and fodder in the archaeobotanical record is challenging, however, as it is shaped by culturally prescribed notions around each of these two categories (e.g., Jones 1998). Ethnographic work is crucial in this context, revealing the actions, the gestures of people and their tools, when processing plants, as the composition of the products and by-products in different parts of the world has shown (e.g., Hillman 1984, 1985; Jones 1983; Capparelli et al. 2011).

This fascinating journey is made possible thanks to the transforming powers of fire: prehistoric flames that captured ancient grains, seeds, fruits, and nuts, moments of prehistoric lives marked by a cooking accident, the burning of a house, a healing ritual, or an offering to ancestors. The very moment ancient plant food ingredients become charred, their journey toward us begins, well protected in the soil. They acquire a "voice" when the archaeologist's trowel brings to light the archaeological deposits that contain them. Sampling the soil and processing the samples secures the retrieval of "ancient grains" with the flow of water carrying them toward specialized sieves that capture ingredients of a distant culinary past, which are then scrutinized with the aid of a stereomicroscope. This work can reveal bulgur fragments 4,000 years old, grapes that were pressed 7,000 ago.

For a period for which written sources are lacking, the charred plant remains retrieved from archaeological sites are the main "text," our main means to enter the realms of prehistoric culinary practices on rare occasions by remains too tiny to be visible, such as phytoliths and starches trapped in grinding tools and cooking vessels, occasionally aided by chemical analyses. Of course we are unable to evoke the specific cultural contexts in which these practices were experienced and the tastes and senses generated during the process. Even when various lines of evidence may suggest special contexts of use, it is impossible to approach the specifics such as the sequence of events, who participated, what they wore, how they were positioned, the sounds, smells, and overall atmosphere of a daily meal or a ritual/healing event. Food preparation is not only about the ingredients but also about perceptions of place, technologies, and equipment as well as a network of social relations mobilized within and between communities (Goody 1982; Hastorf 2017).

The outcome of this fascinating dialogue between people and plant food ingredients would have contributed toward identity expression, through shared or excluded ingredients, recipes, and cooking facilities (see also Valamoti 2017). Besides plant ingredients and actual food preparations, the association of plant foods to the equipment used for processing plant ingredients and transforming them into foods allows valuable insights of past culinary practices. Archaeobotany, when combined with those installations and artifacts related to culinary practice such as hearths, pots, and grinding tools can offer the means to investigate the rich cultural heritage of food preparations, the shaping of past cuisines in the region and their change over time.

The word "cuisine" has been used widely in prehistoric archaeology, with Andrew Sherratt (1991a) being among the first to emphasize the need for an archaeology of food and its social dimension in his article "Palaeoethnobotany: From Crops to Cuisine." The degree to which we can approach the cuisines of prehistoric societies depends partly at least on how the term is defined. According to Parker-Pearson (2003), cuisine is a form of art that requires specialists, inner knowledge, a variety of dishes that are embedded in complicated interrelationships and are associated with ritual. Goody (1982) recognizes three levels of interpretation of the term "cuisine": (1) the product that results from cooking in a kitchen, (2) a culturally differentiated cuisine, (3) a particularly elaborate form of cooking that is encountered in certain societies of the world, China and the Middle East, for example. Which of these notions is most suited to the archaeobotanical remains depends, partly, on the level of approach. Emphasis may be placed on the ingredients that form the basis of the cuisine's food preparations or, on a synthesis of these elements spatially and temporally, exploring thus cultural preferences, rare culinary ingredients, and possible elaboration in food preparation procedures. Obviously, those less tangible elements of cuisine than the remains of the ingredients themselves are more difficult to analyze, and they are touched on here when the evidence allows it.

Behind the meal itself and the steps involved in its preparation are human relationships, connecting people within a settlement or broader regions. Daily or festive meals form an expression of networks and relationships as well as the rules regulating them and boundaries and margins for negotiating and altering them (see Hastorf 2017). The preparation of the plant component of food, as can be gleaned from the anthropological record worldwide, seems to fall into the hands and the realm of women, requiring a considerable part of their daily time and effort (e.g., Stanley 1981; D'Andrea and Mitiku Haile 2002; Lyons and D'Andrea 2003; Hastorf 2017). Cooking food is not only about transformation of ingredients but also society, and, through cooking, various arenas of gender and power associations come into play. Is cooking plants for food always a female business? It is beyond the scope of this book to explore this fascinating aspect of food preparation, yet some caution regarding overgeneralizations is necessary: types of cooking in certain periods of the year may be performed by men, such as meat roasting in Kabylia during the summer (Bourdieu 1977). Through the demands of plant food preparation, solidarity within the group is developed as it requires collective effort and sharing of workload (see, for example, D'Andrea and Mitiku Haile 2002; Counihan 1999).

Beyond the framework of daily food preparation within the household and the community, meals of a collective nature might have been organized for special occasions, such as birth, death, or the completion of a ditch or a house (see examples in Dietler and Hayden 2001a) or on specific days of the year, signified by special celebrations and comprising particular dishes prepared especially for the occasion (e.g., Psilakis and Psilaki 2001 on cereal-based festive foods; also Voutsina 2000; Skouteri-Didaskalou and Voutsina 2005; Goody 1998; Wiessner 2001). Feasts in the literature are classified depending on what is intended to be achieved through them, for example, solidarity among the members of a community or reinforcement of a position of power held by a member (Dietler and Hayden 2001b; Hayden 2001; Wiessner 2001). Such collective meals, known well in modern Greece in the context of the numerous *panygyria* (religious festivals with a feast), could have taken place in specially formed places and spaces that would have been clearly defined in the landscape or within a settlement, in the form of a natural feature such as a water source, a rock, or a cave, or a special building.

Although in most cases these special meals are characterized by the consumption of meat, especially in large quantities, plant-derived foods might have also been important in such contexts. Plant foods often hold a main position in feasts, as is the case with coarsely ground wheat that is cooked in the broth of meat derived from animals especially sacrificed for this purpose at the various Greek *panygyria*. Some feasts, usually religious festivals, are closely interwoven with the consumption of plant foods, as is the case, for example, during fasting, on religious festivals such as that of Saint Barbara on the fourth of December, during which a food called *Varvara* is prepared, consisting of a combination of cereal grain, pulse seeds, fruits, and nuts as well as the *polysporia* (multiseeds) prepared on the twenty-first of November (Megas 1988; Psilakis and Psilakis 2001), or *kollyva*, the boiled wheat grain that is prepared for the dead (see einkorn *kollyva* recipe in Chapter 11). In certain contexts pulse dishes are often the meal served at funerals (known as *makaria*, *parigoria*) probably because meat consumption on that occasion forms a taboo for a specific duration of time (Danforth 1982:44). Pulses transformed into a special soup form the meal for the feast in different parts of the Mediterranean, as encountered in the Aegean island of Amorgos where a chickpea soup is communally eaten after the church service at the festival of St. John Theologian, on the twenty-sixth of September (Figure I.1), and the festival of St. Anton of Gamonal in Burgos, Spain, where traditionally large cauldrons are used to prepare a grass pea or a chickpea soup that is served to the numerous participants on January 17 each year (Valamoti et al. 2011).

Figure I.1. Consuming chickpea soup, festival of St. John the Theologian on the island of Amorgos, September 2007. Counterclockwise from top right: church of St. John Theologian, priests blessing food in a special kitchen area, and chickpea soup being eaten communally. (Photographs courtesy of Nikos Nikitidis.)

Alcoholic beverages, derived from plant food, are often associated with feasting and special contexts of consumption. For example, millet beer that is prepared and consumed in Africa (Goody 1998; Arthur 2003; Dietler 2001). For this reason plant domestication has been linked to the desire to produce alcohol and hence invent agriculture in the Old World (see, however, Chapter 1 for a more cautious approach to the issue). Plants and the foods prepared from them, daily or special, are connected with the material world, providing energy, yet, at the same time they constitute a reference to an immaterial world and a medium to communicate with it, loaded with meanings and symbolisms.

In antiquity, the adoption of specific plants as foodstuffs was connected to gods, goddesses, and mythical creatures. In our study region, people explained the origins of agriculture, wine making, and the exploitation of the olive with myths about Demeter, Triptolemus, Dionysus, and Athena. In modern times, archaeobotanical research forms the tool with which the origins of these activities are explored in prehistoric times, a tool deprived of the magic of myth but able to generate modern mythologies about prehistoric diet. What follows in the chapters of this book are based on this archaeobotanical research.

The plant remains we find in archaeological deposits, and on which this book is based, inevitably constitute the humble material leftovers of human relationships, senses and thoughts, expression of feelings, elements of their memories. These remains of the day, of ordinary days and of special days marking the passage of time, birth, reproduction, and death, hold the key to understanding aspects of past human lives captured in archaeological deposits. It is obvious that a considerable distance separates the burned seeds of a prehistoric house from the actual meals and contexts of their consumption. The task of interpreting long lists of plant species compiled from archaeobotanical studies is often challenging but sometimes rewarding.

The farming communities that emerged in Greece with the onset of the Neolithic had a background of millennia of accumulated knowledge regarding the processing and cooking of food. Little has survived in the archaeological record in regard to cooking practices prior to the emergence of these first farmers in the region. The "works and days" of prehistoric farmers inhabiting Greece would have involved long hours of preparing the land for sowing and selecting the right time to sow, tilling the crops, watching them mature, and, in the hot summer months, harvesting the fruits of their labor, storing a wide variety of ingredients that secured the daily provision of meals as well of special dishes for celebrations and rituals. Sometimes these were further processed, resulting in special ingredients and foods. Rich, pure caches of grain in houses that burned down by fire, such as at Late Neolithic Dikili Tash, or in pits, such as those found at Toumba Kremastis Koiladas, offer snapshots of a culinary past, revealing this annual, repetitive labor that secured food and connected people with the land they worked, their community, and their ancestors. Prehistoric plants encountered in the archaeobotanical record of Greece served as nourishment but also must have been elements of cultural classification systems and perceptions of self and community in a discourse between nature and culture. What constituted food as opposed to the "inedible" largely depends on worldviews, the cultural structuring of identity and otherness (e.g., Hamilakis 1999; see Hastorf 2017 for a review of the relevant literature) as well as the available ingredients and microbial and bacterial fauna and flora. Food taboos would have constituted identity signifiers reflecting at the same time locally available ingredients, yeasts and bacteria contributing toward generating specific "cultures" of taste and smell in both senses of the word, cultural and microbial (see, for example, Brumberg-Kraus and Dyer 2011).

This book opens a window into the daily lives of the prehistoric inhabitants of Greece, through which we may peer into their kitchens and explore their plant ingredients and recipes. Based on a comprehensive overview of the archaeobotanical evidence from prehistoric sites in Greece, the plant component of prehistoric meals in the region is explored, focusing on themes such as the beginnings of agriculture, dietary preferences, and regional differentiation, ways of processing plants, cooking, and consuming a wide range of cultivated and wild plants.

Archaeobotanical indications for the special treatment of plants provide a strong argument regarding their use, and at times, they provide the ability to explore the "recipes" that generated them, leaving little doubt that they were intended for human consumption. This fragile material that has been kept in the earth for thousands of years until unearthed by the archaeologist is sometimes full of surprises, allowing us to more closely observe prehistoric cooking in Greece. Beyond the daily or festive meal, through the humble archaeobotanical remains, we can observe human choice, preferences, and their changes through time, indicative of contacts with neighboring or more distant regions. We can guess the liquid that filled the cups, the foodstuffs that were being cooked in the ovens and hearths, forming through this preparation, consumption, and discard a cycle of activities that defined, and likewise were defined by time and space, in settlements and beyond.

Plant Foods of Greece is based on archaeobotanical remains preserved mainly through charring, the primary means of preservation in our study area, retrieved from prehistoric sites of Greece. Among the flames, while food was being cooked, or when fire embraced a whole building (Figures I.2 to I.5), turning it into burned "destruction layers," fragments of the daily strife of people became transformed into these tiny, charred plant remains that constitute the archaeobotanical record, in other words, the archaeological "ingredients" that make up the chapters of this book.

Charred plant remains retrieved from Neolithic and Bronze Age settlements of Greece by flotation form the basis of the reconstruction of diet in the region. These "archaeobotanical" remains have been found in archaeological deposits containing rubbish, such as chaff and dung, as well as among the remains of prehistoric houses and storerooms destroyed by fire. The available evidence provides a general picture of cultivated plants as well as those harvested from the wild. We should bear in mind, however, that this picture is fragmentary, based only on those plant species and parts that survived contact with fire. A wealth of plant food ingredients consumed as roots, flowers, and leaves, fresh or dried, have limited chances of being preserved at Greek sites (Figures I.6 to I.9). In this group of plants we can include greens,

Figure I.2. Katerina Karatasaki sampling during excavation at Dikili Tash, July 2012. (Photograph by Soultana Maria Valamoti. © Dikili Tash Research Project.)

Figure I.3. Concentrations of different crops visible (here *Lathyrus sativus* and on top *Pyrus amygdaliformis*), allowing their separate sampling during excavation, Dikili Tash, House 1. (Photograph by Pascal Darcque. © Dikili Tash Research Project.)

Figure I.4. Charred wheat grains during excavation at the four lakes region, Limnochori II, TRENCH IV, 2008. (Photograph courtesy of Panagiotis Chrysostomou.)

Figure I.5. PlantCult project experimental cooking: grains and food can get accidentally charred during cooking. (Photograph by Georgina Prats Ferrando. Courtesy of PlantCult.)

Figure I.6. Green pods of pulses harvested to be eaten as a snack, near the village of Peritheia, Corfu, July 2016. (Photograph by Georgios Vily Kapetanakis. Courtesy of PlantCult.)

Figure I.7. Wild mushrooms are a seasonal delicacy or, if dried, a food source for the rest of the year. No such finds have been encountered in the archaeobotanical record of Greece as their chances of preservation are very slim. (Photograph by Soultana Maria Valamoti. Courtesy of PlantCult.)

Figure I.8. St. John's wort (*Hypericum perforatum*) flowers used in medicinal preparations. (Photography by Soultana Maria Valamoti. Courtesy of PlantCult.)

Figure I.9. Sophia Papageorgiou shows Soultana Maria Valamoti her dried greens, village of Kosmati, September 2017. Dried greens (*lachana*) such as *Sonchus*, *Chenopodium*, and *Rumex* can be kept for use throughout the year. (Photograph by Nefeli Stevi. Courtesy of PlantCult.)

eaten like salad, edible wild mushrooms or medicinal and aromatic plants, sometimes cultivated but probably mostly harvested from the wild.

The archaeobotanical record of the area considered here consists of published archaeobotanical reports from nearly a hundred settlements (Appendix A:Table I.1) (Figure I.10), which offer the basis for an investigation of plant food ingredients. At least twenty-six plant food species were cultivated or harvested from the wild and used in the Neolithic and the Bronze Age. These are represented either by rich concentrations of grain or by-products of their processing. The appendix contains the tables, with the archaeobotanical data from each settlement of the Neolithic and the Bronze Age where species have been identified or are present in pure, dense concentrations.

The collection of the available archaeobotanical evidence from Greece used in this book has greatly benefited from work carried out by fellow archaeobotanists who have presented summarized tables of archaeobotanical work carried out in Greece (e.g., Megaloudi 2006; Kroll 1991; Hansen 2000; Sarpaki 2000) as well as my own compilation published in Greek (Valamoti 2009). Although the retrieval of plant remains from archaeological sites in Greece is not a mainstream procedure, for the prehistoric periods a large number of settlements (sixty-one for the Neolithic period and eighty for the Bronze Age) have been investigated archaeobotanically. For the Mesolithic period (8300–7000 BC), only briefly discussed here in Chapter 2, as well as for the Early and Middle Neolithic (approximately 7000–5300 BC), rather few sites are represented by archaeobotanical material; for the subsequent periods more sites have been investigated archaeobotanically (Late Neolithic: approximately 5300–4500 BC; Final Neolithic: approximately 4500–3400 BC; Early Bronze Age: approximately 3400–2100 BC; Middle Bronze Age: approximately 2100–1700 BC; Late Bronze Age: approximately 1700–1100 BC).

The book explores the Neolithic and Bronze Age plant foods of Greece. The Iron Age, from 1100 to 800 BC for most parts of Greece and much later for northern Greece, is beyond the scope of this book. For this period, written sources with rich information for various plant foods are also available. An approach of ancient plant foods from the Iron Age onward would require a thorough consideration of the written sources as well, the subject of a book from PlantCult project (Valamoti et al. 2022a). The book investigates the archaeobotanical record retrieved from Greece since the first archaeobotanical expeditions of pioneers such as Maria Hopf, Jane Renfrew, Wilhelm van Zeist, and later Glynis Jones, Helmut Kroll, Anaya Sarpaki, Julie Hansen, and James Greig, to more recent investigations conducted by foreign

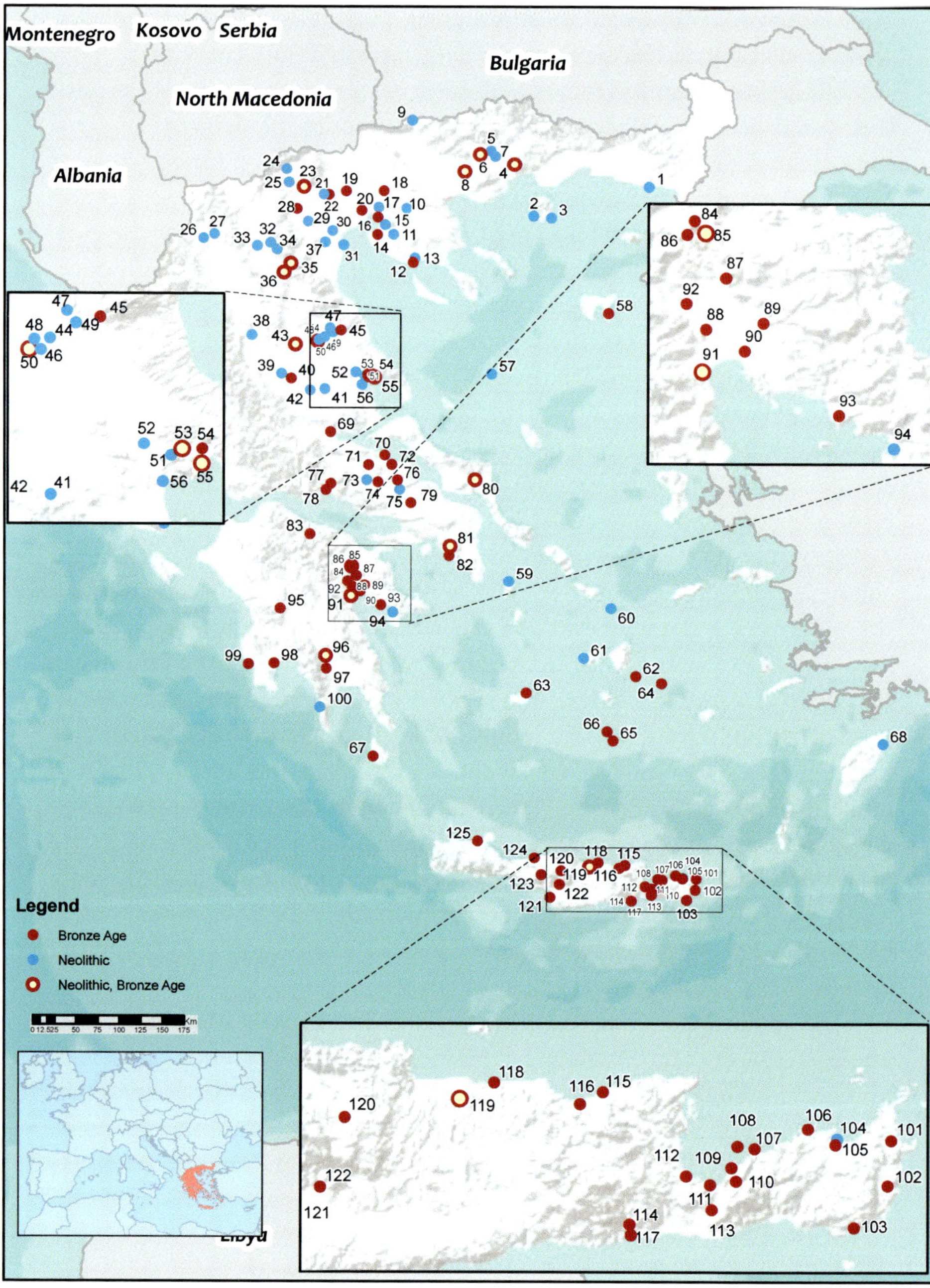

Figure I.10. Map of Greece showing sites that have been investigated archaebotanically (for site numbers, see Table I.1). (Basemap sources: Esri, US Geological Society, and National Oceanic and Atmospheric Administration. Map by Themis Roustanis. Courtesy of PlantCult.)

and native specialists (for a review see, e.g., Valamoti and Bittmann 2015). This exploration focuses on plants used to nourish the body and the senses, plants harvested from the wild or from laboriously tended fields. The minuscule plant remains, often invisible in the excavated earth, become eloquent witnesses of past lives, daily moments, and special occasions.

Not all sites have been sampled in the same thorough way. For example, some have large numbers of samples and a fine mesh was used (Appendix A:Table I.1). This affects comparability between sites, but where an adequate dataset is available, useful insights are possible to obtain and trends to infer. A Greek database of prehistoric plant remains is currently underway within the PlantCult project, something that will allow more fine-grained analyses in the future. For Late Neolithic Greek sites where such evidence is already available, comparisons between cereal and pulse crops are attempted in Chapters 2 and 3. Sampling in the field is of crucial importance, as this can later affect analysis and discussion of the archaeobotanical results. Although ideally we might have wished to know what each inch of excavated area may have contained, in reality this would translate to endless hours of processing of archaeobotanical samples and study, with the strong likelihood to drastically reduce the number of sites an archaeobotanist could study in his/her active career: archaeobotanical work is extremely time consuming. As a result, it is imperative that the soil samples taken for archaeobotanical analysis are as representative as possible. It is also very important to know the exact location of each sample and not its general attribution to an archaeological layer. By taking samples from different areas within a house floor, for example, the archaeobotanist can spot differentiations in space in regard to archaeobotanical composition either in the field and/or later during analysis of the data (e.g., Valamoti and Jones 2003; Valamoti 2015a:Figure 3; Valamoti 2004:Figures 7.35–7.37). A detailed sampling as that performed at Bronze Age Assiros Toumba allowed the reconstruction of the stored concentrations of grain inside the burned down storeroom (Figure I.11) (Jones et al. 1986). When excavating negative features such as pits (Figure I.12), samples that are taken from different levels, following the stratigraphy or else, at regular intervals, allow observations regarding changes over time or different episodes of use as is the case with an Iron Age pit from Karabournaki (Tiverios et al. 2013).

At dry archaeological sites such as those encountered in Greece, where preservation is predominantly by charring, soil samples need to be of a large volume, preferably at least 10 liters (ca. 2.64 gallons), unless the excavated unit is in itself smaller in volume, for example, when the content of a

Figure I.11. Bronze Age burnt destruction layer from a Late Bronze Age storeroom at Assiros Toumba, approx. 1350 BC. (Photograph by Paul Halstead. Courtesy of Ken Wardle.)

Figure I.12. Numerous soil samples retrieved from a pit excavated at the site of Panakton, summer 1991. Each bucket corresponds to a different layer of the pit's fill. (Soultana Maria Valamoti)

medium- or small-sized pot is taken as a sample. Where the volume of soil processed is recorded, it is later possible to estimate the density of plant remains in the sampled feature; in this way a better understanding of human activities in relation to plants can be achieved, working through the predepositional, depositional, and postdepositional processes, starting from the sown field and ending with people's interaction with plant foods at the site, those events that led to the formation of the archaeobotanical record (see, for example, Valamoti and Jones 2003).

Soil samples from Greek sites have been processed with different methods, depending on conditions of preservation and facilities available at different excavation projects (for general principles of retrieval and processing see Greig 1989; Buxó and Pique-Huerta 2008; Jacomet and Kreuz 1999; Pearsall 1989). At dry sites, where carbonization is the main means of preservation, as is the case in the vast majority of archaeological sites in Greece, flotation is mostly employed using a flotation machine (see French 1971). The machine is operated with a barrel specially designed so that water coming out from small holes flows toward test sieves of a smaller aperture of 300 or 250 μm, which collect the "flot," the material that floats (Figure I.13). When water is short, flotation can take place by recycling the water (Figure I.14). The heavy residue is kept in a nylon mesh, usually of 1 mm aperture. When all fractions are dry, they are sorted and then identified, with the aid of a stereomicroscope, to species genus or family level (Figure I.15).

In this book much emphasis is placed on indications of storage or processing of plants and not on the mere presence of a species or the frequent, but in very low percentages, presence in the samples. In cases where even rare finds have their significance, special mention is made. Thus, all edible plants (cultivated and wild), for which there are indications of intentional storage or processing by the inhabitants of the Neolithic and Bronze Age settlements of Greece, are considered as human food, although it is acknowledged that depending on time period, area, and social context what may have been perceived as food or fodder may have varied. Still, as seen in the following chapters, archaeobotanical remains considered do in several cases provide strong evidence for the use of a wide range of plants, including toxic ones that at first sight may appear as unsuited for human food.

More than 120 plant genera and species have come to light from prehistoric sites in Greece, many belonging to cultivated plants but most of them forming part of the wild vegetation (e.g., Valamoti 2009). Which of these species, however, were indeed consumed by the prehistoric inhabitants of the region? Is there a way to "penetrate" Neolithic and Bronze Age kitchens

Figure I.13. Flotation machine in operation at Dikili Tash (2012) with test sieves retaining the charred remains of wild pears. (Soultana Maria Valamoti)

Figure I.14. Flotation machine processing samples from Panakton excavation at the schoolyard of Pyli, August 1991. The machine was operated with a system for recycling water. (Soultana Maria Valamoti)

Figure I.15. Chryssa Petridou sorting a flot for archaeobotanical remains with the aid of a stereomicroscope at the Dikili Tash excavations laboratory. (Soultana Maria Valamoti)

in search of the plant components of daily meals and feasts? This question brings about various issues related to the interpretation of plant remains from archaeological sites. Plant lists at the end of archaeological publications are not the ultimate goal of the archaeobotanist, as sometimes archaeologists may wrongly assume. The tedious, time consuming, and at times even boring job that leads to these long plant lists acquires meaning when this data is translated in terms of the needs of prehistoric people: indications for the storage or processing of a plant, for example, suggest its intentional use by people (Dennell 1976; Halstead 1994). This "translation" is often based on the properties of plants and their potential uses, as these are recorded in ethnographic and ethnobotanical studies as well as agronomy manuals. Extrapolating this information to prehistoric times and also later periods should nevertheless be done with caution. Sometimes this can even be misleading, especially when the presence of toxic substances in pulses is mentioned in the literature: such plants should not be excluded from the food ingredients of a past society; human experience has devised ways of removing plant toxicity.

Projecting modern practice to the past obviously encompasses elements of anachronism. Barley, for example, could be considered as suitable for fodder, based on recent dietary habits or classifications among traditional farming societies of Greece, a concept that does not seem to be the same at all times and regions (Halstead and Jones 1989; Jones and Halstead 1995; Jones 1998). Moreover, plants not reported as consumed by people in recent ethnographic accounts may well have been consumed in prehistoric times, and vice versa (e.g., Ertuğ 2004 for the preparation of bulgur from einkorn) (see also Chapters 2 and 3). Ethnographic observation provides alternative potential ways to interpret archaeobotanical remains rather than a direct image of practices remaining constant over time. By the same token, the presence of a plant species in the archaeobotanical record does not necessarily imply its use as food or medicine in prehistoric times (cf. De Boeck 1994; Hamilakis 2000). Classifications about what is suitable and what not for human consumption are culturally defined, thus ethnographic observations on the use of certain species within a specific society could be misleading if projected on prehistoric societies: plant classifications can be transformed even within the same society, depending on circumstances and social context.

The material on which the book is based consists of samples taken at differing degrees of detail at different sites, which is of course taken into consideration when comparisons among sites, regions, and periods are made here. Despite the challenges noted, this book offers a reliable picture of food ingredients and preparations for the nearly 7,000 years of prehistoric culinary practice in the region. Moving from potential cooking plant ingredients to "prehistoric cuisine" can be an exercise in speculation, but as will be seen, the archaeobotanical record from prehistoric Greece is full of exciting surprises that preserve actual processed food remains that allow a decipherment of at least fragments of past recipes and valuable insights of the culinary past of the area. Such recipes continue to be produced in Greece to the present, and Chapters 10 and 11 explore current plant-food-based traditions and modern trends relevant to archaeobotanical remains as well as a hands-on attempt to "taste" the past.

The book is organized in chapters that follow our modern classifications of groups of ingredients: cereals (Chapter 2), pulses (Chapter 3), oil plants (Chapter 4), fruits and nuts (Chapter 5), fermented brews (Chapter 6), and aromatic and healing plants that would enhance the texture and taste of culinary preparations (Chapter 7). Of course, this is an arbitrary classification, corresponding to modern perceptions of plants in terms of the botanical families they belong to (e.g., cereals and pulses) or their characteristic properties. Our journey to past culinary traditions starts from the inconspicuous Mesolithic hunter-gatherers and the first farmers who shaped the wooded landscapes of early Holocene Greece to suit their needs, then moves to the Neolithic and Bronze Age communities of farmers organizing their settlement space in different ways. The ways plant food ingredients were transformed into dishes and drinks offer fascinating insights into the potential variability of the dishes that could have been prepared and the variable identities that might have been expressed through cuisine (Chapters 8 and 9). The book's journey follows this prehistoric plant-based cuisine in the present, exploring resilience and losses of a culinary legacy with millennia-deep roots (Chapter 10).

Our meals prepared with electric- or gas-operated cooking equipment, often consisting of fast food eaten in haste in front of the television, have nothing to do with the Neolithic and Bronze Age kitchens where food was being prepared, inside houses, or outside, in yards, smells spreading in the village, "alerting" to food being cooked or the feast being prepared. We do share, however, similar ingredients and possibly recipes, despite the millennia separating us from those distant prehistoric people. These ingredients form the basis for the recipes included in Chapter 11; these have been inspired by the ingredients available in prehistoric times and a wide array of traditional recipes gleaned from regional cookbooks, my mother's and grandmother's instructions, as well as my experimentation.

1

The Dawn of a New Culinary Landscape

Plant Foods of Europe's First Farmers

From Asia to Europe: The Emergence of New Ingredients

The domestication of some plant species from the rich wild flora that emerged with the onset of the Holocene led to major changes, with long-term effects on the socioeconomic organization of communities of Western Asia and Europe. In the wider region of the Fertile Crescent, a close relationship between certain plants and people emerged, leading, in some cases, to the transformation of wild plants to cultivated crops (Harlan and Zohary 1966). It is in this area that agriculture, the cultivation of domesticated crops, first appeared, opening up the potential for food surplus accumulation and manipulation; it is also where the wild progenitors of cereals and pulses that were introduced to Europe grow today (see, for example, Smith 1995; Colledge et al. 2004; Colledge and Conolly 2007; Willcox 1999, 2005; Willcox et al. 2012; Zeder 2011; Abbo and Gopher 2017). Agriculture also led to new food ingredients and recipes, thus transforming gatherer-hunter cuisines.

Discussions about the origins of agriculture focus on a very important question: why did gatherer-hunters start cultivating? The factors underlying the emergence of agriculture and domesticated plant food ingredients in the wider region of the Fertile Crescent is a matter of a long, ongoing investigation and debate; factors such as environmental change and population pressure have often been at the heart of interpretative models (e.g., Smith 1995; Thorpe 2003; Zeder 2015; Asouti and Fuller 2013; Abbo et al. 2012, Gopher et al. 2021, for reviews of the history of research and different views on the origins of agriculture and plant domestication).

Social parameters that might have led to agriculture received very little attention in this discourse for many decades with the exception of Barbara

Bender (1978) who, as early as 1978, emphasized social processes as being the major underlying factor that transformed "gatherer-hunters" (Bender 1978:204) to farmers. Although recent evidence for Near Eastern and Anatolian megasites (Belfer-Cohen and Goring-Morris 2011) and "ritual centers" such as Göbekli Tepe (Dietrich et al. 2012) did not exist at the time, Bender (1978) was very perceptive in hypothesizing, on the basis of limited evidence from burials and obsidian trade networks, that the need for surplus food to sustain gatherer-hunter hierarchies and social networks would have led some gatherer-hunters to food production and thus to the domestication of cereals and pulses. In a very important but largely ignored statement she argued that Late Pleistocene gatherer-hunters of the Levant and Anatolia were involved in social networks with regional hierarchies based on control over the economy as well as ceremonial institutions (Bender 1978:215). As social competition will act as an incentive for surplus production, in areas where potential domesticates were available, the need for food surplus could have led to food production (Bender 1978:214). The first cultivators did not set out to become farmers of domesticated species; this was an unforseen outcome with great potential (Bender 1978:204).

Prior to the emergence of what in retrospect is named as the "Neolithic crop assemblage" (Zohary 1989:359) or "crop package" (Zohary 1996:156), different ways of cultivating were probably taking place simultaneously, reflecting the continnuum in human-plant interactions (Harris 1989, 1996; Hillman and Davies 1992; Kislev 1989a; Willcox 2000): gatherers of wild cereals and pulses, cultivators of wild cereals and pulses (Kislev 1989a; Willcox et al. 2008), and cultivators of domesticated cereals and pulses who may have also harvested wild stands. It has been suggested that the new plant food ingredients, the domesticated crops, were the outcome of a long-term interaction between plants and people, and both wild and domesticated plants were growing together in the fields of the first farmers (Willcox et al. 2009:156). An alternative model for 'a core area and a one event' model has been proposed at the antipode of the protracted millennia long model of domestication whereby plant domestication occurred in southeastern Turkey and north Syria, also referred to as the north Levant, where all the wild progenitors of the domesticated Near Eastern crops occur nowadays (Gopher et al. 2021). We can envisage a period during which those growing cultivated crops were perhaps more influential during social gatherings with everybody wanting to have some of those "special" seeds. The first plants cultivated by these early farming communities of the Near East were elements of the wild vegetation, their seed selected to be sown. Wild stands of cereals and pulses might have been domesticated through a process of experimentation and tending of plants, while their diffusion to neighboring communities of people would have been carried out through mating networks and the transportation of "family crops" to the new foci (see Hastorf 1998:778). Agriculture was the result of a long-term transformation of the relationship between plants and people through specific ways of interaction and interdependence of both parties involved (cf. Hillman and Davies 1992; Harlan and Zohary 1966; Hastorf 1998; Harlan 1992; Harlan and de Wet 1965; Zeder 2015). Not all wild plants become domesticated when managed by humans, and not all human societies will inevitably become farmers once they start managing wild plants. On these grounds, management of wild plant food resources in the form of cultivation is considered as neither a prelude nor a prerequisite for plant domestication (Gopher et al. 2021); yet, an intimate knowledge of the biology and nutritional qualities are, the outcome of a close relationship, knowledge of life cycles, harvesting, and consumption.

Both a monophyletic origin of domesticated crops and multiple origins in the wider area of the Fertile Crescent have been proposed on the basis of the same archaeobotanical, botanical, and genetic evidence (e.g., Zohary 1996; Fuller et al. 2010, 2011). Lack of clear definitions regarding the domestication episode and evolution of crops under domestication contribute to contradictory narratives about the emergence of agriculture in the region (Abbo et al. 2012:247). According to Abbo and Gopher, "initial domestication" and crop evolution under domestication need to be differentiated (Abbo et al. 2012) while the domestication episode (initial domestication) took place in a restricted area, was intentional and selective, completely reversing the order in interaction between human societies and nature, the former intervening in the natural course of natural processes in vegetation (Abbo et al. 2012; Abbo and Gopher 2020).

Despite the lack of a consensus regarding the processes leading to domestication, morphologically wild and domesticated cereals were being harvested together, over several millennia, before fully domesticated crops, growing in fields, became dominant (Tanno and Willcox 2006a). This coexistence could have signified either fields where both wild and domesticated plants grew or where three possible practices of (1) harvesting wild cereals, (2) cultivating wild cereals, and (3) cultivating domesticated ones coexisted for some time in the same wider region. The details of these processes, however, are the matter of discussion and lack of overall consensus among archaeologists and archaeobotanists (see for example Fuller and Allaby 2018; Fuller et al. 2018; Abbo et al. 2021).

The emergence of societies cultivating domesticated plants in the Fertile Crescent was the outcome of specific choices made by some groups of people who inhabited the region. At the end of the eighth and the beginning of the seventh millennium BC, some sites such as Halula and Ras Samra show no evidence for cultivation (Nesbitt 2002), which suggests that societies using domesticated cereals and societies who insisted on continuing with the use of wild plants coexisted in the Near East. Such a coexistence has been pointed out very early on in discussions on the emergence of agriculture in the Near East (Harlan and Zohary 1966). One could picture similar processes to those observed in the Near East being underway in southeastern Europe: farming and gatherer-hunter communities coexisted in the landscape, and new, local plants could have been brought into cultivation in addition to the components of the "package"; unfortunately, the evidence for this part of the story is probably lost together with the Mesolithic gatherer-hunter base camps in the now-submerged coastal plains of the Aegean.

Whether a slow or quick process, first occurring in one place or in different locations in the wider area of the Fertile Crescent, domesticated crops were grown by the eighth millennium BC in large parts of the Levant and Anatolia. In the seventh millennium BC these crops are found on both sides of the Aegean Sea. The dynamic in the adoption of new foodstuffs and a new way of life, which is temporally defined as the beginning of the Neolithic, is underlined by differentiations in the elements of the so-called package that appeared in Greece and elsewhere in Europe that may suggest variable itineraries in the diffusion of domesticated plants from the Middle East to Europe.

New Plant Foods in the Aegean: Local or Imported?

The cultivation of domesticated plants for the preparation of food has been considered an essential element of the Neolithic in the Aegean. In the beginning of the seventh millennium BC in Thessaly, as well as in other parts of Greece such as Crete and the Peloponnese, communities organized in "permanent" villages, knowing how to make pots, grow domesticated plants, and keep domesticated animals, appeared. Few but eloquent archaeobotanical finds of domesticated cereal and pulse crops have been identified in a handful of archaeological sites on the island of Crete as well as the Peloponnese, Thessaly, and Macedonia, and on central mainland Greece (Appendix A:Table 1.1).

The new way of life in the early seventh millennium BC replaced or modified the ways of acquiring food through the hunting and gathering of plants in the preceding Paleolithic and Mesolithic periods. Cereals and pulses from cultivated fields emerged as new food ingredients, which, alongside ceramic cooking technology, must have contributed, in the long run, to a considerable change in the prehistoric cuisines of the region. Understanding the processes through which these new food ingredients and perhaps recipes, too, appeared in prehistoric Greece with the onset of agriculture requires an examination of the processes that transformed subsistence on wild plant foods to subsistence on cultivated crops.

Several archaeologists have emphasized the contribution of native elements in the process of cultural change and the adoption of agriculture in Greece (Theocharis 1967, 1973; Dennell 1983; Barker 1985; Kotsakis 1992, 2000, 2001, 2003; Séfériadès 1993; Halstead 1996), and some have gone as far as to hypothesize that the transition to the Neolithic way of life might have been a local development of the area (Hansen and Renfrew 1978; Kyparissi-Apostolika 2000, 2003). Other researchers attribute the emergence of Neolithic societies in Greece to the introduction/imposition of a fully developed new way of life, either through colonization (Ammerman and Cavalli-Sforza 1971, 1973, 1984; Hansen 1992; Perlès 2001; Runnels 2003) or through cultural diffusion or both (Dennell 1992; Runnels and Van Andel 1988).

Scholars have sought the arguments for or against an indigenous development of agriculture in the Aegean in the Mesolithic background of Greece. In this region, the Mesolithic that precedes the Neolithic has been, until recently, very poorly investigated. In addition to the emblematic site of Franchthi Cave (Figure 1.1), investigations in the last two decades have drastically increased the number of Mesolithic sites on the mainland as well the islands (Kyparissi-Apostolika 2000, 2003; Sampson et al. 2003; Runnels and Van Andel 2003; Sampson 2006; Galanidou and Perlès 2003; Strasser et al. 2010; Tourloukis and Palli 2009), radically altering the picture of what earlier scholarship described as an almost uninhabited area prior to the arrival of the first farmers. Mesolithic (10,000 years ago) societies of mainland Greece and the Aegean islands are situated in caves as well as open sites on the coast and farther inland (Jacobsen 1976; Kyparissi-Apostolika 2000; Perlès 2001; Runnels 2001; Galanidou and Perlès 2003). Stone tool production seems to differentiate the region from other contemporary lithic industries of other parts of Europe and the Near East, continuing a tradition already known from the Upper Paleolithic, based on flakes (Perlès 2001,

Figure 1.1. View of Franchthi Cave. (Photograph courtesy of Nicholas Thompson.)

2003). We know that, during this period, people were hunting and exploiting marine resources as well as collecting wild plants (Hansen 1991; Payne 1982; Shackleton 1988; Kotzamani and Livarda 2014; Perlès 2016). Rather than a uniform Mesolithic, however, new evidence and approaches argue for a complex palimpsest of populations with varying subsistence and cultural practices (Galanidou 2011; but see Tourloukis and Palli 2009:34).

The first farmers of Greece have been perceived as small groups of colonizers of a nearly empty landscape, bringing along the Neolithic "crop package" (Runnels 2003:124–126). This crop package might have also been diffused via various maritime and overland routes by seafarers of the Aegean and the east Mediterranean, connecting the eastern and western shores of the Aegean and the east Mediterranean during the end of the Pleistocene and early Holocene (Runnels and Van Andel 1988; Hansen 1992; Runnels 2001, 2003; Galanidou and Perlès 2003, Galanidou 2011; Asouti et al. 2018; Horejs et al. 2015).

The way plants were used within a framework of food production is a focal point in the construction of the term "Neolithic" and the association of Early Neolithic societies of the Aegean with the cultivation of cereals and pulses. To understand the processes of introduction of the new food ingredients coming from fields, we need to explore the archaeobotanical record of Mesolithic and Early Neolithic sites from Greece. The archaeobotanical remains available for the early period of agriculture in Greece are fragmentary and at present insufficient for shedding light on the processes through which the transition from harvesting wild plants to cultivating domesticated plants occurred (Valamoti and Kotsakis 2007; Valamoti 2009; Kotzamani and Livarda 2018). Of the Mesolithic sites that have been excavated, only three, Franchthi, Theopetra, and Cave of the Cyclops, have so far yielded archaeobotanical material. Of these, the remains from Franchthi and Theopetra are fully published (Hansen 1991, 1992; Kotzamani 2010), whereas a general report is only available from Cave of the Cyclops (Sampson 2006; Sarpaki 2011). The preliminary publication of archaeobotanical remains from Theopetra (Mangafa 2000) should be treated with caution as among the Mesolithic finds six-row barley is included, without, however, any evidence for the presence of earlier forms of barley, that is, wild or two-row.

The identification of emmer wheat in Upper Paleolithic layers from Theopetra (Kotzamani 2010) further highlights the problematic nature of the available pre-Neolithic archaeobotanical finds. It is very likely that the Theopetra emmer glume bases are intrusive from later deposits, as wild emmer, a species preceding domesticated emmer, has never been found at the site and is not an element of the Greek flora.

In the early days of investigating Europe's first farmers, the spotlight was on Franchthi Cave and its long stratigraphic sequence spanning the Mesolithic and the Early Neolithic together with archaeobotanical evidence for early plant harvesting. Wild barley and wild lentils are listed among wild plants harvested at Franchthi during the Mesolithic (Hansen 1992). They imply harvesting of wild plants as part of food gathering; therefore, some form of relationship between plants and people was already established in Mesolithic Greece. A similar picture has been obtained from Theopetra (Kotzamani and Livarda 2018). Seed size, however, on which inferences regarding the presence of wild plants that are later found domesticated (barley and lentils), is not a reliable criterion (see, for example, Fuller 2007 on pulses and Abbo et al. 2014). Moreover, stratigraphic sequences at cave sites such as Francthi and Theopetra are often disturbed, and seed size is not a safe identification criterion for distinguishing wild from domesticated crops; hence, these Mesolithic "wild" cereals and pulse seeds should be directly dated. The Mesolithic archaeobotanical evidence needs to be treated with caution, an approach adopted in a review of the available Mesolithic evidence for plant exploitation at Greek sites (Livarda and Kotzamani

2014; Kotzamani and Livarda 2018). The finds of wild barley among the Mesolithic Franchthi Cave deposits led to a publication in *Nature*, raising the possibility for a local process of domestication of crops such as barley and lentils (Hansen and Renfrew 1978). This suggestion was subsequently dropped in favor of an introduction of the Neolithic crop package from the Near East (Hansen 1991). Were the Mesolithic inhabitants of Franchthi harvesting barley and lentils from wild stands or from fields where morphologically wild plants were cultivated? Can we entirely rule out the possibility that some of the grains retrieved from the Mesolithic layers are intrusive? At present, the nature of the relationship between plants and people in Mesolithic Greece is open to investigation, discussion, and speculation and in need of more data.

The archaeobotanical evidence that would allow a good understanding of the origins of cultivated cereals and pulses in prehistoric Greece is very scarce, not only for the Mesolithic sites of Greece but also for Early Neolithic ones (Valamoti and Kotsakis 2007; Livarda and Kotzamani 2014; Kotzamani and Livarda 2018). Despite the relatively large number of Early Neolithic sites, few have been investigated archaeobotanically (Appendix A:Table I.1 and 1.1) and the number of samples and charred seeds of cultivated plants is too small or difficult to evaluate due to the summarized form in which the data has been presented (e.g., Kroll 1983; Valamoti 1995; Valamoti 2011a). From the south of Greece, Franchthi and Knossos are relatively well represented by a number of samples (Franchthi: Hansen 1991, 1992; Knossos: Evans 1964, 1968; Sarpaki 2013), whereas in the north, Nea Nikomedeia (Zeist and Bottema 1971) and more recently Revenia, Paliambela, and Mavropigi in northern Greece (Kotzamani 2010; Kotzamani and Livarda 2018; Valamoti 2011a) confirm the presence of cereal and pulse crops and thus the existence of early farming communities in Europe in the seventh millennium BC. The domesticated species identified at these sites are einkorn, emmer wheat, the new glume wheat, recently identified as *Triticum timopheevii* (Czajkowska et al. 2020), free-threshing wheat, two-row and six-row barley, lentils, and peas (Appendix A:Table 1.1). Thus, the available data is adequate to identify the cultivation of domesticated plants as an element of the way of living in Greece during the Early Neolithic but inadequate for investigating the processes through which the domesticated species were introduced, and perhaps selected, at each site. Where sampling has been systematic such as at Franchthi and Mavropigi-Fillotsairi, both wild and cultivated plant species occur at the sites. The alleged underrepresentation of wild plants at Early Neolithic sites (Perlès 2001) is most likely the outcome of taphonomic factors (Antolín and Jacomet 2015). Chickpea, an element of the "Neolithic crop package," has not been found among Early and Middle Neolithic deposits from Greece (Appendix A:Table 1.1) (see also Chapter 3).

Although the botanic distinction between wild and domesticated cereals and pulses is usually clear and straightforward, on an ideological level, such a distinction is culturally prescribed and variable in space and time (Ingold 1996). Nevertheless, such a dichotomy has often been assumed in archaeological approaches of early farming communities of Europe, for example, in the "Domus-Agrios" distinction put forward by Hodder (1990) and applied by Perlès (2001) to Early Neolithic archaeobotanical material from Greece, lacking wild plant remains. Crucial turning points in this relationship in regard to Old World cereals and pulses that became domesticated concern (1) the transition from depending on wild plants, as they grow wild in nature, to controlling their reproduction by keeping seed and sowing it and (2) the sowing of this seed in regions where these were not indigenous (Harlan and Zohary 1966; Jones et al. 1998).

In this discourse, the idea of colonizers bringing with them a set crop package from the east led to a conceptual dichotomy between gatherer-hunters and farmers (see Valamoti 2015a for a critical discussion). The debate on the processes through which agriculture emerged in Greece is still lively and ongoing and relevant to the nature of interaction between plants and people in the region.

Consuming domesticated crops has often been considered the opposite of harvesting and consumption of wild plant foods, corresponding to two well-defined and distinct worldviews, that of farmers on one hand and gatherer-hunters on the other. The roots of such a dichotomy can be seen in ancient Greek thought (see Vidal-Naquet 1983) as well as narratives about the emergence of agriculture in the Near East (e.g., Ammerman and Cavalli-Sforza 1971; Cauvin 1994; Hodder 1990). The relationship between plants and people, however, is more nuanced than the wild-domesticated dichotomy. The continuum of this relationship ranges from harvesting wild plants to cultivating domesticated plant species (Jarman 1972; Jarman et al. 1982; Harris 1989, 1996). Tending wild plants becomes a form of management, with plants being nursed like children: archaeological and anthropological literature offers examples of farmers who do not perceive fields as opposed to wild vegetation (Hastorf 1998; Ingold 1996). Such a scenario has recently been questioned with agriculture being regarded as a drastically different way of perceiving and interacting with nature, a conscious choice rather than the unconscious outcome of millennia-long cultivation of wild cereals and pulses (Abbo and Gopher 2020).

Identifying incipient cultivation of cereals and pulses cannot be achieved on the basis of specific, morphologically diagnostic features preserved on archaeobotanical remains. As there is no way to distinguish between the harvest of wild or the cultivation of these wild plants (Hillman and Davies 1992; Hansen 1992), it is not easy to recognize archaeobotanically human societies that cultivate morphologically wild cereals. A means of identifying this stage is the coexistence of wild and domesticated forms in the same archaeological contexts, in proportions different from those encountered in natural wild stands (Hillman and Davies 1992; Nesbitt 2002; Tanno and Willcox 2006a). The presence of weed seeds in archaeological deposits containing morphologically wild cereals and pulses is another means of identifying incipient cultivation (Hillman 2003; Colledge et al. 2004; Willcox et al. 2008). Some scholars have regarded both criteria with great skepticism (Abbo et al. 2012). None, however, of these criteria established for detecting incipient cultivation for the Near East have been observed in the archaeobotanical record from Mesolithic Greece. Nor are there indications for the presence of wild progenitors of domesticated wheats in Greece deriving from the archaeobotanical record or modern plant distribution (Harlan 1976) (Figure 1.2). Barley and some pulses, however, could have been locally brought into cultivation. Barley grows wild in Greece and so do some pulse species such as *Lens nigricans* and various *Lathyrus* species.

Grass pea (*Lathyrus sativus*) is rarely encountered among Neolithic finds from the Near and Middle East (Zohary and Hopf 2000), and although not among the crops of the so-called Neolithic package, it is very common among Neolithic finds from Greece and Bulgaria (Halstead and Jones 1980; Zohary and Hopf 2000; Valamoti and Kotsakis 2007; Marinova and Valamoti 2014). This pulse species may have been domesticated in Europe, the Aegean in particular (Kislev and Bar-Yosef 1988; Kislev 1989b). This suggests divergences from the "crop package" as chickpea is nearly absent not only among early sites but throughout the Neolithic of Greece, whereas grass pea is very common (Valamoti and Kotsakis 2007). The contribution of elements from local vegetation in Greece and the Mediterranean in general in the shaping of the range of cultivated species of the Neolithic seems to be the case.

According to Hansen (1992), fully domesticated plants appear suddenly at Franchthi, after an abandonment prior to the Neolithic, and thus these are introduced as a package from the Near East. The evidence presented as proof

Figure 1.2. Wild einkorn growing in the region of Antalya, Turkey, July 2019. The dehiscent ear leads to spikelets with a distinct clean scar that allows the archaeobotanical identification of wild cereals. (Photograph courtesy of Shahal Abbo.)

for an introduced way of life could be used, however, as evidence for indigenous processes, too. Some form of wild lentil, locally available such as *L. nigricans*, a wild lentil species distributed around the Mediterranean (Ladizinsky et al. 1983; Ladizinsky 1989), could have been domesticated and the same could have applied to wild barley. Both a monophyletic (e.g., Badr et al. 2000) and a polyphyletic (e.g., Molina-Cano et al. 2005; Pankin et al. 2018; Jones et al. 2013) origin of domesticated barley have been suggested on the basis of DNA analyses performed on modern barley accessions, including wild barley and various cultivated. According to one scenario based on DNA studies conducted on modern barley landraces, there are at least three different centers of origin of barley cultivation in western Asia, indicating that barley spread to southeastern Europe along different diffusion paths (Jones et al. 2013). If the wild barley and lentils found at Franchthi are not intrusive from later layers, the harvesting of seeds from the wild plants that later appear as domesticated is attested; this raises the possibility that local wild barley populations might have played a role in barley crop evolution under domestication and diversity in barley landraces that would have developed in these parts of Europe where agriculture spread first from the Near East. One might even hypothesize that the wild barley present in Greece might have contributed to the genetic stock of the domesticated barleys of Europe. Yet, the coastal plains where base camps and contact points with the eastern coasts of the Aegean might have been located, of key importance for exploring the origins of agriculture in Greece, are missing from the available records due to the rise of sea level taking place in the first millennia of the Holocene (Van Andel and Shackleton 1982; Perissoratis and Conispoliatis 2003).

The role of local Mesolithic groups in the process of transition to agriculture in the Aegean and southeastern Europe appeared in narratives arguing that local populations had good knowledge of the landscape and its resources (Dennell 1983; Kotsakis 2001). The role of local gatherer-hunters' interacting with the new-coming, farming populations, something that contributed to the characteristics of the Greek Neolithic, has also been recognized by Perlès, who saw a connecting thread between farmers and gatherer-hunters in the domain of the hunting apparatus and local knowledge of lithic raw materials (Perlès 2001:50–51): the Mesolithic tradition of making projectile heads and preferring the use of obsidian from Melos for tool making is also encountered in Early Neolithic settlements. Local hunter-gatherers would have been familiar with the wild flora of Greece. Thus, when contacts with farmers were established, some of the wild-growing local plants could have been domesticated as an outcome of this gatherer-farmer interaction; this would have generated complex histories at the crossroads of cultural and natural processes, merging elements from both sides of the Aegean Sea.

Further arguments for continuity between the Mesolithic and the Early Neolithic at Franchthi have been inferred on the basis of the continuous use of the mollusk *Cerithium vulgatum* from the Paleolithic through to the Neolithic (Dennell 1983; Halstead 1996; Dilcock 2001). Visits to the cave or continuous habitation over millennia by human groups, at times for the burial of ancestors (Cullen 1995), might have attributed to Franchthi Cave the character of a point of reference in space, a natural feature of the landscape acquiring symbolic significance and monumentality (Dilcock 2001) transcending our cultural periodizations, underlining further a suspected continuity between gatherer-hunters and first farmers.

Recent literature coming from different lines of investigation highlights the shared elements and contacts between the east and western shores of the Aegean before the emergence of farming settlements in Greece, which are evidenced in vegetation cover, networks of obsidian circulation, and lithic technologies (Asouti et al. 2018; Reingruber 2018; Kaczanowska and Kozłowski 2018). Moreover, recent more nuanced approaches of architecture and artifactual evidence suggest that in northern Greece elements of gatherer-hunter societies characterize the first farmers in the Aegean (Kotsakis 2019) while variable trajectories in the socioeconomic organization and choices of these early farming communities of Greece are beginning to emerge from more contextual approaches to settlement and landscape interaction during the Early Neolithic (Souvatzi et al. 2021). This ongoing research confirms earlier suggestions that "patterning will probably be more diverse and mosaic-like than monolithic and package-like" (Watson 2003:37), unlike former assumptions that a "highly structured package" arrived in a nearly empty landscape (Runnels 2003:124, 126). Yet, even Runnels acknowledges a more complex picture in certain areas where his surveys revealed Mesolithic habitation such as the Argolid and Epirus (Runnels 2003:126).

One or Several Packages?

Inferences about the processes leading to the emergence of agriculture in the Aegean rely heavily on Mesolithic and Early Neolithic archaeobotanical remains and comparisons with the Near East. Perlès argues that domesticated crops appeared all together as a "package" and thus could only have been brought to Greece with the establishment of farming villages by farming populations arriving from the east; at the same time, she recognized that different itineraries could have been followed by different groups moving from Anatolia and the Levant to Greece (Perlès 2001). This would imply variations within the "package." The notion of a "package" of domesticates arriving in Greece has been questioned on other grounds: divergences from the Levantine situation in terms of crop species have been pointed out (Valamoti and Kotsakis 2007) as well as of regional variability in preferred cereal crops visible in the archaeobotanical record of Greece (Valamoti 2001a, 2003, 2004, 2015a, 2017). Divergences along the path from Asia to Europe have been attributed primarily to environmental factors (Coward et al. 2008 and more recently Ivanova et al. 2018). Yet, more complex processes were likely underway related perhaps to different origins of the first farmers (Valamoti 2001a, 2004, 2017) as well as to different trajectories in the establishment of agriculture (see Perlès et al. 2013).

Of the crops reaching Greece in the Early Neolithic, the new glume wheat type now identified as *T. timopheevii* is especially interesting as it highlights the strong possibility of different paths being followed in the spread of agriculture to Europe. *T. timopheevii* identified from Early Neolithic deposits in Greece, such as Giannitsa B and Mavropigi-Fillotsairi, must have been introduced along paths that connected areas of western Asia and the east Mediterranean with southeastern Europe. Connections between northwest Anatolia and northern Greece have been established on the basis of genetic analyses of Early Neolithic populations, and there are indications that such genetic connections may already have existed between Greek Mesolithic and Anatolian early farming populations (Hofmanová et al. 2016). Further analyses suggest a genetic input from European gatherer-hunters to early farming populations from southeastern Europe (Mathieson et al. 2018). Archaeobotanical finds from Yeni Kapi near Constantinople (6400–5400 BC, Ulaş and Fiorentino 2020) show that this wheat species dominates those assemblages as is also the case with some Early and Late Neolithic sites from northern Greece. As the geographical distribution of its most likely wild progenitor, *Triticum araraticum*, lies to the east of the Fertile Crescent, its introduction to the Aegean might have followed different paths within the wider framework of the spread of agriculture. The Iranian Neolithic and Caucasus gatherer-hunter gene flow detected in Neolithic and Early Bronze Age populations of Greece (Clemente et al. 2021) shows that variable paths and complex networks were involved in the spread of crop plants and human genes with the spread of agriculture to Europe: *T. timopheevii* might have moved westward with women carrying their ancestral crops to new homelands through marriage. Indeed, its variable geographic distribution during the Middle and Late Neolithic of Greece could be pointing to different cultural groups of agriculturalists bringing along their ancestral cereals.

If the selection of cereal species reflects identity and origins of the first groups of farmers arriving to southeastern Europe as previously suggested (Valamoti 2004, 2017; also Chapter 10), the dominance of *T. timopheevii* at certain Neolithic sites in northern Greece might indeed reflect diverging culinary identities among early farming groups: northwestern Anatolian groups included this wheat in their "crop package," unlike populations of the south where emmer and free-threshing wheat may have been preferred as they are prominent in archaeobotanical assemblages. Similar trajectories were probably followed in different parts of Europe. In Spain, for example, it is the free-threshing wheats that appear to dominate together with emmer wheat (Buxó 2007; Peña-Chocarro 2007; Zapata-Peña 2007) and not einkorn, which is very prominent in the north of Greece and parts of Bulgaria (Marinova 2007; Valamoti and Kotsakis 2007). Relevant to this suggestion inferred from Early Neolithic cereal species are recent ^{14}C dates from Greece that also suggest different paths that might have been followed by the first farmers establishing themselves on mainland Greece (Maniatis 2014; Perlès et al. 2013). In the future when earlier Neolithic archaeobotanical data, directly dated, become available, it might be possible to investigate more closely this hypothesis, that is, that different groups of farmers arrived from the east with different origins reflected in the selection and combination of the cereal and other crops they brought along with them. This would require a rich archaeobotanical dataset from both sides of the Aegean Sea.

The new way of farming life involved a shift from consuming only wild plant foods to also including the consumption of cultivated domesticated plants. Some of the species that today, and possibly in the Neolithic, were native in regions of the Mediterranean, could have been locally domesticated and incorporated in the new way of life we describe by the term "Neolithic," transforming and enriching agriculture and its products in its journey from Asia to Europe. After all, foods, among other goods, may well have circulated between communities of the Aegean and the Near East as is indicated by a range of artifactual evidence.

Until the discussion is enriched with new archaeological (including archaeobotanical) data and their publication, both from Mesolithic and Early Neolithic sites, a need emphasized by various scholars (e.g., Efstratiou 1996; Kotsakis 2003, 2007; Efstratiou 2007; Reingruber 2018; Galanidou 2011), questions on the shaping of the basic elements of Neolithic cuisine will remain open. Despite the paucity of archaeobotanical data for the crucial centuries marking the transition to agriculture in the Aegean, recent evidence from northern Greece and Turkey as well as radiocarbon dates from early sites in different regions of Greece lend support to the idea that different trajectories and itineraries were followed in the spread of cereal and pulse domesticates in our study area. The main questions in the discussion are why the prevailing gathering-hunting lifestyle changed, what caused this change, and through which processes a new way of life was established in the region.

Wild plant foods, however, unlike stereotypical narratives about the Early Neolithic, were collected, and, contrary to what is generally believed, local gatherer populations, familiar with wild plants growing in the region, could have contributed with knowledge on wild plant food resources. Some of these later became domesticated, such as fruit trees such as the grapevine and the almond. It is also possible that this interaction would have led to the domestication of crops different from the elements of the crop package, such as grass pea. Exploring the range of crop species in greater detail in the future might also reveal possible connections between Early Neolithic farming villages of Greece and populations in the Near East and western Asia as indicated by the geographic distribution of the various wheat crops of Neolithic Greece.

Thus, archaeological discourse for the emergence of the Greek Neolithic has increasingly become more nuanced, recognizing the contribution of local gatherer-hunter communities, elusive though they may be in regard to the plants they used (see Valamoti and Kotsakis 2007 and Kotzamani and Livarda 2018 for a recent review of the flimsy data). What was already put forward by Halstead (1996), on rather limited evidence, for complex processes of establishing domesticated crops in prehistoric Greece with different trajectories being followed in different parts of the country, seems to still hold.

2

Prehistoric Cereal Foods

Cereals involving different species in different regions have formed staple foods for human populations around the globe. Going back in times for which written sources are available, "cereals" and "bread" are terms equivalent to civilized life, a gift of the Gods in ancient Greek texts (Valamoti et al. 2022a). Ancient recipes offer a wide range of food preparations (Valamoti et al. 2022a) into which cereals could have been transformed, and our only means to approach them for prehistoric times lie in the ashes of hearths, the burned storerooms where grain was kept in containers, occasionally preserving intriguing remains of specific cereal foodstuffs. These charred remains of cereal food preparations potentially hold the key to decipher prehistoric recipes. Cereals, a staple food in modern and prehistoric times alike, are the starting point in the exploration of plant foods from prehistoric Greece. The following sections examine the cereal species that fed societies of the Neolithic and Bronze Age and, through special finds corresponding to actual food remains, offer glimpses of past culinary practices related to cereal foods.

Ancestral Cereals, Cereals of Choice, and Newcomers

The cereal genera identified from Neolithic sites are wheat and barley. Both glume wheat and free-threshing cereals have been identified, the former having their grain tightly surrounded by glumes, the latter loosely (see further on for a detailed description). Wheat is present in nearly all sites investigated archaeobotanically (Appendix A:Table 2.1) evidenced through a variety of wheat species: einkorn wheat (*T. monococcum* L.) (Figure 2.1), emmer wheat (*T. dicoccum* Schübl.) (Figure 2.2 and Figure 2.3), the new glume wheat type recently identified as *T. timopheevii* Zhuk (Figure 2.4 and Figure 2.5) (Czajkowska et al. 2020), and bread or/and durum wheat

(*T. aestivum/durum*) (Figure 2.6 and Figure 2.7). The new glume wheat, now *T. timopheevii*, was first identified in Greek Neolithic and Bronze Age material from northern Greece (Jones et al. 2000) and Bronze Age Austria (Kolher-Schneider 2003).

Free-threshing wheat, that is, durum and/or bread wheat, is present in samples from Greece but does not dominate Neolithic deposits. Rachises of durum/bread wheat have been found at sites of the Early Neolithic, for example, at Giannitsa B in the region of Macedonia, as well as of Late Neolithic (e.g., Makri in Thrace and Arkadikos in Drama plain), whereas few grains among thousands of einkorn are reported from Sitagroi (Sitagroi II) (Appendix A:Table 2.1). A rich concentration of grains is mentioned from

Figure 2.1. Einkorn wheat (*Triticum monococcum*) charred grains from Dikili Tash, 4300 BC. (Photograph by Pavlos Lathiras. Courtesy of PlantCult.)

Figure 2.2. Emmer wheat (*Triticum dicoccum*) charred grains, overview, from Mavropigi-Fillotsairi, Early Neolithic. (Photograph by Stavros Dimakopoulos. Courtesy of PlantCult.)

Figure 2.3. Emmer wheat (*Triticum dicoccum*) charred grains, detail, from Mavropigi-Fillotsairi, Early Neolithic. (Photograph by Soultana Maria Valamoti. Courtesy of PlantCult.)

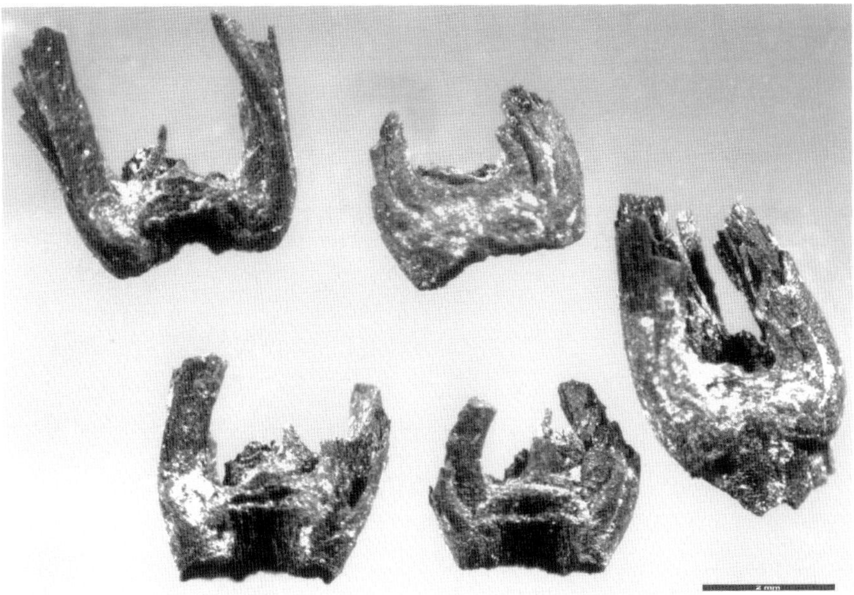

Figure 2.4. *Triticum timopheevii* from Neolithic Greece: spikelet forks from Vasilika Kyparissi, mid-sixth millennium BC. (Photograph by Pavlos Lathiras. Courtesy of PlantCult.)

Figure 2.5. *Triticum timopheevii*, probable grains from Late Neolithic Kleitos, Greece. (Photograph by Stavros Dimakopoulos. Courtesy of PlantCult.)

Figure 2.6. Free-threshing wheat (*Triticum aestivum/durum*) grain from Late Neolithic Kyparissi. (Photograph by Stavros Dimakopoulos. Courtesy of PlantCult.)

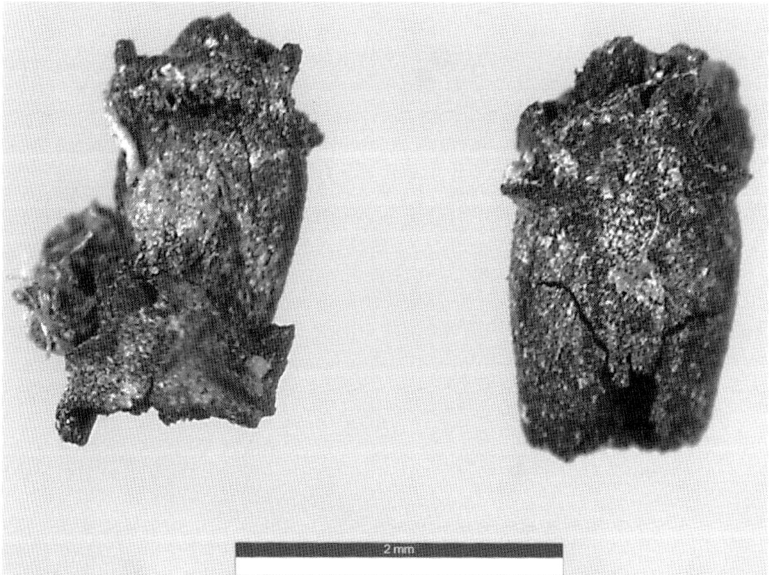

Figure 2.7. Free-threshing wheat (*Triticum aestivum/durum*) rachis from Late Neolithic Kyparissi. (Photograph by Pavlos Lathiras. Courtesy of PlantCult.)

the Neolithic layers of Knossos in Crete (Evans 1964). Likewise, from Makri there are indications of processing free-threshing wheat (Valamoti 2004). Leaving the finds from Knossos, Koufovouno, and Makri aside, the presence of free-threshing cereals in the Neolithic could be described as limited, especially in the north. Another free-threshing cereal, barley (*Hordeum vulgare*), is nearly omnipresent (Figure 2.8 and Figure 2.9) (Appendix A:Table 2.1). At sites with adequate archaeobotanical preservation, both two-row and six-row barley have been identified among grains and rachis remains (e.g., at Arkadikos Dramas) (Valamoti 2004). Both hulled and naked varieties of barley have been identified at prehistoric sites from Greece.

Based on species composition of archaeobotanical samples from the Neolithic in northern Greece, the glume wheats (einkorn, emmer, and new glume wheat *T. timopheevi*) dominate the assemblages as they are found in rich concentrations of seeds and/or glume bases of these cereals, often unmixed (Valamoti 2004, 2006a). Free-threshing wheat, *T. durum/turgidum* or *T. aestivum*, is present in few cases only (Appendix A:Table 2.1). A recent synthesis of the evidence based on Late Neolithic material from northern Greece, derived from extensively and systematically sampled settlements, confirms this picture both on the basis of absolute counts and frequencies (Valamoti

Figure 2.8. Barley (*Hordeum vulgare*) grain, naked variety, Final Neolithic Dikili Tash. (Soultana Maria Valamoti)

Figure 2.9. Barley (*Hordeum vulgare* grain), hulled variety, Middle Bronze Age Thessaloniki Toumba. (Soultana Maria Valamoti)

et al. 2022b). Taphonomic factors that could have influenced the picture obtained from the archaeobotanical material include crop processing, the ways crops were handled and used at the site, and the depositional and postdepositional processes that influence the chances of survival of different plant species in archaeological deposits, such as differential preservation when in contact with fire, different burial environments, and variable retrieval methods (see, e.g., Dennell 1976; Jones 1983; Boardman and Jones 1990; Hillman 1981). Thus, a skeptic might have argued that this prevalence of glume wheats is the result of taphonomic biases influencing the components likely to survive in the archaeological deposits, compared with the entire plant spectrum used by the inhabitants of a site (see Dennell 1976).

The taphonomic factors that could have influenced the presence of glume wheats in the archaeobotanical record differ from those affecting free-threshing cereals, as the processing steps required for obtaining clean grain from the plants are different (see detailed description further on). The limited presence of free-threshing cereals, that is, bread or durum wheat, could be attributed to different ways of processing: the glumes of glume wheats tend to be removed at the very last stage of crop processing, usually stored in spikelets (see, for example, Valamoti 2004), and therefore have more chances to reach household fires on a regular basis. Moreover, free-threshing cereal by-products are more vulnerable to fire and therefore have limited chances of preservation in the archaeobotanical record (Boardman and Jones 1990). There is, however, no rule of thumb for which wheat species have a better chance of preservation. Free-threshing chaff, which is removed early in the processing sequence, if mixed with dung to prepare dung cakes used as fuel, would be equally likely to be preserved in deposits of spent fuel.

There are several reasons why the near absence of bread/durum wheat during the Neolithic in the north of Greece is not an artifact of differential predepositional treatment of the crops: the paucity of the free-threshing wheat remains is encountered both in "refuse" contexts and storage areas where crops have been charred during destruction by fire (e.g., Mandalo and Dikili Tash) both among grain and chaff-rich finds. It is unlikely, therefore, that differences in processing have introduced a bias against the representation of free-threshing cereals. Free-threshing wheat in prehistoric Greece is generally underrepresented both among stored products and processing waste by-products; therefore, taphonomic biases related to differential processing of these groups of cereals (see Dennell 1976) are unlikely to be the cause of the limited presence of free-threshing wheat that has been observed for the Neolithic (see also Valamoti 2004). Moreover, there seems to be a geographical gradient with free-threshing wheat being more common in the south of Greece, as it has been found at Crete at the site of Knossos dated to the Early Neolithic and at Kouphovouno in the Peloponnese, dated to the Middle Neolithic (Vaiglova et al. 2014).

Among the glume wheats it has been argued that einkorn wheat is a dominant crop at many Neolithic sites of northern Greece as well as at sites farther south (Valamoti 2001a, 2003, 2004). This picture is more clear-cut at sites where a large number of samples have been retrieved from different parts of an excavated area, and thus less likely to result from a partial representation of a site. This is especially impressive at Makriyalos, Makri, and Apsalos in northern Greece, where nearly all the samples are dominated by einkorn glume bases and spikelet forks (Valamoti 2004). At Dikili Tash, einkorn wheat was regarded as the sole wheat species stored among the Neolithic houses of phase II, which were destroyed by fire around 4300 BC (Valamoti 2004). Nevertheless, more recent finds suggest that emmer was also in store (Valamoti, work in progress) as seems to have been the case at earlier phases of the site (Matterne 1993; Kokkidou 2018). At Sitagroi (Renfrew 2003) and Toumba Balomenou (Sarpaki 1995) einkorn again prevails.

Recently published data from the region of western Macedonia at sites such as Apsalos (Valamoti 2006a), Toumba Kremastis Koiladas (Karathanou and Valamoti 2011), and Kleitos (Valamoti and Stylianakou 2015) indicate the dominance of *T. timopheevii* together with einkorn among certain settlements or households in the Middle Neolithic of northern Greece, whereas it is present in low proportions in Early Neolithic Giannitsa B and Mavropigi-Fillotsairi (see also Chapters 1 and 10). At the recently excavated site of Vasilika Kyparissi (Pappa and Nanoglou 2019), *T. timopheevii* appears to dominate the assemblages where glume wheat chaff is present, while grains of this species may also be present (Lathiras 2020). This recent evidence from Greece is in accordance with further new archaeobotanical data from sites in Serbia (Filipović and Obradović 2019) where rich concentrations of *T. timopheevii,* dated to the Late Neolithic, occur. Early Neolithic finds from neighboring regions come also from the site of Yeni Kapi in Turkish Thrace (Ulaş and Fiorentino 2020).

Thus, although the archaeobotanical record of northern Greece shows a predominance of einkorn at several sites, more recent evidence allows for a more nuanced picture of different species dominating different sites or small regional catchments. If glume wheat species' distribution is considered on an intrasite level, again spatial differences become visible. This can be observed at the site of Makriyalos where different glume wheat species dominate the assemblage. In most cases, einkorn dominates, but, in others, a different cereal species, either emmer or *T. timopheevii,* dominates (Valamoti 2004:Figures 7.24, 7.25, 7.28).

The evidence from the north of Greece alters the picture generated by Hansen who considered emmer wheat and barley as the dominant species throughout the Neolithic and the Bronze Age (Hansen 2000). This picture is probably resulting from the fact that most of the sites Hansen takes into consideration (47 out of 54) (Hansen 2000) originate from areas south of Macedonia and Thrace and consist of few and mainly poor samples. Megaloudi (2006), in her book, points out a prevalence of emmer in the south of Greece. Indeed, this is obvious at sites, especially in Thessaly, and it may well reflect different preferences in cereals pertaining to the south. Recent finds from the Peloponnese and Crete seem to support a predominance of emmer and free-threshing wheat in the south, at Koufovouno (Bogaard et al. 2013; Vaiglova et al. 2014) and Knossos (Nitsch et al. 2019). Thus, although at present the evidence from the south is not as rich as that from the north, some patterns begin to emerge. Work in progress within the project PlantCult aims to complete the database for prehistoric Greece and fully explore such patterns that may relate to regional crop variation, inevitable choices and/or preferences.

The dominance of einkorn in northern Greece at several Neolithic and Bronze Age settlements as well as at sites farther south (as is the case from Toumba Balomenou, something Sarpaki calls "cultural traditionalism" (Sarpaki 1995) undermines the emphasis encountered in the literature on emmer wheat as part of the Neolithic package and its arrival in Greece and elsewhere in Europe. This is probably an indication of internal variability in the "package" as it "exited" the Fertile Crescent. In fact, it seems more likely, in light of recent archaeobotanical research, that several "packages" may have spread westward. A similar situation may hold true for the Near East. Despite the emphasis placed on emmer for Neolithic societies of the Middle East and for the spread of agriculture in Europe (Zohary 1996), it seems that the role of einkorn had been underestimated in discussions concerning the origins of agriculture. In tables provided by Nesbitt (2001, 2002) based on archaeobotanical remains of the Middle East, wild einkorn appeared to be more frequent than wild emmer, and the predominance of domesticated emmer was overestimated at the time (nineteen sites have domesticated einkorn and twenty-four emmer) (Nesbitt 2001). Further research in the Near East over the last two decades has emphasized the complexity of the domestication process and the multilocal domestication of what has been described as the "Neolithic package," a term that was widely used and unquestioned for decades (see Valamoti and Kotsakis [2007] for a critical approach of the term).

So, if einkorn and *T. timopheevii* dominate in absolute numbers the Neolithic of northern Greece, and emmer and bread wheat that of the south, what are the possible underlying factors? Since emmer managed to adapt and become a basic food species in Britain, in a climate much less suited to its cultivation than that of Greece (Van der Veen 1992), it is unlikely that environmental reasons determined this specific preference observed at the Greek sites. It is possible that cultural, culinary preferences, perhaps relevant to the origins of the inhabitants of each region, might underlie this pattern, perhaps corresponding to the wheat of ancestors, culturally defining different regions of the Aegean (Valamoti 2001a, 2003, 2017).

This pattern that begins to emerge could be related to different places of origin of the first farmers of northern and southern Greece (see Chapters 1 and 10 for further discussion). As more evidence on *T. timopheevii* from other parts of Europe (e.g., Rottoli and Pessina 2007; Toulemonde et al. 2015; Kenéz et al. 2014) and more recently Serbia (Filipović and Obradović 2019) and northwest Turkey (Ulaş and Fiorentino 2020) is gradually accumulating, this species emerges as a crop in its own right. Future syntheses tracking its occurrences in space and time might prove an interesting line of investigation in the wider exploration of the spread of agriculture to Europe and the formation of culinary identities (see Valamoti 2017).

The prevalence of einkorn is regionally visible in the Neolithic of northern Greece and parts of Bulgaria (Marinova and Valamoti 2014), while emmer and free-threshing wheats are preferred in other parts of the Mediterranean (Valamoti 2004; Marinova 2007; Buxó 2007; Peña-Chocarro 2007). This long "einkorn tradition" in parts of prehistoric Greece may be reflecting origins from ancestors who domesticated or brought einkorn to Greece. This hypothesis can only be tested when the archaeobotanical record from Early Neolithic sites from western Asia and Europe is significantly enriched. Different itineraries were probably followed by early farmers, and through these itineraries, different combinations of domesticated cereals moved across southeastern Europe and the wider Mediterranean basin (see Valamoti 2001a:187–188). Thus, the suggestion by Perlès of "multicultural pioneer groups . . . of (adventurous) individuals, . . . having followed different pathways from their original ancestral homes up to Greece," based on possible sea routes and a coarsely drawn picture of Early Neolithic cereal crops of Crete and mainland Greece (Perlès 2001:62), is now further enhanced by the available archaeobotanical data of the last twenty years (see Appendix A:Table 2.1).

If we consider einkorn, based on the above, as a "group identity signifier" (see Hamilakis 2000) for the inhabitants of broad regions such as northern Greece, emmer wheat might represent something different: at Mandalo (Valamoti and Jones 2003) there is a lot more emmer wheat than einkorn, while at Makriyalos and Makri emmer corresponds to a very small proportion among the rubbish that contains remains of glume wheat de-husking by-products (glume bases and spikelet forks), while at Apsalos *T. timopheevii* wheat is prominent. This differentiation seems unlikely to be the result of different microclimates at these sites, favoring different species. After all, both einkorn and emmer are very hardy wheat species in regard to soil conditions and climate (Christides 1963). Emmer wheat as well as new glume wheat type might have characterized food choices of certain settlements of Greece as part of cultural differentiations that are difficult to disentangle for the time being. Such generalizations, however, require a systematic quantitative analysis of a large body of data, as a different picture can be obtained depending on how crop choice is assessed from the archaeobotanical record: absolute numbers, concentration values, or ubiquity (see Valamoti et al. 2022b for a regional overview of Neolithic northern Greece using ArboDat).

A similar differentiation is detected in the Neolithic of Bulgaria: einkorn wheat dominates the Struma valley and north Bulgaria while emmer the Thracian plain (Marinova and Valamoti 2014). Perhaps einkorn was the traditional wheat of northern Greece, which was initially cultivated by the first farmers in the region who continued to grow it for millennia, while emmer was the wheat of the "others," a species that might have been accumulated and exchanged for status, or a species that was consumed in specific circumstances or by certain members of the community or for the preparation of specific dishes (see also Chapter 10 for a discussion). It becomes evident that archaeobotanical research has the potential to ask questions about preferred foodstuffs or foods to which access was limited, and the ongoing assembling of many, good-quality archaeobotanical data will soon enable us to map "culinary" regions of prehistoric Greece.

Glume wheats and einkorn, in particular, continue to be the dominant cereals during the Bronze Age (Appendix A:Table 2.2), in storage contexts, at sites in northern Greece (Kroll 1983; Jones 1983; Halstead 1994) with impressive caches of stored einkorn spikelets encountered at Archondiko, Assiros Toumba, and Kastanas. At Assiros Toumba, despite the wide range of cereal crops stored in the central storeroom, it is einkorn that predominates (Jones et al. 1986). The evidence from southern Greece is more fragmentary; it is therefore not so easy to distinguish a regional dominance of a specific species. The dominance of einkorn wheat, mainly in Late Neolithic deposits, both in storage and rubbish contexts, can be used to rule out, on safe grounds, a primary use as fodder. As the three wheats are all glume wheats, their reaction to charring conditions would have been more or less the same, thus, it is unlikely that the predominance of einkorn is due to biases introduced from crop processing for consumption and processes of deposition of plant remains in archaeological deposits. Thus, for various reasons (archaeobotanical composition, archaeological context, and spatial distribution), it is unlikely that the prevalence of einkorn is the outcome of taphonomic biases (Valamoti 2004). Einkorn was probably the staple cereal food par excellence for the inhabitants of several settlements of Greece, the north in particular, both in Neolithic and Bronze Age times, yet, inter- and intrasite variability, although difficult to detect, is in fact present and potentially very informative regarding cultural choices, access to specific crops and landraces, social differentiation, and identity.

During the course of the Bronze Age and increasingly so in later periods, free-threshing wheat is encountered more frequently in rich concentrations (Figure 2.10). Among the Early Bronze Age layers at Sitagroi, a considerable number of free-threshing wheat rachis, indicating processing for grain extraction, is reported. In the Late Bronze Age storerooms at Assiros, stored grains of naked wheat have been found (Figure 2.11), and aDNA analysis showed that they correspond to a hexaploid wheat (Brown et al. 1998).

Figure 2.10. Free-threshing wheat (*Triticum aestivum/durum*) grains from Archondiko Giannitson, end of third millennium BC. (Photograph by Stavros Dimakopoulos. Courtesy of PlantCult.)

Figure 2.11. Free-threshing wheat (*Triticum aestivum/durum*) grains from Assiros Toumba, Late Bronze Age. (Photograph by Glynis Jones. Courtesy of Glynis Jones and Ken Wardle.)

Free-threshing wheat seems to take over in the historical periods, perhaps as more suitable to sustain the increasing populations of towns and cities (Livarda 2012; Valamoti et al. 2018).

During the Bronze Age two more cereals appear, spelt wheat (Figure 2.12 and Figure 2.13) and millet (Figure 2.14 and Figure 2.15). Spelt wheat, also a glume wheat, has been found farther north than Greece at sites in central Europe dated to the fourth millennium BC (Zohary and Hopf 2000; Blatter et al. 2004; Akeret 2005; Kohler-Schneider and Caneppele 2009), in contexts earlier than the northern Greek finds. Thus, suggestions that spelt wheat appears in northern Greece around the time it occurs as an established crop in central Europe (Valamoti 2007a and 2007b; Valamoti et al. 2008a) are no longer valid. The presence of spelt grains and glume bases (the latter being far more reliable for species identification) leaves little doubt that spelt was cultivated at Archondiko in the end of the third millennium BC. The Late Bronze Age Assiros storerooms, destroyed by a conflagration episode around 1350 BC, have yielded both spelt grain and glume bases in rich concentrations, pointing toward storage in spikelets (Jones et al. 1986). There is very sparse evidence for the presence of spelt wheat in the south of

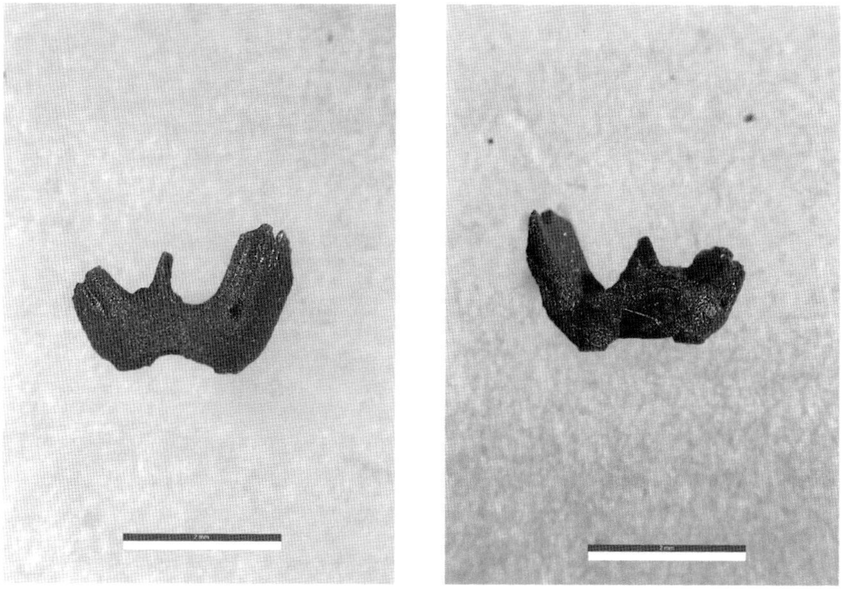

Figure 2.12. Spelt wheat (*Triticum spelta*) spikelet fork from Archondiko Giannitson. (Photographs by Pavlos Lathiras. Courtesy of PlantCult.)

Figure 2.13. Spelt wheat (*Triticum spelta*) grains from Archondiko Giannitson. (Photograph by Stavros Dimakopoulos. Courtesy of PlantCult.)

Figure 2.14. Millet (*Panicum miliaceum*) seeds, perhaps the oldest in Europe, from Early Bronze Age Skala Sotiros, Thassos island. (Soultana Maria Valamoti)

Figure 2.15. Millet (*Panicum miliaceum*) seed agglomeration from Angelochori, Late Bronze Age. (Soultana Maria Valamoti)

Greece, and these are temporally restricted to the Late Bronze Age, consisting of very few seeds or glume bases at sites such as Tiryns and the Athenian Agora (Appendix A:Table 2.2).

At sites such as Nea Nikomedia, Servia, Sitagroi, Arkadikos Dramas, Dikili Tash, and Franchthi where storage contexts or rich, dense grain concentrations have been found, barley is well represented throughout the Neolithic (Appendix A:Table 2.1). Barley is ubiquitous as it is found at nearly all sites. Compared with the glume wheats, however, it has a more limited presence in the Neolithic, especially at sites where glume bases dominate the assemblages. Its limited representation among refuse deposits rich in glume wheat chaff (e.g., Makriyalos and Apsalos) might be attributed to a difference in handling the by-products of barley compared with the glume wheats, as well as to their comparatively limited chances of preservation when in contact with fire (Boardman and Jones 1990). During the Bronze Age barley is a very common find, encountered in rich grain concentrations at Kastanas, Archondiko, Angellochori, Agios Athanasios, Mesimeriani Toumba, Agios Mamas in the north, and Agios Vasileios in the Peloponnese. Kroll points out this dynamic presence of barley in the archaeobotanical material of Bronze Age Greece (Kroll 1983, 2003), and as will be shown, it must have constituted an important element of this period's "cuisine."

A different kind of cereal in relation to those mentioned earlier is millet. It appears in the Bronze Age, and its use as a crop seems to be confined to the Late Bronze Age. Regarding its spatial distribution, it is mainly encountered at northern Greek sites on the basis of current evidence. The presence of isolated millet seeds is known from archaeological deposits dated to the Early Bronze Age. Until now there is only one case where millet has been found in a pure, dense concentration in Neolithic layers. The find comes from Olynthus and is dated to the end of the Neolithic period (Mylonas 1929). Yet, this is an old find, and in light of the results obtained from dating of millets in Europe (Filipović et al. 2020; Motuzaite-Matuzeviciute et al. 2013), it would need to be directly dated prior to any discussion of this potentially early appearance. Millet is present in deposits of the Bronze Age at Skala Sotiros on the island of Thassos dated to 2200–2000 BC on the basis of ^{14}C dates from adjacent Celtic bean seeds found in the same destruction layer (Valamoti 2016). These millet seeds have recently been dated with ^{14}C AMS in the Poznań Radiocarbon Laboratory (Goslar et al. 2004), and their date falls around 1450–1500 BC with a probability of 95%; the same date has been obtained for millet seeds from Archondiko (Appendix B). These two cases of millet from northern Greece are the earliest finds from southeastern

Europe so far, although not as early as previously assumed on the basis of the layer where the Skala Sotiros seeds were found. They are among the earliest millet occurrences in Europe (Filipović et al. 2020). Millet occurs in nearly all Late Bronze Age sites of northern Greece where systematic sampling for plant remains has been applied, such as at Angellochori, Agios Mamas, Kastanas, and Assiros. The southernmost site where millet has been found is Kalapodi in Phthiotis. Recent evidence shows that it is absent in east-central mainland Greece at sites such as Agia Paraskevi, Kynos, and Mitrou (Ntinou et al. 2022), in an area that was part of the Mycenaean world in the Late Bronze Age. It is, however, present among archaeobotanical finds from the northernmost limit of the Mycenaean world at Mount Olympus (Clemence Pagnoux, personal communication). The relationship of millet to culinary identities during Mycenaean times in the Late Bronze Age is a challenging and exciting matter in this respect, as millet's geographic occurrences evidenced so far indicate a northern distribution, something that appears to continue in ancient Greece through the textual evidence. Although this geographical distribution may reflect the limited archaeobotanical research in the south of Greece, given its "exotic" origin it might reflect complex cultural

Figure 2.16. Oat (*Avena* sp.) grains from Late Neolithic Kyparissi Vasilika. (Photograph by Stavros Dimakopoulos. Courtesy of PlantCult.)

processes of new crop introductions and their culinary transformation (Valamoti 2016).

In regard to oats (*Avena* sp.) although sporadic seeds are occasionally mentioned in site reports, until now it has not been identified as a crop in its own right with a possible exception currently under investigation from Late Neolithic Vasilika Kyparissi in central Macedonia (Lathiras 2020; Valamoti et al. 2022b) (Figure 2.16).

Cereals as Food: Cleaning the Grain

The wheat species of the Neolithic, einkorn, emmer, and *T. timopheevii*, belong, as mentioned, to the glume wheats (Nesbitt and Samuel 1996). The grain in glume wheats is surrounded by glumes that remain intact even after threshing, unlike those of free-threshing cereals that free their grain on threshing (Figures 2.17 to 2.19). Barley, in its hulled variety, although it is a free-threshing cereal, requires extra processing for the removal of the hulls that surround the grain. The harvested crop of the glume wheats, after threshing, instead of corresponding to clean grain, as happens with the

Figure 2.17. Glume wheat. From left to right: einkorn ear, spikelets, clean grain (*top*) and chaff (*bottom*). (Soultana Maria Valamoti)

Figure 2.18. Glume wheat: einkorn spikelets and ear. (Soultana Maria Valamoti)

Figure 2.19. Free-threshing wheat ear components. Left to right: spikelets, clean grain (*top*), chaff (*bottom*). (Soultana Maria Valamoti)

free-threshing wheats we are familiar with nowadays, corresponds to grain enveloped by its glumes (see Figure 2.18). To remove the glumes from glume wheat grain, people would have had to make a considerable effort, which was most certainly also time consuming, as is observed in ethnographic accounts of glume wheat processing in Turkey (Hillman 1981, 1984), Romania (Hajnalová and Dreslerová 2010), Spain (Peña-Chocarro 1999), Ethiopia (D'Andrea and Mitiku Haile 2002; Lyons and D'Andrea 2003), and Morocco (Peña-Chocarro et al. 2009).

Hillman (1981, 1984) offered a thorough and detailed recording of the steps of glume wheat processing in Turkey. There, the removal of glumes required pounding of the spikelets in wooden mortars, subsequent winnowing for the removal of the light parts, and successive sieving for the separation of glumes from the grain. A wooden mortar is used for pounding spikelets in Ethiopia by the few families that still cultivate emmer wheat (Figure 2.20)

Figure 2.20. Pounding glume wheats with wooden pestle and mortar, Ethiopia. (From Valamoti 2009, Figure 4.21. Photograph courtesy of Catherine D'Andrea.)

(D'Andrea and Mitiku Haile 2002). In Morocco, pounding of glume wheat spikelets is carried out in a mortar that is dug out of the soil (Figure 2.21) and is lined in its interior with clay (Peña-Chocarro et al. 2009). A similar structure has been excavated at Late Neolithic Thermi B (Figure 2.22) (Valamoti 2009), perhaps used for similar purposes, whereas, on the basis of phytolith analysis a similar function has been attributed recently to pits dug out from bedrock at the site of Paliambela in northern Greece, dated to the Early Neolithic (Tsartsidou and Kotsakis 2020). The removal of glumes from the grain of the glume wheats is done in Spain using grindstones, although the use of mortars is mentioned for the past (Peña-Chocarro 1996, 1999). Winnowing of the pounded spikelets is carried out by using big "sieves" without holes in Spain and Ethiopia (D'Andrea and Mitiku-Haile 2002).

Experimental investigations have shown that pounding in mortars is much more efficient than grinding on a grinding stone for the removal of glumes from glume wheats (Meurers-Balke and Lüning 1992). The ethnographic examples do not mention any special treatment of spikelets prior to pounding. Experimental investigations, however, suggest that parching of spikelets (Hillman 1984) or, their light wetting, facilitated the removal of glumes: wetting of spikelets was applied to the experimental removal of glumes from emmer grain, carried out as part of an investigation of ancient Egyptian breadmaking using emmer grain (Samuel 1993, 1996, 2000). Samuel's experiments showed that moistening of the spikelets facilitates the removal of glumes, and I realized the same when I applied Samuel's "tip" on einkorn spikelets during einkorn de-husking for experimentation (Valamoti and Anastasaki 2007).

Deposits dominated by glume bases in Greece have been interpreted as refuse, containing among other things the by-products of processing glume wheats for cleaning grain from the surrounding glumes. This process is unnecessary when the glume wheats are used as fodder (see Nesbitt et al. 1996). Moreover, ethnographic evidence from Ethiopia suggests that the removal of glumes from emmer spikelets was such a tedious job that demanded a great deal of effort invested by women, to the extent that they complained specifically about this aspect of crop processing (D'Andrea and Mitiku Haile 2002). This observation might imply that the archaeobotanical remains of glume bases correspond to the remains of processing glume wheats as human food consumed by the Neolithic inhabitants of the settlements (Valamoti 1995).

Experimental research has shown, however, that if animals (goats in the specific experiment) consume whole glume wheat spikelets, these come out

Figure 2.21. Pounding glume wheats. In Morocco, a dug-out hole lined with plaster is used. (Photograph courtesy of Leonor Peña-Chocarro.)

Figure 2.22. Hole for possible pounding of glume wheats, from Neolithic Thermi B. (From Valamoti 2009, Figure 4.23. Photograph courtesy of Maria Pappa.)

in dung as badly preserved glume bases (Valamoti and Charles 2005; Valamoti 2013a). Therefore, the connection between glume bases and human food is not as straightforward as previously thought. It is nevertheless unlikely that the majority of the available evidence from prehistoric Greece consisting of glume bases corresponds primarily or exclusively to the remains of fodder ending in animal dung.

Transforming Cereals into Dishes: Some Insights from Ethnography and Ancient Texts

A cereal grain, be it of wheat or barley, can have different properties more suited to specific preparations than others, as one could argue based on modern perceptions about flour qualities for different products like pasta or bread. A glance into a modern bakery reveals the wide variety in which cereal flour can be transformed into bread with the multitude of types of bread available. Exploring the diversity of how cereals might have been transformed into food in prehistoric cuisines is facilitated by ethnographic accounts and ancient texts. The transformation of cereals into food with traditional technologies as well as ancient "recipes" from textual evidence can offer a heuristic device for understanding the prehistoric ways into which cereals might have been prepared for food, their additional ingredients, properties, and context of preparation and consumption, that is, whether for daily meals, religious festivals, special rituals, or meals for special occasions. Insights from periods later than prehistory and different cultural contexts need of course to be used with caution, more as a source of inspiration for broadening the range of possible ways into which cereals might have been converted into food in prehistoric times.

The ways of transforming cereal grain into food are many and correspond to different levels of processing of the clean grain ranging from hardly any transformation (e.g., clean grain) to a sophisticated preprocessed ingredient, the preparation of which needs a series of steps, prior to its cooking proper as part of a recipe (e.g., bulgur). Ancient Greek texts show a great variability in the ways cereals were transformed into food (Valamoti et al. 2022a): specific recipes required different levels of maturity of the grain or different degrees of processing of the grain by mechanical means (fine or coarse grinding and sieving) or heat (roasting, parboiling). Thus, although cereals can be mainly transformed into some form of porridge or bread, the multitude of potential recipes can be impressive for each of the cereal species identified in prehistoric Greece. Of course, these insights correspond to specific social and economic conditions that generated or favored a specific recipe, and the observed variability in ancient texts and ethnographic accounts cannot be assumed for the prehistoric past but investigated through actual food remains.

Experimentation, ethnography, and speculation are the main means to approach the ways in which einkorn and the other glume wheats were eaten in prehistoric Greece. Nowadays, einkorn is used as fodder in Turkey and Spain, or for its straw, for the production of objects like hats, pot lids, or as roof cover in Spain and Morocco (Peña-Chocarro et al. 2009, Hajnalová and Dreslerová 2010). Its consumption in Spain is connected to periods of hunger or poor families (Peña-Chocarro 1999), and the same applies to Morocco (Peña-Chocarro et al. 2009). Nevertheless, together with other cereals it is used for special ceremonial foods, prepared for festivals (Peña-Chocarro 1999). In the Rif area of Morocco, where the cultivation of einkorn still survives, it is consumed in the form of whole, boiled seeds or coarsely ground grain used in various recipes as well as in the form of flour mixed with other cereal flour in breadmaking, (Peña-Chocarro et al. 2009). A wealth of recipes based on einkorn grain are also reported from Turkey (Ertuğ 2004).

Most researchers classify einkorn wheat among these cereals unsuited for breadmaking (e.g., Le Clerc et al. 1918; Abrams 1978; McGee 1984; Vallega 1992; Christides 1963; Dalianis 1983). We should, however, be cautious in projecting our modern tastes and criteria on what is suitable or not for breadmaking, as well as on what constitutes "bread" in a prehistoric context. This wheat species is rich in protein and carotene, which gives einkorn bread a strong, dark-yellow color (D'Egidio et al. 1993; Vallega 1992; Hidalgo et al. 2006) while analyses and comparisons of einkorn to other wheat species have shown that it is superior in its nutritional properties and content of antioxidants (Brandolini et al. 2008). Einkorn breadmaking trials showed that the bread has a dark color, is tasty, but usually does not rise, unlike bread wheat, widely used in breadmaking nowadays. As these recent studies show, however, some einkorn varieties yield well-risen breads (Borghi et al. 1996; Corbellini et al. 1999). My experimentation with a mixture of einkorn and barley flour using homemade sourdough was very successful, although the bread was very different from the fluffy breads we are nowadays used to (see Chapter 10 and Figure 10.2). Bulgur, that is, parboiled, de-branned cereal grain, can be made of einkorn, as is in parts of Turkey, such as on the Black Sea coasts (Ertuğ 2004).

Emmer wheat nowadays is used in parts of Spain and Italy as fodder. In Asturias, however, the inhabitants use it for breadmaking, although the dough of this bread is considered difficult to knead and the bread is dark and flatter than other breads (Peña-Chocarro 1999). It is cultivated and sold as farro in the area of Tuscany, as a local delicacy with roots in the past. It is the cereal species par excellence for bulgur making in Turkey, both on the level of traditional agricultural production for family consumption and in industrial production (Bayram 2000). Bulgur, which traditionally is made of emmer wheat, consists of parboiled wheat that is then dried, de-branned, and ground and is used in many ways in Anatolian, east Mediterranean, Aegean, and North African cuisines (e.g., Valamoti 2011b). Nowadays, emmer is cultivated especially for making bulgur in Sinope and Kastamonu by poor farmers, on a low scale, and because of the extra processing involved in the removal of glumes, its cultivation is considered of no financial interest (Giuliani et al. 2009). Bulgur (*chondros* in ancient Greek) is also a very popular traditional food prepared in Greece today. It is prepared in the summer, as was done in antiquity, during the extremely hot "dog days," as this is necessary for properly drying the produce to avoid spoiling. Today, it is made with durum wheat (Figure 2.23), and in antiquity the use of some species of

Figure 2.23. Bulgur making at Kosmati, Grevena, September 2017. Grain is spread to dry under the sun after parboiling but before grinding. (Photograph by Georgios-Vily Kapetanakis. Courtesy of PlantCult.)

a two-seeded wheat (perhaps emmer) is reported in *Geoponika* (see Valamoti 2011b; Valamoti and Anastasaki 2007 for detailed reviews of ancient and ethnographic information on bulgur preparation and consumption).

Emmer could also be used for other recipes, for which nowadays durum wheat is considered appropriate: κολλυβόζουμο/*kollyvozoumo* is like a "cream" made with the liquid resulting from boiling wheat during κόλλυβα/*kollyva* production (the grain for the souls of the dead; see the recipe in Chapter 11), a foodstuff known as *aşure* in parts of Turkey and Palestine as well as in Greece (e.g., Hillman 1985; Psilakis and Psilakis 2001). Pulses and honey can be added as well. The tradition of boiling cereal grain to offer as food for the dead is widely seen throughout the Balkans (Mesnil and Popova 2002; Megaloudi 2004, 2006). It is thus possible that emmer or other wheat grains might have been consumed as whole, boiled grain, although whether in a daily or ritual context reserved for special occasions is impossible to tell on present evidence. The charred cereal grains found in an Early Neolithic burial at Mavropigi-Fillotsairi correspond to their ritual offering in a funerary context, yet, besides them being fully processed and clean, no traces of cooking could be observed.

In addition to modern culinary traditions, recipes for glume wheats can be researched in ancient texts. The researcher is faced with the problem of identifying the exact botanical species referred to in the ancient texts: the terms *zea*, *tiphe*, and *olyra* could generally be considered as glume wheats, yet, attempts to identify the exact species, that is, whether it was einkorn, emmer, or spelt, result in different conclusions (e.g., Amouretti 1986; Gennadios 1997; Christides 1963; Valamoti et al. 2022a).

In regard to the new cereal species used during the Bronze Age, spelt wheat and millet, new recipes might have appeared alongside their arrival in prehistoric Greece. Alternatively, the new species might have been adapted to already existing ones. Spelt wheat survives today as a crop in countries of Central Europe such as Austria and parts of southern Germany, in regions where, according to available archaeobotanical evidence, its cultivation may have begun in prehistoric times (Berihuete-Azorín et al. 2020; Mielke and Rodemann 2007). Spelt wheat has been also traditionally cultivated in regions of Spain (Peña-Chocarro 1999) and Belgium (Mol 2002). Nowadays, it is sold in supermarkets in packages either as ripe grain known as *dinkel* or as grain resulting from unripe ears that have been smoked and dried, known as *Grünkern* (Mielke and Rodemann 2007). Traditionally, it is cooked as a soup, but more recently it has been included in various recipes in an effort to promote this wheat species (Puumalainen et al. 2002). As Micha-Lampaki

(1984) mentions, *hidra* was a kind of gruel, a soup, made of unripe wheat, a wheat referred to by Aristophanes as *hidrias* (Valamoti et al. 2022a). Although this practice appears similar to the use of *Grünkern* in soups, it is impossible to determine the cereal species Aristophanes had in mind, as the use of unripe wheat in recipes could be relevant to other cereal species such as durum and bread wheat, of which *frikeh* or *farik* or *freekeh* is prepared in many regions of the east Mediterranean (Hillman 1985; Palmer 2002; Bayram 2008): ears are harvested unripe and parched, acquiring a delicious "nutty" taste (Figure 2.24 and Figure 2.25).

Durum and bread wheat nowadays constitute the cereals par excellence for making pasta and bread, respectively, the main forms in which cereals are consumed today in Greece, followed by bulgur and *trachanas* (Psilakis and Psilakis 2001). In prehistoric Greece, on the basis of available evidence, they had a limited presence, especially during the Neolithic. A free-threshing wheat might have been among the preferred wheats of the south, together with emmer. Their limited presence, however, does not preclude their use on a reduced scale also in the north. Durum wheat resembles emmer wheat when regarding its culinary properties, and we could imagine similar uses

Figure 2.24. Unripe wheat harvested for *freekeh* production, Altınözü-Hatay, Turkey, June 2019. (Photograph courtesy of Mustafa Bayram.)

Figure 2.25. Unripe wheat scorched, Altınözü-Hatay, Turkey, June 2019. (Photograph courtesy of Mustafa Bayram.)

for both species. Bread wheat, however, due to its high gluten content, might have been used in prehistoric times for the preparation of special foods.

Barley may be known today from the barley rusks available in supermarkets and local delicacy shops where traditional products from the islands of Crete and Lemnos are available. A culinary use of barley is typical of Tibet where roasted barley flour is used for the preparation of *tsampa,* a food reminiscent of ancient Greek *maza,* consumed both in daily and ritual meals (Weckerle et al. 2005; Dalby 2003:47). One can find recipes for barley bread in special cookbooks safeguarding Greek regional culinary traditions or in ethnographic accounts aimed to facilitate the comprehension of archaeological materials (Procopiou 2003) or from bread recipes with unusual ingredients (Voutsina 2000; Psilakis and Psilakis 2001). Yet, despite its marginal role as food in modern Greece, barley has played an important role as human food in different cultural contexts of the Old World (Newman and Newman 2006; Baik and Ullrich 2008). In prehistoric Greece, barley might have made a much more significant contribution to people's diet. Bear in mind that barley was used in rituals and was the main ingredient of the famous *maza* of ancient Greeks (meaning "kneaded" and made of ground barley kneaded with a liquid), a common foodstuff that in some contexts was used as an identity signifier of Greeks versus Romans who were farro (emmer) eaters (Braun 1995; for variations in *maza* preparation see Micha-Lampaki 1984 and Valamoti et al. 2022a). Hillman (1984, 1985) provides references to the use of barley in bread-baking, references that exist in Pliny, and for the use of some form of sourdough made of bitter vetch or chickpea for the preparation of barley bread.

Barley and wheat are also basic ingredients of beer-making from prehistoric times until today. The grain needs to be left to sprout, then this process is interrupted by heating, the grain is dried and ground, resulting in ground malt, the main raw ingredient for brewing. The malt is heated in water and left to ferment. Alterations induced by such processes can be recognizable on charred archaeobotanical remains that demonstrate a sucked-out surface and traces of the sprouting embryo on the barley surface (Stika 2011) as well as other features recently observed with the aid of scanning electron microscopy (SEM).

Millet, the new species introduced to the prehistoric cuisine of northern Greece during the Bronze Age, could have been used either as clean grain, boiled slightly in the form of a gruel, or ground and kneaded in some form of bread, or in the form of a fermented drink (Wells 1990). Different ways of consuming millet can be found in ancient Greek texts with tips aimed at rendering this not very popular cereal more palatable (Valamoti et al. 2022a). Although today in our study region millet is known mainly as bird feed, *Panicum miliaceum* and other *Panicum* species are still used as a staple food in different parts of the world, for example, in Africa (D'Andrea and Mitiku Haile 2002; Hamon and Le Galle 2013), in India (Jaffrey 1985), and in the Ukraine, China, and Japan. In some parts of Europe millet cultivation survives as a marginal crop, as is the case in the Iberian Peninsula where it is cooked mainly as porridge, with water or milk, as well as in breads and a variety of dishes (Moreno-Larrazabal et al. 2015). Recently, the "bio" movement has made millet seeds available in special shops in Greece where it is consumed as an alternative food by people allergic to gluten. In India, millet flour can also include chilli pepper and spinach in the preparation of pies (Jaffrey 1985), in Iran it is cooked with lentils after it has been soaked (Wells 1990), while in parts of Africa a millet beer is prepared (Goody 1998; Arthur 2003, 2014).

The preparation and consumption of a fermented millet drink are known from the east Mediterranean (Hillman 1985) as well as from the Balkans. In the 1990s one could still buy this drink (*boza*) in the pastry shop of Hatzis in Thessaloniki, on Venizelou Street. *Boza* is still prepared, from lactic acid fermentation, and sold in countries such as Albania and Turkey (Hancioğlu and Karapinar 1997). In Romania it is known as *baska*. References to *boza* between the fifteenth and nineteenth centuries are available for Bulgaria (Pavlov 2001). Millet flour (or other cereals, occasionally coarsely ground) is roasted and briefly boiled in water. After it has cooled, it is mixed with sugar or honey and yeast from old *boza* and covered for two days until fermentation is complete (Petrov et al. 1978).

A wealth of recent finds that might correspond to breads, porridges, or other foodstuffs offer the potential for more detailed investigations of ancient cereal food preparations. Such work has been pursued by teams working in different parts of Europe, yielding promising results (González-Carretero et al. 2017; Valamoti et al. 2019; Heiss et al. 2015).

In most cases, however, recipes based on prehistoric cereals are more of a speculative nature rather than based on hard archaeological evidence. Grinding of the grain is a step toward its conversion into food, and different recipes employ cereal grain ground in different "grades," that is, degree of grinding, of meal. Thus, the cereal grain could have been used whole, coarsely ground, or ground into small particles or very finely into meal or flour. As the next section shows, occasionally some of these ground cereals may be preserved in the archaebotanical record, offering actual insights of prehistoric culinary practice.

Cereal Foods of Prehistoric Greece: Deciphering Ancient Recipes

Little is known about prehistoric bread in Greece. In our study area, no information on breads, whole loaves or parts of loaves, of the type encountered in Egypt (Samuel 2000), Pompei (Renfrew 1973a), France (Lannoy et al. 2002; Heiss et al. 2015), Germany, Austria, Switzerland (Währen 2002; Heiss et al. 2017, 2019), Slovakia (Hajnalova 1989), Romania (Monah 2002), and Sweden (Hjelmqvist 1984, 1990; Hansson 1994, 2002), of dates ranging from prehistoric to historic times, has been published.

Research on the preparation of beer and bread in Egypt relied on, among other lines of evidence, the experimental reproduction of possible recipes and the use of scanning electron microscopy (SEM) for the investigation of ancient food remains from Amarna (dated to the New Kingdom in Egypt). These finds allowed for the identification of fermentation of barley for the preparation of beer: the micrographs of starch granules demonstrated a characteristic porous structure that is created when starch is attacked by yeast enzymes (Samuel 1993, 1996, 2000). More recently, observations on experimental and archaeological material from ancient Egypt have led to a proposal of an additional potential criterion for the identification of beer, that of aleurone thinning, which might occur during fermentation (Heiss et al. 2020). Pilot investigations in charred cereals, experimental and archaeological, have indicated that different starch structures may be preserved in charred specimens (when charring occurs under low charring temperatures), opening up the potential to detect the processing of cereals in order to make beer and bread (Valamoti et al. 2008b). A series of experiments using SEM have aimed to understand past culinary practices based on cereal ingredients (cf. González-Carretero et al. 2017; Valamoti 2002a; Valamoti et al. 2008b; Valamoti 2011b). Such an investigation is currently underway thanks to the support of the European Research Council and project PlantCult (Valamoti et al. 2017; Valamoti et al. 2019; Berihuete-Azorín et al. 2020; Heiss et al. 2020; Valamoti et al. 2021). This aspect of past culinary investigations is very promising, and despite the limitations, the analysis and interpretation of such finds have made considerable progress.

The main ingredient of bread, flour, is difficult to detect and retrieve archaeologically because, in the best case, it would appear as a concentration of ash. Under specific circumstances, however, it may be preserved, a situation reported by Sarpaki on certain finds of what she interprets as cereal flour from Akrotiri in Santorini (Sarpaki 1987a, 2001a). As the coarseness of ground cereal particles increases, so do the chances of their preservation from various sites of prehistoric Greece (see below discussion of Bronze Age bulgur/*trahanas*). Coarsely ground cereals that have been generated through specific recipes can, at the same time, constitute ingredients themselves. Thus, coarsely ground wheat can be left untreated or can be parboiled in water or milk, ground in different grades, each corresponding to different ingredients, and used in different recipes (see Valamoti 2011b, especially Figure 7). A look at traditional forms of transforming cereal grain into fragments reveals variations in what can be described as ground cereal grain, often superficially grouped in the archaeobotanical literature under the term "bulgur." Coarsely ground wheat without any prior treatment is known as *kourkouta* in Lesvos island, as *korkoto* among the Pontic Greeks, and as *chondros* in Crete and ancient Greece. It is used in soups and as an ingredient of *trachanas* (Valamoti and Anastasaki 2007; Valamoti 2011b).

The term *pligouri* (bulgur) refers to a cooking ingredient that at the same time results from a recipe, as cooking of the grain is required for its production

(Abdalla 1990; Bayram 2000). Wheat grain, preferably emmer or durum wheat, is parboiled, then the grain is sun dried, the bran removed through pounding, and then the dried grain is ground (Valamoti 2011b:Figure 8; Valamoti et al. 2021:Figure 2). This boiled and ground grain is preserved in a dry form and is used as an ingredient of various meals ranging from simple *pilavs* to sophisticated stuffed vine leaves. With a mere soaking, very brief simmering with a little water, bulgur is then ready for consumption. Today it is widely used in Mediterranean cuisine (Rivera-Nuñez and Obón de Castro 1989; Aubaile-Sallenave 1994) and used instead of rice in many recipes (Abdalla 1990; Scott 1983; Kiziridou 2002).

What criteria could one use to identify bulgur in cereal fragments retrieved from archaeological sites? Could archaeologically retrieved cereal fragments from prehistoric Greece be distinguished as *chondros*, bulgur, or other ground cereal products such as ground malt or *trachanas*? Experimental archaeology provides some insights. Based on two concentrations of charred cereal fragments from two sites in central Macedonia dated to the Early Bronze Age, Mesimeriani Toumba (Grammenos and Kotsos 2002) and Archondiko Giannitson (Papaefthymiou-Papanthimou et al. 2002), a series of experiments were carried out to investigate the conditions under which these fragments were generated in the past (Valamoti 2002a). First, this experimental investigation intended to find a means to distinguish fragments generated after charring, due to postdepositional breakage, from those generated through grain grinding. Thus, the fracture surfaces of modern einkorn grain that was (1) charred and then broken; (2) broken and then charred; and (3) soaked in boiling hot water, dried, broken, and then charred, were examined.

This first attempt demonstrated that seeds broken prior to charring feature a characteristic fracture surface with a "swollen" endosperm (Figure 2.26). By contrast, grain that has first been charred, then broken, demonstrated a porous surface and no sign of bulging (Valamoti 2002a; Valamoti et al. 2021). Moreover, grain that has been soaked in water, ground, and then charred at low temperatures demonstrated a characteristic shiny surface (Figure 2.27). Based on these first experiments, the archaeological fragments from Mesimeriani (Figure 2.28) were interpreted as the remains of a type of einkorn bulgur, as the grains demonstrated both a bulging endosperm and a shiny fracture surface (Valamoti 2002a:Figure 2). Alternatively, if treatment of the grain at Mesimeriani had taken place with milk instead of water, as is the case with modern *trachanas* prepared in various parts of Greece (Valamoti and Anastasaki 2007; Valamoti 2011b), the Mesimeriani

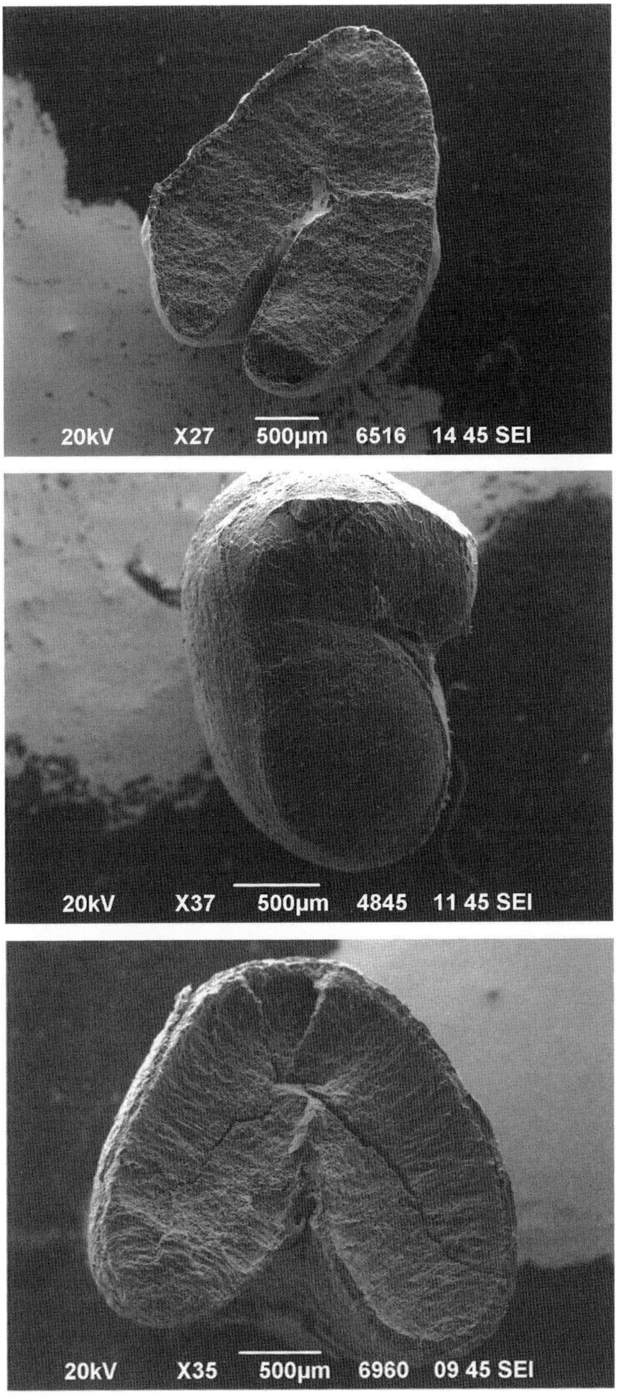

Figure 2.26. Experimental/ethnographic specimens of fracture surface of cereal fragments. *Top to bottom*: einkorn charred dry bearing a post-charring fracture, bulgur fragment generated prior to charring and bulgur charred soaked, and fragmented post-charring. SEM=Scanning Electron Microscopy (From Valamoti et al. 2021, detail of Figure 16. Photographs by Chryssa Petridou and Lambrini Papadopoulou. Courtesy of the *Journal of Archaeological Science*.)

Figure 2.27. Shiny surfaces on experimental/ethnographic cereal fragments, induced by starch gelatinization. (From Valamoti et al. 2021, detail of Figure 14. Photograph by Chryssa Petridou. Courtesy of the *Journal of Archaeological Science*.)

Figure 2.28. Ground wheat grain from Mesimeriani Toumba. Based on experimental observations, the grain was most likely boiled in some liquid prior to charring. (Photograph by Stavros Dimakopoulos. Courtesy of PlantCult.)

fragments could suggest perhaps the earliest known *trachanas* remains from Greece. Yet, for the time being, it is impossible to distinguish parboiling in milk (*trachanas*) from boiling in water (i.e., bulgur). Various aspects of foods made of coarsely ground cereal grain are extensively discussed in a recent integrated study that brings together ethnography, experimentation, and archaeobotany (Valamoti et al. 2021).

The barley fragments from Archondiko, although bearing clear indications for grinding prior to charring (i.e., a bulging endosperm), did not demonstrate the characteristic shiny surface that could be linked to treatment with some liquid such as water or milk (Valamoti 2002a:Figure 2; Valamoti 2011b:Figure 2). This could be either due to lack of such treatment of the grain or due to high charring temperatures.

This first series of experiments generated new questions that needed a more systematic experimental investigation and the collaboration of several researchers. The new series of experiments were carried out in collaboration with Mustafa Bayram from Gaziantep University in Turkey, a food scientist specializing in bulgur, and Delwen Samuel, a specialist in Egyptian breads and beer who had used SEM for the investigation of such remains from Deir el Medina and Amarna in Egypt. In this experiment we attempted to explore the effect of boiling on cereal starch granules, namely, those of einkorn and barley grain, using SEM.

The research demonstrated that when cereal grain has been boiled, dried, and ground, the starch granules are subject to drastic morphological changes, and some of the characteristics of these changes may be preserved after charring at low temperatures (Valamoti et al. 2008b). Starch granules in the grain endosperm are distorted due to boiling, and depending on the distance of the granule from the grain's outer surface, it may be distorted in the form of a red blood cell shape or totally distorted into an amorphous matrix. Further research on endosperm morphology, recently completed and also in progress, will help to decipher recipes from prehistoric cereal foods (Valamoti et al. 2019; Valamoti et al. 2021).

When the archaeological fragments were compared with the experimental material, it was possible to observe on the endosperm of the Mesimeriani fragments a structure similar to those present on the experimental einkorn seeds that had been boiled in water, dried, and then ground. Thus, the initial interpretation of the fragments as a form of bulgur, based on the oozing endosperm and the shiny fracture surface, was confirmed by the later pilot analysis using experimentation and SEM (Valamoti et al. 2008b). More recent analyses using a wide range of experimental specimens generated under different cooking conditions simulating prehistoric processes (e.g., grinding with Neolithic replica tools) have further confirmed the initial interpretation of the archaeological finds (Valamoti et al. 2021). Turning to the material from Archondiko, the SEM micrographs were, again, not particularly illuminating, as the formations recorded on the micrographs taken from the experimental material did not match the formations observed in the archaeological material.

Further examination of the Archondiko finds in a systematic way is in progress, in regard to the food remains per se, the processes that led to their formation, and also to their spatial associations within the house complexes at the site (Valamoti and Petridou 2021). The Archondiko finds are especially interesting as they may be linked to brewing (Valamoti 2018). A more thorough examination of these remains, through a combined examination of morphological characteristics visible with the stereomicroscope and micrographs obtained with the aid of SEM, will hopefully contribute to a better understanding of the processes that led to the formation of the Archondiko cereal fragments (Chryssa Petridou, PhD, in progress). Such approaches could contribute to the study of other remains of processed cereal food, such as flour and bulgur reported from Akrotiri in Santorini (Sarpaki 2001a).

It is very likely that the two finds from Archondiko and Mesimeriani Toumba correspond to two different recipes, something perhaps already implied by the different size range of fragmented cereal kernels and the use of two different cereal species (Valamoti 2002a). The Archondiko fragments have been found as loose fragments and as masses of fragments (Figure 2.29). These formations are reminiscent of modern *trachanas* prepared in Greece and *xinochondros* in Cyprus (Valamoti and Anastasaki 2007), with variations in other circum-Mediterranean countries (see, for example, Palmer 2002; Valamoti 2001a).

Nevertheless, before connecting the prehistoric cereal fragments and masses from Archondiko with a form of *trachanas*, the question of whether conglomerations of fragments resulted from high temperatures or represent an intentional form of a specific food preparation needs to be resolved. An alternative interpretation currently being investigated is whether these fragments and agglomerations of fragments represent ground malt or "malt cakes" that could have been used as starters for beer fermentation (Valamoti 2018). More cereal finds from Archondiko are currently being investigated in light of the possibility that beer may have been brewed at the site, a suggestion put forward on the basis of finds of sprouted cereals (Valamoti 2018;

Figure 2.29. Lump of cereal grains fused into a mass, Archondiko Giannitson, end of third millennium BC. (Soultana Maria Valamoti)

Petridou 2023). The Archondiko masses and fragments could correspond to ground malt or boiled cereals reduced into fragments used for brewing or even other foodstuffs containing malt. Ongoing experimentation and the use of SEM conducted within the framework of PlantCult could contribute to deciphering these ancient barley fragments.

These ground cereal remains, dating to the end of the third millennium BC, from Mesimeriani and Archondiko, correspond to foodstuffs prepared for human consumption. Despite ethnographic evidence for the grinding of cereals and other grains for the preparation of fodder (e.g., Peña-Chocarro 1999; Halstead and Jones 1989), a practice also advised in animal husbandry handbooks (e.g., Morrison 1957), these specific archaeobotanical remains are unlikely to represent animal fodder. They were found in large quantities, in storage contexts devoid of other remains of plants such as straw or other plant species that could correspond to fodder. If these grains were intended as animal fodder, there would have been no reason to remove the einkorn chaff from the spikelets or the barley hulls. Moreover, it is difficult to envisage that the stone-grinding equipment of the time would render grain grinding for fodder a worthwhile operation.

More recent finds of processed foods based on ground cereals have been retrieved from sites such as Late Neolithic Dikili Tash (Kokkidou 2018), Early Bronze Age Armenochori, Late Bronze Age Angelochori, and many sites in the four lakes region near Amyndeon in northern Greece (Valamoti et al. 2019). Some correspond to bulgur/*chondros*/*trachanas* type, that is, coarse fragments, others to lumps of ground cereals in different grades of grinding. These finds are currently being studied and together with those from Archondiko and Mesimeriani indicate a variety in the ways of converting cereals into food among different settlements of northern Greece and perhaps different recipes used within each settlement, too (Valamoti et al. 2019). The charred cereal fragments present a research challenge, and a solid methodology for deciphering the ingredients of these foodstuffs is much needed.

The Bronze Age finds of coarsely ground cereals, parboiled at Mesimeriani, may at present be the oldest ones known from prehistoric Greece and perhaps a much wider region in southeastern Europe and the east Mediterranean. These prehistoric ground cereals probably correspond to preprocessed cereals of the type of bulgur and *trachanas*, both well known from southeastern Europe and the east Mediterranean in recent times as well as from the Greek world in antiquity (Valamoti and Anastasaki 2007). Earlier finds have been reported from neighboring Bulgaria, from various Neolithic and Bronze Age sites (Marinova 2006; Valamoti et al. 2019), yet, in light of the recent experimental studies, the Bulgarian finds need to be reexamined for a more accurate interpretation. It may well be the case that common food preparation methods were shared throughout the Balkans (Valamoti et al. 2019), yet, variations on the same basic recipes or different recipes may well have existed to underline regional or intrasettlement differences related to identities. The ongoing archaeobotanical research of such food remains is expected to bring to light more evidence on cereal recipes and variability in culinary practice in prehistoric Greece.

3

Pulses

Adding Protein, Color, and Variety

Despite the well-embedded notion of a stereotypical Mediterranean triad consisting of bread, wine, and olive oil, pulse dishes hold a dominant position in modern cuisines of Greece, and as will be shown here, the same was true for prehistoric times. This chapter provides an exploration of the pulse ingredients identified in the archaeobotanical record of prehistoric Greece and of the possible ways these might have been consumed as food. The significance of pulses for prehistoric communities of Greece has been stressed by various experts who have highlighted the nutritional properties of pulse seeds and their ability to improve soil fertility through their cultivation (e.g., Halstead 1981, 1994; Hansen 2000; Jones and Halstead 1995; Sarpaki 1992a, 2000). Pulses, an important plant protein source for prehistoric people relying heavily on plant foods for their subsistence (Halstead 1981; C. Renfrew 1972), would have formed a significant dietary element.

A wide range of pulse species appear to have been cultivated since the beginning of the seventh millennium BC in Greece (Appendix A:Table 3.1), and it is very likely that in prehistoric times they added variety in cereal-based meals. Lentils (*Lens culinaris* Medik.) occur at almost every site and in some cases are the dominant pulse species, appearing at nearly all sites during the Neolithic (Figures 3.1 to 3.3). They are often present in pure, dense concentrations, for example, at sites such as Nea Nikomedia (Zeist and Bottema 1971), Arkadikos (Valamoti 2004), and Dikili Tash (Kokkidou 2018). Other species that are perhaps less frequently occurring but have been retrieved in pure dense concentrations are pea (*Pisum sativum* L.) (Figure 3.4), grass pea (*Lathyrus sativus* L./*cicera* L.) (Figure 3.5), and bitter vetch (*Vicia ervilia* [L.] Willd.) (Figure 3.6).

Figure 3.1. Charred lentil seeds (*Lens* sp.) from Early Neolithic Sossandra. (From Valamoti 2015b, Figure 2. Soultana Maria Valamoti)

Figure 3.2. Charred lentil seeds (*Lens* sp.) from Final Neolithic Mandalo. (Photograph by Stavros Dimakopoulos. Courtesy of PlantCult.)

Figure 3.3. Charred lentil seeds (*Lens* sp.) from Early Bronze Age Skala Sotiros Thassou. (Photo by Stavros Dimakopoulos. Courtesy of PlantCult.)

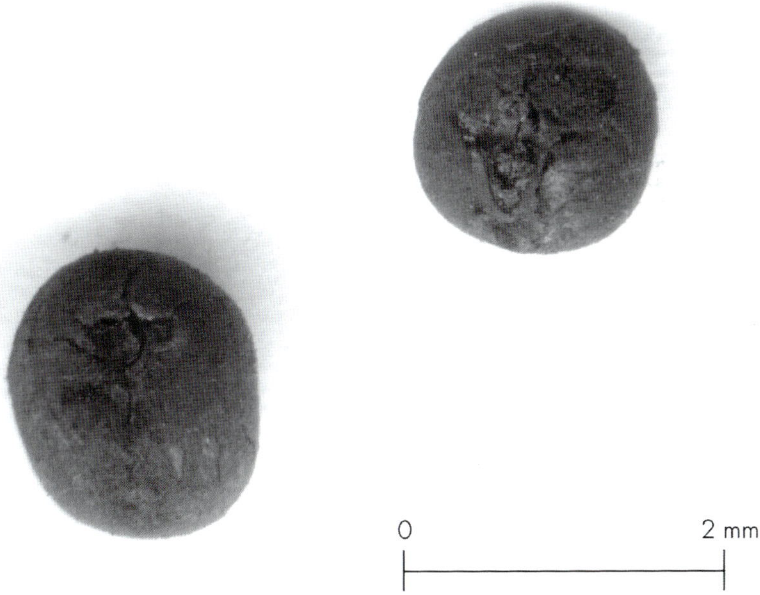

Figure 3.4. Charred pea (*Pisum sativum*) seeds from Arkadikos, Late Neolithic. (Photograph courtesy of Georgios-Vily Kapetanakis.)

Figure 3.5. Charred seeds of grass pea (*Lathyrus sativus*) from Toumba Kremastis Koiladas. (Soultana Maria Valamoti)

Figure 3.6. Bitter vetch (*Vicia ervilia*) is often found in pure, rich concentrations of seeds in prehistoric Greece (inset) (Soultana Maria Valamoti). Charred seeds from Dikili Tash, Late Neolithic. (Photograph by Pascal Darcque. © Dikili Tash Research Project.)

Chickpea (*Cicer arietinum* L.) is a pulse that, although known in the Near East 10,000 years ago (Tanno and Willcox 2006b), is nearly absent from the meals of prehistoric Greece. From the nearly 6,000 years of Neolithic and Bronze Age habitation in this part of Europe, chickpea is represented by a total of forty-eight seeds. The only chickpea concentration reported for the whole of the Neolithic of Greece is thirty-two chickpea seeds that were found at Dimini in Thessaly, dated to the Late Neolithic (Kroll 1979). Seeds are also reported from Neolithic Otzaki Magoula without, however, any mention of their number (Kroll 1983), and one chickpea seed is reported from Sitagroi (Renfrew 2003). Thus, with the exception of these scarce finds, chickpea is extremely rare in prehistoric Greece and cannot be considered as a regular plant food ingredient then, unlike ancient Greece in historic times (Dalby 1997; Valamoti et al. 2022a).

A similar situation has been observed in neighboring Bulgaria where chickpea seeds have been identified at several Neolithic sites such as Kovacevo, Balgarcevo, Kapitan Dimitrievo, and Galabnik (Marinova and Popova 2008). There, as in Greece, the Bulgarian finds are characterized by a small number of seeds, fewer than ten for the total of the Neolithic sites, except for one rich concentration of approximately three hundred chickpea seeds that were found charred at Galabnik (Marinova and Popova 2008). The geographic distribution of Bulgarian sites where chickpea has been found is limited to the southwest, and their occurrence is considered an outcome of contact networks with Anatolia, especially during certain periods of the Neolithic, representing, according to Marinova and Popova, a contaminant of other crops circulating among different communities rather than a crop in its own right. This hypothesis remains to be tested when more evidence from Anatolia is made available and is compared with the southeastern European finds.

Like chickpea, Celtic bean (*Vicia faba* L.) does not appear to have been grown and consumed in the Neolithic of southeastern Europe in general (Valamoti 2003, 2009; Marinova and Valamoti 2014). This, despite the fact that Celtic bean is also an early component of the crops that were domesticated in the Fertile Crescent: more than fifty seeds of this plant were found at the site of Tell el-Kerkh in northwestern Syria and more than two thousand seeds at Ahihut in Israel, both dated approximately 10,000 years ago (Tanno and Willcox 2006b and Caracuta et al. 2017, respectively). Thus, chickpea, together with Celtic bean, although being important components of the Near Eastern Neolithic, seem to have been "dropped off" from the "crop package" of domesticates grown by the first farmers of southeastern Europe.

The rarity of chickpea and the absence of Celtic bean have some significance: they allow a glimpse into regional differences among the pulses that became established as crops in Europe once the agrarian way of life spread from the Near East to new lands. Some plant foods that were important elements of Near Eastern early farmers' diets failed to reach new lands. As farming spread westward alongside contact networks through which the new way of living with/from nature spread to Greece, complex natural and cultural processes influenced crop choice. Different origins of farmers, preferences prescribed by culture and taste, and failure of the initial attempts to cultivate some of the crops in these new lands are possible reasons contributing to this picture. Some of these plants that did not establish themselves in new areas, may, on rare occasions, have been exchanged, brought in as gifts and consumed, or left unused in store.

Although Celtic bean and chickpea are absent from the farming "crop package" that reached Greece, another pulse species, a species that occurs regularly in archaeobotanical samples from Greece, is often found in pure, dense, rich concentrations from the north mainland of the country to the Aegean islands of the south. This species is grass pea. An impressive concentration of grass pea is known from the Late Neolithic site of Kremastis Koiladas, in the region of western Macedonia, near the modern city of Kozani. There, at least 66,000 seeds of grass pea were stored in the form of complete seeds, not split in cotyledons (see Figure 3.5). They were found in a pit on the margins of the settlement, inside a large, partly preserved pot (Karathanou and Valamoti 2011; Valamoti et al. 2011). Grass pea is also found stored inside the Neolithic House 1 at Dikili Tash (see Figure I.2). Another impressive concentration of approximately 6,500 seeds of *Lathyrus cicera* has been found on the island of Mykonos at the site of Phtelia, dated to the Final Neolithic (Megaloudi and Marinval 2002).

It has been argued that pulses became more important dietary elements toward the end of the Neolithic, associated with changes in agricultural practice induced by soil exhaustion, on the basis of evidence primarily from Sitagroi (Blouet 1986). Such diachronic comparisons, however, are faced with problems such as the paucity of adequate data: archaeobotanical evidence from earlier periods of the Neolithic is limited and fragmentary in most cases. Moreover, it is unlikely that pulses were doing better than cereals on depleted soils: certain pulse species, like pea and lentil, have greater requirements in terms of soil properties (fertility, moisture) compared with einkorn and barley, for example (see, for example, Jones et al. 1999 for pulses grown on manure on the island of Evvoia, Greece).

What is clear from the archaeobotanical record is that new pulse species make their appearance in the crop repertoire of prehistoric Greece during the Bronze Age (Appendix A:Table 3.2). These are Celtic bean (*Vicia faba* L. var. *minor*), Spanish vetchling (*Lathyrus clymenum* L.), and Cyprus vetch (*Lathyrus ochrus* [L.] DC).

Celtic bean is absent from Neolithic contexts as noted, yet it is encountered in rich concentrations dated to the Bronze Age. It has been found as a clean, stored crop, at sites such as Mandalo (Valamoti and Jones 2003), Skala Sotiros (Figure 3.7) (Valamoti, unpublished material), Angelochori (Valamoti 2010a), and Thebes (Jones and Halstead 1993a). Roughly around the same time period, Celtic bean appears in central Europe (Karg 1996). This picture is further confirmed by recent reviews of the Neolithic evidence from central European sites whereby Celtic bean is not reported as a crop found in dense, pure concentrations before the Bronze Age (see, for example, a comprehensive regional review of the data for the Late Neolithic Michelsberg and Bischeim cultures dated to fifth/fourth millennia BC, Kreuz et al. 2014); the few seeds mentioned in publications from Neolithic contexts may be intrusive. Thus, Celtic bean corresponds to a pulse species that becomes established in Europe as a crop during the Bronze Age, despite its very early use in the Near East where it was found in pure rich concentrations

Figure 3.7. Charred Celtic bean (*Vicia faba* minor) seeds from Skala Sotiros, Early Bronze Age. Celtic beans feature from the Bronze Age in archaeobotanical assemblages from Greece. (Soultana Maria Valamoti)

indicating its cultivation and use 10,000 years ago at sites such as Ahihud in Israel (Caracuta et al. 2017) and at Tell el-Kerkh in Syria (Tanno and Willcox 2006b).

Celtic bean, although not dangerous to everybody, can prove extremely dangerous, even lethal, causing a hemolytic anemia known as favism, if consumed by people with a genetic deficiency of the erythrocyte (red blood cell) enzyme glucose-6-phosphate dehydrogenase (G6PD). However, this deficiency as well as the consumption of Celtic bean can be advantageous against malaria, especially for those people with this genetic deficiency (Davies 1961; Katz 1987; Hansen 2000; Dalianis 1993; Crépon et al. 2010; Katz 1987; Hansen 2000; Jones and Halstead 1993a). Although the consumption of Celtic bean to counteract malaria in prehistoric Greece, as is argued by Hansen (2000), cannot be excluded, malaria symptoms are not visible in skeletal remains of Bronze Age people when Celtic bean consumption is attested (Triantaphyllou 2001). The late adoption of this pulse species in prehistoric Greece, despite it being part of the Neolithic crop package, may be related to food taboos established early on concerning consumption of Celtic bean seeds, perhaps related to health issues. A genetic vulnerability of local gatherer-hunters who eventually became farmers, under influences from the east or by becoming integrated with farming populations that moved northward and westward to Greece from the Near East and Anatolia, might have contributed to the late adoption of the crop in Europe and the food taboos related to it in ancient Greece. Celtic beans, in antiquity, were considered responsible for symptoms such as unease, dreams, and increased male sexuality, which were described by the ancient Greek writer Plutarch (Katz 1987).

The two new Bronze Age *Lathyrus* additions to the plant foods of prehistoric Greece, *L. clymenum* L. and *L. ochrus* (L.) DC, have a distribution in the Aegean islands of the south as well as the northwest (Yenibademli site on the island of Imbroz; Oybak Dönmez 2005). Finds of *L. clymenum* and *L. ochrus* have been recently retrieved from the eastern coast of the Aegean sea, at the site of Liman Tepe and Bakla Tepe, dated to the end of the 4th millennium B.C. which corresponds to the Early Bronze Age of Greece; *Lathyrus clymenum* in particular was represented at Liman Tepe by a total of more than 10000 seeds though both species were components of otherwise cereal comminated samples (Maltas et al. 2022). The seeds of *L. sativus*, known and grown throughout the Neolithic, are different from those of *L. clymenum*. Those of *L. ochrus*, however, can be easily confused with those of pea. Thus, a range of criteria has been established for the identification of the different *Lathyrus* species in archaeobotanical assemblages (Jones 1992).

L. clymenum, known today as *arakas* on the islands of Santorini, Anafi, and Karpathos, has been found in a rich concentration of seeds at the Bronze Age site of Akrotiri on Santorini as well as a contaminant in a concentration of *L. ochrus* seeds at Knossos (Sarpaki and Jones 1990). The second *Lathyrus* species, *L. ochrus*, was found at Knossos stored in the "Unexplored Mansion" and was initially identified as pea (Jones 1984, 1992). Recently, finds of *L. ochrus* were reported from excavations at Knossos-Gypsades (Bennett 2016).

These three *Lathyrus* species, documented as distinct crops grown on separate fields in prehistoric Greece, are nowadays sometimes confused in the modern agronomic literature, probably because they represent minor crops of limited commercial value, unsuited to large-scale exploitation. Dalianis (1993), for example, acknowledges the presence of all species, yet he mentions the fava of Santorini in the *L. sativus* section while in the *L. clymenum* section makes no reference to the cultivation of this species for human food on the island (Dalianis 1993:135–136). Often these three species and especially *L. ochrus* and *L. clymenum* are treated as exclusive fodder crops, ignored for their value as human food, partly also due to their toxicity (López-Bellido 1994; Hanbury et al. 2000; Getahun et al. 2002). Both species are traditionally grown as human food as well as fodder in many of the Greek islands (Jones 1992). *L. ochrus* is nowadays a minor crop cultivated in parts of Greece and on Cyprus. It is known as *afkos* on the island of Lemnos where it was grown on a large scale in the recent past and exported as seed to Crete where *L. ochrus* is consumed as greens in a salad known as *papoules* (Valamoti field notes). In the Aegean islands it is also known as *lathouri* or *spika* (Jones 1992). On Cyprus it is known under the name of *louvana* where it represents a pulse crop widely grown but cultivated on a small scale (Jones 1992; Photiades and Alexandrou 1979).

The significance of these two Bronze Age *Lathyrus* species as culinary elements seems to be confined to the archaeobotanical record and ancient Greek sources (Valamoti et al. 2022a) as well as in their survival in traditional small-scale cultivation and the initiatives of few modern farmers who continue their cultivation on some Greek islands and pockets of mainland Greece (the recent revival of these crops has taken place in the context of modern culinary trends and the ancient grains movement) (see Figure 10.8).

Besides these three pulses (*Vicia faba, Lathyrus clymenum, Lathyrus ochrus*) that appear for the first time in Greece during the Bronze Age, all remaining pulse species, known since the Neolithic, continue to be used during the Bronze Age. The seed of the pulses, cleaned of their pods and/or straw, found in rich, pure concentrations among the debris of burned destruction layers eloquently attests their importance for prehistoric communities of the region. Pulses feature prominently not only in the Neolithic but also during the Bronze Age (Appendix A:Table 3.2). Sarpaki (1992b) mentions that at Akrotiri, on the island of Santorini, pulses were stored, and she underlines their significance for prehistoric communities of Greece as foodstuff, buffer against crop failure, and fertilizer. Certainly, by the end of the Bronze Age a wide range of pulse seeds harvested from crop fields could have constituted ingredients of prehistoric meals prepared on the Greek mainland and the islands.

The variety observed in the different pulse species identified stored in houses that were destroyed by fire is impressive. Growing of a wide range of pulses is observed not only on a site level but also within individual houses throughout the Neolithic and the Bronze Age. Inside two Neolithic houses of Dikili Tash, dated to end of the fifth millennium BC and explored in detail, were three pulse species kept in store: lentils, grass pea, and bitter vetch (Valamoti 2004:Figure 7.36). At Arkadikos, a variety of pulse species (grass pea, pea, lentils) stored together is also observed (Valamoti 2004:Figure 7.17). At destruction layers of third millennium BC Mandalo, four species of pulses have been identified in what were probably storage contexts: Celtic bean, grass pea, bitter vetch, and lentil (Valamoti and Jones 2003). At Knossos, during the Late Bronze Age, grass pea, pea, and Celtic bean were stored at the "Unexplored Mansion" (Jones 1984). In addition to these Bronze Age cases, recent research from the south offers further evidence for a wide variety in stored pulses: at Late Bronze Age Knossos-Gypsades (Bennet 2016) Amy Bogaard has identified concentrations of charred bitter vetch, grass pea, lentil, Celtic bean, and *Lathyrus ochrus*, suggesting different plots dedicated to each pulse and their storage for consumption.

In northern Greece, however, such a variety in pulse species is not encountered everywhere during the Bronze Age. At Kastanas (Kroll 1983) and Agios Athanasios (Moniaki 2009), only bitter vetch has been identified as a stored crop, while at Assiros, lentil and bitter vetch were stored, with the latter being the most prominent species (Jones et al. 1986). At Middle Bronze Age Skala Sotiros, Celtic bean was stored in the same destruction layer as grass pea and lentils, all three found in pure dense concentrations (Valamoti, under study). At Archondiko, throughout its Bronze Age occupation layers spanning the end of the third millennium BC, through to the end of the second millennium BC, archaeobotanical research has failed to yield stored pulses in pure dense concentrations at the site despite rich finds associated to cereal and oil plant storage (Telioridou 2014; Petridou 2016; Valamoti et al. 2008; Papaefthymiou-Papanthimou et al. 2013). The limited presence of pulses in the Archondiko samples is unlikely to be an artifact of sampling of the excavated area of the site, as sampling was very detailed; it is possible that other areas of the settlement, not yet excavated, were reserved for the storage of pulses.

At Late Bronze Age Assiros, the dominance of bitter vetch in the large storerooms may be related to their function and the intended use of the stored products. Perhaps at Assiros storage was based on basic, reliable staple foodstuffs like the glume wheats and also fodder crops like bitter vetch, though either can be used for both purposes. The other pulses might have represented crops grown by individual households and therefore are unrepresented in the central, large storerooms of the site. A narrow range of pulse crops is encountered also at Kastanas, a site demonstrating a different organization of space compared with that of Assiros during the Late Bronze Age, lacking central storage facilities. Although Kastanas may have occupied a different position in the networks of local hierarchies of Bronze Age communities of the region (Andreou et al. 1996), there again glume wheats and bitter vetch are identified as the main stored crops. If we consider the species range of pulses available for the Bronze Age, a shift of emphasis toward the cultivation of bitter vetch may be the case in the north of Greece, compared with the diversity of species stored inside the Neolithic houses from the area.

The cultivation of a hardy pulse, more suitable for fodder, especially for animals used for agricultural labor (Valamoti field notes, Lemnos 2009, Crete 2009), may be related to changes in the goals of agricultural production. Cattle could get a boost from eating bitter vetch seeds (Valamoti field notes, Lemnos island). Bitter vetch would have been a nutritious fodder for horses as well. Horses make their appearance in the end of the third millennium BC in Greece as rare individuals, becoming more prominent in the second millennium BC (Cantuel et al. 2010). Besides the potential association of millet with the arrival of the horse in Greece (Valamoti 2016), a link between this animal and an increased cultivation of bitter vetch in the north seems plausible.

Although garden cultivation regimes have been proposed for Late Bronze Age northern Greece on the basis of evidence from weed seeds (Jones 1992), variation in the intensity of manuring has been observed even among different houses of the same site during the Early and Late Bronze Age at the sites of Archondiko Giannitson and Toumba Thessalonikis, respectively (Nitsch et al. 2017). Extensive cultivated plots may have been more widespread than generally thought during the Late Bronze Age as is indicated by pollen diagrams suggesting a considerable reduction of the forest cover that dominated the north regions of Greece throughout the Neolithic (Bottema and Woldring 1990).

Thus, the prominence of bitter vetch at the northern Greek sites may indirectly reflect a tendency, in some Bronze Age contexts, toward an emphasis in certain fodder crops (e.g., *Medicago* at Kastanas and Melilotus/*Trifolium* at Agios Athanasios) or/and more extensive methods of cultivation. Also, samples during the Bronze Age tend to be richer in weeds, compared with the Neolithic (Valamoti 2004), which could also be a hint toward less intensive and more extensive methods of cultivation, larger fields being less carefully weeded. Animal power, necessary in extensive cultivation regimes, would have had high requirements for fodder, and bitter vetch could have been such a suitable fodder crop. The recent finds from Knossos-Gypsades, however, where a wide variety of stored pulses has been identified, suggest a more complex situation and the need for cautious interpretations of the available archaeobotanical evidence.

Different regions, different areas within settlements, and different excavated contexts yield different pieces of information about available pulse ingredients. These are the pieces of a puzzle composed of differences in crop choices, settlement types, and functions and perhaps also different archaeobotanical recovery methods. It is therefore difficult for the time being to make more generalized statements about the role of pulses in human diet during the Bronze Age in Greece. What is reflected through the archaeobotanical record is probably different scales of production, one geared toward generating, storing, and mobilizing surplus and another operating on the small scale of subsistence where intensive farming aimed to provide food for the household or household groups.

Despite the importance that pulses appear to hold throughout the Neolithic and the Bronze Age in Greece, due to their nutritional value and their ability to enhance soil fertility (C. Renfrew 1972; Halstead 1981), they are absent from Linear B tablets, as Halstead and Jones point out (Halstead 1992, 1994; Jones and Halstead 1993a). Linear B tablets, made of clay and inscribed in Linear B script, recorded goods and personnel of the Mycenaean palaces of southern Greece in the fourteenth to thirteenth centuries BC. These archives mention cereals but not pulses. The latter, however, have been unearthed at locations outside the big palatial complexes, as is the case for the "Unexplored Mansion" at Knossos (Jones 1984) and the recent Knossos-Gypsades finds (Bennet 2016). This divergence observed between the written sources and the archaeobotanical material should be unsurprising. Daily agricultural strife and meals of ordinary people of the Late Bronze Age were probably not the subject of concern in the archives of the Mycenaean centers (Halstead 1992, 1994; Valamoti 2009:76; Valamoti et al. 2011:392). This point is further underlined by Livarda in a recent overview of the archaebotanical data from Crete (Livarda and Kotzamani 2013). Similar "contradictions" between the written archival records and the archaeobotanical record have been observed in other areas and periods (see, for example, Bakels 2005 for a similar case from medieval northern Europe). Archival evidence may focus more on what constitutes the taxed product paid to the "elites" rather than the elements of daily staples for ordinary peasants, in this northern European case, bread wheat and rye, respectively.

The variety in stored pulse ingredients during the Neolithic and the Bronze Age in Greece connected prehistoric kitchens to a colorful landscape with different plots of pulses, especially in the south during the Bronze Age, with yellow, blue, and purple flowers corresponding to the fields of the different pulse crops (Figure 3.8). In light of this discussion, Anaya Sarpaki is

Figure 3.8. Colorful fields in spring would mark the location of pulse crops in the landscape of prehistoric Greece. (Collage by Kostas Prokos. Photographs by Yannis Gavalas [*top left and center*]; Stergios Salamousas [*bottom center*]; Soultana Maria Valamoti [*top right and bottom left and right*]. Courtesy of PlantCult.)

correct in noting the "Mediterranean tetrad," adding pulses to the three elements, wheat, wine, and olive oil, that have closely been connected to the nutritional tradition of the Mediterranean (Sarpaki 1992a).

Processing Pulses for Food

Despite ample information on the range of pulse species potentially used as food by the prehistoric inhabitants of Greece, our knowledge on the ways of their consumption is rather limited. The ethnographic record of Greece and other places where pulses constitute important food elements offers useful insights into their culinary use and the ways pulse crops might have been consumed as food in prehistoric times. Further information about the particular properties of certain pulse species can be found in agronomic books and the works of ancient written sources.

After the Harvest

Ethnographic studies of crop processing reveal the steps involved in cleaning pulse seed after harvesting of the crop (Jones 1983, 1984; Fuller and Harvey 2006; Butler et al. 1999). The basic steps for separating the seed from the rest of the harvested crop are similar to those followed for free-threshing cereals. The crop is left to dry, in the field or the threshing floor, then threshed. Small quantities of a harvested crop, like those observed on a grass pea harvest on the island of Corfu (Figure 3.9) or a *L. ochrus* harvest at Lemnos (Figure 3.10), require mere beating with a stick or trampling by feet; similar practices are reported by Hillman for glume wheats and free-threshing cereals (Hillman 1984:121–122, 1985). Larger quantities require a threshing floor and the use of animal power as described for cereals. After threshing, the crop had to be winnowed (Figure 3.11 and Figure 3.12), sieved, and hand cleaned for the removal of impurities such as chaff, stones, and weed seeds. In prehistoric Greece, bulky harvests for large groups of people would have required more elaborate means of processing, such as threshing on a threshing floor, perhaps with the use of more sophisticated equipment, a threshing sledge, for example.

Preparing the Seed for Consumption

Pulses can be consumed in various ways that do not necessarily involve processing of the harvested crop or cooking. They can be a delicious snack while still green in their pods in the fields (Figure 3.13), and they are sometimes sold in this way in local markets nowadays in Greece and elsewhere (Asif et

Figure 3.9. Threshing grass pea in the village of Agios Georgios, Corfu, by beating with a stick, July 2016. (Soultana Maria Valamoti)

Figure 3.10. Threshing Cyprus vetch (*L. ochrus*) by trampling, near the village of Moudros on the island of Lemnos, June 2011. (Soultana Maria Valamoti)

Figure 3.11. Winnowing Cyprus vetch after threshing at Moudros, Lemnos, July 2011. (Soultana Maria Valamoti)

Figure 3.12. Second winnowing of the grass pea crop, after threshing and removal of stems in the village of Agios Georgios, Corfu, July 2016. (Photograph by Georgios-Vily Kapetanakis. Courtesy of PlantCult.)

al. 2013). When fully mature, the clean seeds may undergo various kinds of processing prior to cooking, depending on the species and the type of desired dish. Pulses, as the etymology of the word indicates, can be cooked in various forms, one of which gives them their English name: pulses cooked in the form of a mush is the equivalent to the Latin *puls* from which the word *pulse* derives (Asif et al. 2013), a word encountered also in ancient Greek texts in the form of *poltos* (πόλτος) (Valamoti et al. 2022a), meaning grain cooked in the form of a mush.

Some of the pulse species available to the prehistoric inhabitants of Greece, such as various *Lathyrus* species and bitter vetch, are rich in toxic substances (e.g., Dalianis 1993; López-Bellido 1994; Hanbury et al. 2000; Getahun et al. 2002). Grass pea and bitter vetch, two pulses that occur frequently in prehistoric Greece, can be toxic for humans and some animals (see, for example, Dalianis 1993; Enneking 1995). Toxicity in grass pea is widely known, yet people in some regions and under certain circumstances consume these grains in countries such as Greece, Ethiopia, and India. Ancient writers tell of the toxicity of bitter vetch (e.g., Hodkinson 1988; Scarborough 1991; Simoons 1998; Flint-Hamilton 1999); moreover, the dangers of their consumption for prehistoric communities often features in the archaeobotanical and archaeological literature (e.g., Cohn and Kislev 1987; Butler et al. 1999; Halstead 1990; Hansen 2000; Peña-Chocarro and Zapata-Peña 1999).

Figure 3.13. Unripe grass pea from a field near Petaleia, Corfu, July 2016. Eaten raw, they taste mildly sweet and are a seasonal delicacy. (Photograph by Georgios-Vily Kapetanakis. Courtesy of PlantCult.)

The consumption of *Lathyrus* can cause lathyrism. The toxic substance responsible for lathyrism is the amino acid, b-N-oxalylamino alanine (BOAA) (Rao et al. 1964; Murti et al. 1964). This condition, which leads to the paralysis of the lower limbs and affects primarily male populations and less so females (Cohn and Kislev 1987; Hansen 2000; Stockman 1917), is described as "irreversible, nonprogressive spastic paraparesis in humans with a definitive history of *Lathyrus* intake, which is due to poorly understood degenerative changes in the spinal cord" (Mishra et al. 2014:38; Zanicovic 2001).

Bitter vetch, a commonly occurring pulse seed, is yet another toxic pulse. Its frequent occurrence in the archaeological record of prehistoric Near Eastern sites led van Zeist (1988) to suggest that it was probably consumed as human food after being detoxified by means of leaching. Dioscorides refers to the toxicity of this species when it is consumed by humans, underscoring that it causes headaches, bowel problems, and hematuria (Scarborough 1991; Hodkinson 1988). Pliny also mentions that it is dangerous for both humans and animals (HN V:18.139). Cannavanine, the toxic substance contained in bitter vetch seed, can have detrimental effects on human health, if consumed.

Toxic pulse seeds usually require special treatment for the removal of toxic substances, when the seed is intended for human consumption and occasionally for fodder, too (Cohn and Kislev 1987; Dalianis 1993). The testa of the pulse seeds contain the largest proportion of toxic substances, and testa removal therefore results in the reduction of their toxicity (Enneking 1995; Hansen 2000; Butler et al. 1999). In grass pea this process results in the removal of the largest percentage of toxic substances (Hansen 2000), and indeed this is the traditional means for preparing grass pea sold in supermarkets in Greece, that is, as split cotyledons without testa. In some packages of split grass pea, a sequence of treating with water is sometimes included in the cooking instructions: boiling, discarding the water, cooking by bringing to boil, and slowly simmering.

Pulse seeds can be split by grinding without any prior processing by other means. Splitting of *Lathyrus ochrus* is traditionally performed on the island of Lemnos on fully cleaned seeds, using a rotary quern. There are ethnographic accounts whereby soaking of the seeds is performed prior to the removal of the testa. This is the case with the processing of grass pea in Ethiopia whereby the seeds are soaked, subsequently dried, and then ground (Butler et al. 1999). A similar practice was reported to me for villages in the north part of the island of Corfu; there, the seeds of grass pea are soaked in ash water prior to testa removal. Soaking in an alkaline medium is in fact mentioned as a very efficient means of detoxifying *Lathyrus sativus* seeds based on experimental studies (Khokhar et al. 1996). Treatment with water prior to de-husking is reported for Morocco where bitter vetch seeds were soaked in water and then roasted or sun dried prior to the removal of the testa, presumably to facilitate this process when bitter vetch was consumed as human food (Enneking 1995; Enneking et al. 1995). Soaking bitter vetch seeds in water for a certain amount of time renders them suitable for food consumption. Grinding and subsequent soaking of the seeds was referred to as a practice for the removal of bitter vetch toxins on the island of Karpathos during World War II (Halstead and Jones 1989). These traditional processes involving soaking of toxic pulses are also supported by experimental studies (Mohan et al. 1966; Tekle-Haimanot et al. 1993). Ancient Greek sources also refer to the detoxification of bitter vetch whereby the seeds are sprinkled with water, roasted, ground, and sieved (Valamoti et al. 2022a:118).

Once split, the cotyledons have to be winnowed to remove the testa to end up with clean seeds, regardless of whether the seeds have received treatment with water. This step has been observed in the sequence of grass pea processing in Ethiopia (Butler et al. 1999) as well as on the island of Lemnos (Figure 3.14).

During the Late Neolithic and the Early Bronze Age an increased occurrence of cooking pots has been linked by certain researchers to an increased importance of pulses in prehistoric diet in Greece (Hansen 2000; Demoule and Perlès 1993; Blouet 1986). In this context, the higher presence of Neolithic pots bearing indications for contact with fire and therefore cooking (Perlès and Vitelli 1999) has been associated with boiling pulses (Vitelli 2000). This association of pulses with boiling, however, has a certain degree of arbitrariness (Valamoti 2003): pulses could have been left to soak overnight, reducing thus significantly the duration of cooking, while cereals, in contrast, could have equally required boiling or simmering for some time (cf. for bulgur making) or roasting, thus requiring contact with fire of the vessel in which they were being cooked.

There is little doubt that among the pulse ingredients available to the prehistoric inhabitants of Greece, some would have needed some level of detoxification. The negative effects of *Lathyrus* species and bitter vetch consumption must, at some point, have become known to the prehistoric inhabitants of Greece and other areas of the Mediterranean where *Lathyrus* species such as *L. sativus*, *L. clymenum*, and *L. ochrus* were brought to cultivation. The negative effects are sensed when they are consumed in large quantities over long periods of time. The consumption of toxic pulses as food during periods of food shortages is attested from the recent past, with special references in the

Figure 3.14. Split pulse seeds are winnowed to remove the *testa* from the cotyledons. Cyprus vetch *testa* (on the left side mostly) and split cotyledons partly separated, Moudros village, Lemnos, July 2011. (Soultana Maria Valamoti)

literature regarding the negative effects on human health. This is, for example, the case with grass pea in Ethiopia and India, a famine food to which people resort as it sometimes is the only available food resource, hence the high incidence of lathyrism in these parts of the world (Tekle-Haimanot et al. 1990; Rizzotti 1952; Getahun et al. 1999). The problem has also been reported for parts of Europe and the Mediterranean (Stockman 1917). In World War II Greece, when hunger was a grave and even deadly problem, lathyrism affected some individuals (Halstead 1990; Hodkinson 1988; Valia Angelakoudi, personal communication for the village of Theologos on the island of Evvoia).

Recent evidence has shown that lathyrism is the outcome of complex factors, only one of which is grass pea consumption (Mishra et al. 2014). Lathyrism affects people who are poorly fed, on grass pea only, while the seeds contain varied levels of the toxic substance BOAA, depending on environmental factors related to field conditions and overall rainfall regimes (Fikre et al. 2011). Old and recent studies of the disease clearly suggest that a balanced diet that includes grass pea consumption will not lead to lathyrism (McCombie-Young 1927; Tekle-Haimanot et al. 1990). A moderate consumption of grass pea and bitter vetch combined with cereal intake in the same meal would counteract the negative effects of the toxic substances contained in them when consumed as human food (e.g., Lambein 2000; Getahun et al. 2005). Moreover, variation in levels of toxicity among different landraces of *Lathyrus sativus* (Başaran et al. 2011), as well as different ways of consuming the seeds, for example, boiling versus. roasting (Tekle-Haimanot et al. 1993), seem to play a part in the outbreaks of lathyrism among human populations.

Episodes of lathyrism might have occurred in prehistoric Greece in times of food shortage due to environmental or other crises that have been postulated for prehistoric subsistence societies of the region (e.g., Halstead 1981). Episodes of lathyrism may potentially be detected in skeletons of people who might have been affected by the disease in prehistoric times, though at present this has not been reported. It has been argued that a male figurine from Dispilio that shows an absence of the lower limbs may be depicting humans affected by lathyrism (Hourmouziadis 2007:62–63). Nevertheless, grass pea in modern Greece does not correspond to a famine food but to a delicious starter, much appreciated on both sides of the Aegean.

Removing Testa and Processing with Water

Testa removal would have been an efficient way to reduce toxicity in *Lathyrus* species and bitter vetch. My experimental and ethnographic observations on pulse seed splitting on the island of Lemnos suggest that testa removal can be achieved mechanically with light grinding of the pulse seeds, which results in the separation of the cotyledons and in testa removal at the same time (Figure 3.14). The identification of this practice in the archaeobotanical record, especially when in a charred state of preservation, is extremely difficult, in particular, distinguishing between absence of testa due to intentional removal or due to charring.

Splitting can be carried out relatively fast, especially if seeds have been treated with water prior to splitting (soaking then drying of the seed) as has recently been shown by experimental investigations carried out on grass pea seeds using a small-sized grinding stone similar to those used in Neolithic Greece (Valamoti et al. 2013). Further experimental splitting of bitter vetch seeds, this time using replica grinding stone tools, showed that 200 grams of bitter vetch can be freed of their testa in less than 15 minutes (Bofill et al. 2020; PlantCult project unpublished observations). At the same time this light grinding of bitter vetch seeds leads not only to a separation of the testa

Figure 3.15. Experimental grinding of bitter vetch seeds using a replica of a Neolithic pair of grinding tools. The process leads to the splitting of cotyledons and simultaneous removal of *testa*. (Photographs by Maria Bofill. Courtesy of PlantCult.)

but also to the splitting of the seeds in two cotyledons (Figure 3.15). In the experimental removal of the pulse seed testa, the seeds did not receive any prior processing to facilitate the process.

Removing the testa of pulse seeds could have been easily performed by using the grinding stones available in the Neolithic and the Bronze Age, as was observed during experimental splitting of both bitter vetch and grass pea seeds in the context of project PlantCult. I am unaware of a traditional mechanical method that would result in testa removal without the concurrent splitting of the seed in cotyledons. This process, described as "polishing," is reported as being practiced in Turkey and the Near East, without, however, any description of the way this can be achieved without the simultaneous splitting of seeds into cotyledons (Cappers et al. 2016:2:Figures 1224–1228).

A Late Bronze Age Celtic bean concentration from Thebes was retrieved without testa but most importantly was split in the form of modern split pulses to be cooked as *fava* (mush). Although absence of testa is likely to occur due to charring, the splitting of the Celtic bean seeds, carefully stored in a decorated vase, suggests deliberate processing for splitting dry seeds and removing the testa at the same time (Jones and Halstead 1993a).

It is unlikely that pulse seeds will break up into intact cotyledons due to depositional or postdepositional processes. These would most likely generate uneven fragments of seeds rather than intact cotyledons. Pulse seed finds from sites recently investigated in northern Greece consist of whole pulse seeds rather than cotyledons as is the case with Toumba Kremastis Koiladas (Karathanou and Valamoti 2011; Valamoti et al. 2011) and Dikili Tash (Valamoti, under study). Attempts to induce breakage of charred bitter vetch seeds from Dikili Tash into cotyledons through mechanical pressure proved to be very difficult and resulted in uneven fragments.

If soaking of pulse seeds has been performed prior to the removal of testa, this could have resulted in potentially distinct cotelydons recognizable in the archaeobotanical record. Some indications for such a practice come from Early Bronze Age northern Greece. At Agios Athanasios, a site inhabited at the end of the third millennium BC (Mavroidi et al. 2008), the contents of a pit outside a building offer insights into prehistoric pulse processing. In a pure concentration of charred bitter vetch, a total of 917 seeds and 2,985 cotyledons were encountered (Moniaki 2009; Valamoti et al. 2011:Figures 4, 5). The larger proportion of cotyledons to seeds is an unusual find compared with other bitter vetch or *Lathyrus* finds from northern Greece. In addition to this observation, a closer examination of the cotyledon

Figure 3.16. Charred, split cotyledons of bitter vetch from the Early Bronze Age site of Agios Athanasios. (Soultana Maria Valamoti)

inner-surface morphology indicated a pattern that was also unusual: the cotyledons demonstrated a slight depression (Figure 3.16).

To investigate these specimens further, an experiment was designed. Experimentally generated cotyledons of modern bitter vetch seeds from the Lassithi Plateau in Crete were treated with water on the basis of recent ethnographic observations and literary information (Valamoti et al. 2011). The study showed that soaked or boiled bitter vetch seeds, split to cotyledons, differ in their morphology from bitter vetch seeds that were not treated with water (Valamoti et al. 2011). In this study, bitter vetch seeds were (1) soaked, (2) boiled, (3) left untreated, and after charring in controlled conditions, they were subsequently examined with a stereomicroscope and with scanning electron microscopy. This study indicated that pulse seeds treated with water are likely to demonstrate modified inner cotyledon morphologies in an archaeobotanically recognizable way (Valamoti et al. 2011). In specimens treated with water (boiled or soaked in water) and charred in a dry state, a concave inner cotelydon surface seemed to be the main characteristic feature with the concave effect being more pronounced when the experimental specimens were first split then boiled; untreated specimens demonstrated a flat surface.

The splitting of the Agios Athanasios bitter vetch seeds could suggest processing to remove the testa prior to cooking to help reduce the toxicity of the seeds. It is possible that the seeds had been first boiled or soaked, then dried and split, processes that have been described in the ethnographic record already presented. Had the seeds been soaking when the house was destroyed by fire, we would expect to see the voids observed in the experimental specimens (Valamoti et al. 2011:Figures 6a, 6b). In the absence of such a feature, it is unlikely that the bitter vetch seeds could correspond to some mixture of ground seeds in water at the time of destruction (as is done for leaven preparation).

In light of these observations, the large proportion of bitter vetch cotyledons compared with whole seeds retrieved from the site of Agios Athanasios and dated to the end of the third millennium BC was interpreted as a batch of bitter vetch seeds processed with water and ground, probably intended for human consumption (Valamoti et al. 2011). Examination of cotyledon morphology would be a more reliable criterion for detecting such forms of pulse processing, yet, further experimentation and investigation with scanning electron microscopy are much needed to establish a reliable methodology.

Besides the Agios Athanassios finds that offer solid evidence for processing pulse seeds, archaeobotanical finds that have been associated with pulse processing come from the Neolithic site of Servia in northern Greece. Processing by soaking has been suggested as being practiced already in the Neolithic, on the basis of a cache of charred grass pea seeds found among the debris of a burned destruction layer of a house at Servia in northern Greece, dated to the sixth millennium BC (Hubbard 1979). The seeds were found around the sherds of a clay pot, and it was speculated that they were soaking inside the pot during the conflagration, making them spill as the pot was smashed and its contents poured out. Hubbard based his interpretation on the particular arrangement of the seeds around the fragments of the pot. Such an arrangement, however, could have resulted from the mere breaking of the pot during the destruction, the contents spreading and rolling around the broken pot. Therefore, this interpretation of the Servia find still remains a hypothesis to be proven through actual examination of the effects of soaking or boiling pulse seeds for the removal of toxins. A study of their cotyledon morphology could further contribute toward understanding the depositional taphonomy of this cache (Valamoti et al. 2011).

This pilot study on the effects of processing pulses on seed morphology using mechanical pressure and water highlighted the potential in exploring pulse seed processing for food in the archaeobotanical record (Valamoti et al. 2011). For Greece, most of the published records of pulse finds do not distinguish between complete seeds and cotyledons; thus, there are hardly any indications for the ratio of cotyledons to complete seeds. This may well be worth recording in the future. A high percentage of cotyledons in the archaeobotanical record would provide not only an indication of pulse seed processing but also a hint toward the way it was converted into a dish. As discussed, splitting the pulse seed to remove the testa from toxic pulse seeds results in the removal of a considerable amount of the toxic component, and this might be the underlying factor for the widespread way of cooking *Lathyrus* species in the traditional form of *fava* dish in Greece and *dhals* in India (see Singh 1995). This practice need not have been confined to toxic pulses like grass pea and bitter vetch but could apply to other pulses, too, as testa are difficult to digest and consumption in the form of mushes might have been preferred both for reasons of taste and ease of cooking and digestion.

Grinding

Besides splitting, pulse seeds could have been ground in the form of meal or flour. Detecting flour in the archaeobotanical record and retrieving it during excavation are difficult tasks as it is hard to recognize during excavation or even to retrieve if present in a very fine form. Sarpaki has identified ground pulse seeds, presumably flour intended for culinary use (Sarpaki 2001a). The find is dated to the mid-second millennium BC. Sarpaki considers this means of processing very efficient in terms of food preparation (Sarpaki 2001a), yet the use of this particular flour concentration remains elusive. Grinding pulse seed would have required splitting of the seeds into cotyledons and the subsequent removal of testa by winnowing and then grinding proper of the pulse seed. Although testa removal and splitting would have been rather straightforward and fast, grinding into flour, according to preliminary investigations, would have been much slower (Chondrou et al. 2018; Valamoti et al. 2013).

A pulse meal could have been prepared with the available grinding equipment, depending on labor availability and a need for specific dishes for which a finely ground ingredient was required. The study of usewear could potentially provide indications for grinding of pulses for flour, assuming that grinding tools, unlike general consensus, were at times specialized for the processing of certain plant food ingredients. Although this may seem farfetched, food taboos or functional reasons might have prescribed the use of specialized grinding stones for different types of grains, as has been the case, for example, for grinding millet, tef, and maize in Africa (see, for example, Shoemaker et al. 2017; Nixon-Darcus and D'Andrea 2017).

Cooking Prehistoric Pulses: Mushes, Soups, and "Breads"

Pulses in prehistoric Greece could have been consumed as whole dry seeds or in various recipes requiring different levels of processing. Unless pulse seed flour was needed for some special foodstuff, split, testa-free cotyledons would be easy to cook into a mush. A finely ground flour, however, would have reduced cooking time but increased processing time. Therefore, the selection of the level of grinding would have depended on the desired dish and occasion of consumption rather than on time requirements for processing the pulse seeds.

Pulse seeds in the form of split, testa-free cotyledons are a dominant form of obtaining seeds of *Lathyrus* or other pulse genera, ready for cooking in modern Greece, Cyprus, and Turkey. The split seeds are consumed in the form of a mush in Greece, Cyprus, and the Aegean coasts of Turkey. Throughout Greece, nowadays, the most common way of consuming the different *Lathyrus* species is by slowly simmering, split, testa-free pulse seeds into a thick mush, called *fava* (Figure 3.17 and Figure 3.18), which is featured in cookbooks with traditional Greek recipes (Sarpaki and Jones 1990; Delatola-Foskolou 2006). The word *fava* in Greek implies this special way of cooking pulses and not a certain pulse species, as *fava* can be prepared from various pulse seeds such

Figure 3.17. Dish of *afkos fava*, Lemnos island, July 2011. (Soultana Maria Valamoti)

Figure 3.18. Grass pea fava generated experimentally for the PlantCult project. Note the differences in cooking achieved in the different vessels: under the same cooking conditions, the closed-mouth vessel cooked *fava* in less time than the open-mouthed one. (Photograph by Georgina Prats Ferrando. Courtesy of PlantCult.)

as Celtic beans and peas. The dry seeds of *L. clymenum*, *ochrus*, and *sativus* are split with the testa removed and are basic ingredients for fava preparation, *L. clymenum* corresponding to the famous *Fava Santorinis*.

Consuming pulses in the testa-free form has the advantage that toxicity is drastically reduced while the seeds become easier to digest and the absorption of nutrients is enhanced (Jones and Halstead 1993a; Singh 1995; Singh et al. 2000; Patil and Sokhansanj 2003; Amiri et al. 2012). Dehulling and splitting are effective methods for facilitating cooking of pulses, as in this form they cook much faster than as whole seeds in their testas (Aykroyd et al. 1982:45). Testa-free, split pulse seeds are not confined to toxic species. As is well known from the cuisines of India and Pakistan (Reddy 2006; Tiwari and Singh 2012), for example, numerous dishes including dhals are prepared on the basis of testa-free, split pulse seeds (Tharanathan and Mahadevamma 2003). The same is known from the Cyclades in Greece, where seeds of a local, small-seeded variety of pea, known as *katsouni*, are split into cotyledons for consumption, and this is the form in which they can be purchased in modern specialized shops. *Katsouni* seeds are simply sun dried prior to splitting (Yiannis Gavallas, personal communication). A similar way to cook pulses can be found in the region of Puglia in Italy, where a traditional dish, *fava e cicorie*, consists of split *Vicia faba* seeds cooked in a mush and served with boiled *cicorie*, that is, wild radishes (Renna et al. 2015). Mushy peas, often served with fish and chips, is the British contribution to cooking pulses nowadays in a form of a mush, very similar to *fava*.

Despite the toxicity of grass pea and bitter vetch also reported by ancient writers, these pulses were consumed as food in antiquity (Valamoti et al. 2022a), and a mixture of bitter vetch flour and barley for making bread reported recently from Crete (Psilakis and Psilaki 2001) may be related to the use of bitter vetch as the starter for a sourdough used for making barley bread mentioned by Pliny (HN XVIII:103, quoted by Hillman 1985). A similar mixing of the flour of bitter vetch and cereal grain for the preparation of human food is recommended in an Andalusian twelfth-century written source by agriculturalist Ibn Al Awam (Enneking 1995).

The use of pulse seed flour is encountered in several traditional Greek recipes, for example, in the preparation of the επτάζυμο/*eptazymo*, meaning seven times kneaded/fermented bread, as well as for other special breads prepared on Crete and elsewhere in Greece (Psilakis and Psilakis 2001). Pulse seed flour, too, has a culinary use in the preparation of breads or pancakes as is done in various parts of the world (India, Bangladesh, Nepal, Ethiopia) where pulses are a main plant food ingredient (e.g., Aykroyd et al. 1982; Stockman 1917; Campbell 1997). The use of bitter vetch yeast and chickpea is a characteristic example mentioned by several authors (Hillman 1985; Psilakis and Psilakis 2001; Maltezou 2004). Thus, the seeds of pulses can be used for the preparation of yeast for making raised bread. The making of *eptazymo* bread requires a chickpea sourdough: ground chickpeas are left to soak overnight before being mushed in a puree that is left to ferment in lukewarm water in a warm temperature of 30°C (86°F) (Psilakis and Psilakis 2001; Maltezou 2004). The use of bitter vetch seeds in a similar way, reported in various sources, could also potentially result in distinct cotyledons and/or seed fragments. Experimental specimens of this type of material are currently being investigated by the PlantCult team. Pulse flour could be mixed with cereal flour for the preparation of bread, for example, lentil flour, something mentioned in ancient texts (Valamoti et al. 2022a). It could have been also used in its own right for the preparation of pancakes (see the *socca* from Nice, known as *farinata* in Italy) (Albala 2013). Various recipes for breadmaking including pulse flour or a pulse component for the preparation of a special sourdough are well known from rural Greece (Psilakis and Psilaki 2001).

The archaeobotanical material available until now has not provided indications for the preparation of yeast made of pulse seeds, and indeed that might be too difficult to recognize in the archaeobotanical record. Only if a house caught fire while yeast preparation was underway could we hope to find some remains of the soaking grain demonstrating the characteristic cavities observed in experimental bitter vetch seeds (Valamoti et al. 2011).

In prehistoric Greece pulses could have been consumed as part of daily meals accompanying cereals. Traditional ways of cooking pulses are the same as those found in the ethnographic record and ancient texts (Valamoti et al. 2022a). Despite the limited archaeobotanical evidence for cooking practices related to pulses, we can assume that cooking of pulse seeds in various forms would have rendered them easy to digest, while in certain cases it would have further facilitated the removal of toxic substances.

A way of consuming *Lathyrus* seeds, in this case grass pea, reported from the Dervenochoria plateau in Thebes on central mainland Greece, involves cooking of whole seeds, with their testa, which are brought to the boiling point, then the water is discarded, and subsequently they are cooked in a form of a pulse soup as is done with lentils (Niki Chondrou, personal communication). A similar way of cooking grass pea has been reported by a villager from Karya on the island of Lefkada in the Ionian Sea and the villages of Petaleia and Agios Georgios on the island of Corfu (personal observations). In Spain the soaking of grass pea is reported not as part of splitting grass pea seeds but prior to cooking them in stews; seeds are also ground into flour and cooked as gruel (Peña-Chocarro and Zapata-Peña 1999).

Thus, besides the removal of testa, treatment of toxic pulse seeds with water could have helped remove toxic substances in *Lathyrus* and bitter vetch as is indicated by ethnographic studies and food science. The quick boiling and discarding of the water before cooking would have been effective in achieving this (Zeist and Bottema 1971; Dalianis 1993). This would have been a challenging task in a prehistoric context, especially if the cooking for many people needed to be done in the large cooking pots of 4-liter (ca. 1.056 gallons) capacity: they are heavy and difficult to handle as was easily realized during experimental cooking of *Lathyrus sativus* seeds (see experimental pots in Figure 3.17 and Figure 3.18). Soaking and the removal of testa would have been much easier ways in a prehistoric context.

At present no preparations such as a pulse seed mush or seed soup have been identified in the archaeobotanical record, unlike cereal-based preparations such as bulgur or *trachanas* (see Chapter 2) or other cereal-based food preparations (Kohler-Schneider 2001). We can speculate that pulses could have been consumed as mushes prepared with split cotyledons or as soups made of whole seeds. They could also have been ground to flour used for the preparation of leaven, breads, or pancakes. Starch analysis may prove a useful tool for exploring the cooking of pulse seeds, and experimentation in this direction is certainly a means to develop a methodological tool for such analyses (Kasapidou 2020).

Besides corresponding to an important staple food, the cooking and consumption of pulses in rituals and feasts are very likely, despite the flimsy nature of the archaeological evidence when it comes to exploring such aspects of prehistoric life. Although projections to the prehistoric past based on ancient sources and the ethnographic record are inadvisable, they can be used as a heuristic device that aims to move beyond the subsistence role of pulses as food in prehistoric societies.

Pulses have been associated with food taboos and with the dead in the recent and more distant past. Celtic beans were a taboo food among certain groups of people; therefore, their consumption was regulated by restrictions and prohibitions. The Pythagoreans, for example, refrained from eating them (Flint-Hamilton 1999; Scarborough 1982). Various pulses are cooked in the context of religious festivals across the Mediterranean, for example, in Greece and Spain nowadays (Valamoti et al. 2011) and constitute a food often offered in funerals in Greece, while pulses such as lentils and broad beans constituted offerings as part of Roman funerary rites, with Celtic beans in particular being considered as housing the souls of the dead (Flint-Hamilton 1999; Megaloudi 2004). The limited evidence available from prehistoric Greece may be pointing to similar uses in prehistory. Pulse seeds have been found in funerary contexts in prehistoric Greece, for example, at Neolithic Avgi and Toumba Kremastis Koiladas, where a connection with pulses and ancestors may be suggested by the context of the finds (see Valamoti 2011a).

The rich finds of pulses in great variety of species and the few indications of prehistoric processing and special contexts of consumption show that pulses were important food elements in prehistoric Greece. A wealth of ways to transform them into food could have existed, by means of detoxification and through their use in various recipes ranging from liquid consistencies to solid foods in the forms of breads/pancakes and even in fermented foodstuffs.

4

Oil Plants in the Prehistoric Cuisines of the Aegean

Olive oil is an emblematic element of what is widely known as the "Mediterranean diet" and the "Mediterranean triad" consisting of bread, wine, and oil or the tetrad with the addition of pulses (Matalas 2006a; Sarpaki 1992a). The association of olive oil with Greek cuisine may well be justified given the long history of its use in the broader Mediterranean basin, for example, cooking in frying pans with oil, various dough preparations or fish in the ancient Greek world and modern times (e.g., Dalby 1997; Kochilas 2001; Valamoti et al. 2022a). In areas marginal or unsuited to the growth of olive trees, nowadays, alternative sources of culinary fat are used, including plant oils (linseed or sesame in recent times) or animal fat (see Bozi 1997). Yet, the emphasis on prehistoric olive oil consumption as food in prehistoric Greece prevails in the literature (Renfrew 1972; Zohary et al. 2012; Riley 2002; Tzedakis and Martlew 1999; Sarpaki 2009a). There is evidence, however, to suggest that our modern perception of the culinary contribution of olive oil in Mediterranean cuisines should not be projected to the past: olive oil was most likely used in special, ritual contexts while a culinary use was most likely limited and confined to the wealthy, both in ancient and more recent times (Hamilakis 1996; Boulotis 1996; Foxhall 2007; Raftopoulou 1996; Valamoti 2009; Fappas 2008; Matalas 2006a and b). Some potentially oil-yielding crops, mentioned in the Linear B archives (the clay tablets used to record goods in the Late Bronze Age palaces [Melena 1983; Sarpaki 2001b]), such as sesame (*Sesamum indicum*; sa-sa-ma) and *Carthamus tinctorius* (ka-na-ko) have not been recognized among prehistoric archaeobotanical assemblages from Greece. The Linear B word for sesame strongly resembles the Akkadian *sammasammu* and is considered to be borrowed in Mycenaean Greek from an oriental source (Melena 1983). This underscores the contact paths that brought new plant food species, especially condiments and special plants such as oil plants in the Eastern Mediterranean "world system" of the Late Bronze Age. This chapter focuses on the actual archaeological remains of plants that might have been used for oil extraction in Greece in the Neolithic and the Bronze Age and briefly touches on the context of prehistoric oil consumption.

A wide range of plants with oil-bearing fruits and seeds is indicated through the archaeobotanical record of prehistoric Greece, and these plants could potentially have been used for the extraction of oil. Such plants may be underrepresented in charred archaeobotanical assemblages because their content in oil renders them vulnerable during charring (Wilson 1984; Kroll 1993; Märkle and Rösch 2008). Olive stones are probably an exception and can be preserved even when in contact with temperatures of 350°C (662°F), as has been shown through experimental work (Adam-Veleni and Mangafa 1996; Margaritis and Jones 2008; Marinova et al. 2011; Braadbaart et al. 2016). In the case of the olive, the use of *pomace* (pressed olive cake, the by-product of olive oil extraction) as fuel, attested from later periods, could have greatly contributed toward its preservation and representation in the archaeobotanical record (see Rowan 2015).

Figure 4.1. Grinding linseed with replica grinding stones. Traces of usewear and plant microremains could provide insights of processing plants for oil in the past. (Photograph by Maria Bofill. Courtesy of PlantCult.)

In light of the taphonomic limitations on oil plant representation among charred archaeobotanical assemblages, the available rich concentrations of seeds/fruits of oil plants from sites with burned destruction levels are rather impressive and indicate a predominant role of certain oil-yielding plants in prehistoric agriculture, economy, and probably diets of the inhabitants of the region under consideration. Even a quick glance at the different plant species potentially used for oil extraction with such properties (Appendix A:Tables 4.1 and 4.2) is enough to show that their variety increases remarkably during the Bronze Age. At present no solid evidence exists that these plants were indeed used for their oil-rich seed; therefore, oil extraction can only be inferred from the archaeobotanical record rather than proven. An analysis of plant microremains from the surfaces of archaeological grinding stones might be a way forward, opening up a pathway to explore the possibility of oil extraction from various small-seeded oil plants in prehistoric times (Figure 4.1).

Oil Plants Rooted in the Neolithic: Terebinth and Linseed

Terebinth

Terebinth, *Pistacia terebinthus* L., together with other *Pistacia* species, grew in vegetation *refugia* in the Balkan Mountains during the Pleistocene, from where it spread with the onset of the Holocene (Willis 1992, 1994), marking a change in climate conditions favoring the appearance and spread of this plant in the east Mediterranean (Rossignol-Strick 1995) and Greece (Bottema 1974; Lawson et al. 2004; Turner and Sánchez-Goñi 1997; Allen 1990; Turner and Greig 1975; Bottema 1979; Lawson et al. 2005; Wijmstra 1969; Panagiotopoulos et al. 2013; Willis 1992).

In Greece, wood charcoal analysis from late Pleistocene and early Holocene hunter-gatherer sites provides information for the earliest appearance of *Pistacia* in the vegetation of different areas and for the use of its wood for fuel by Late Paleolithic and Mesolithic groups (Maria Ntinou, personal communication). Direct ^{14}C dating of *P. terebinthus* charred wood from the Mesolithic sequence of Theopetra Cave confirms its presence at 8500 cal. BC (Ntinou and Kyparissi 2016). The taxon spread on the mainland and the islands with the onset of the Holocene (Franchthi Cave, Cave of the Cyclops) and was continuously used for fuel thereafter as suggested by finds at numerous sites in the north and south, coastal, and inland (Maria Ntinou, personal communication).

Charred nuts of *Pistacia* from northern Greece have been identified as *P. terebinthus* (Valamoti 2004). Based on the morphological characteristics of the nuts it is unlikely that the finds in the north of Greece correspond to *P. lentiscus*, but *P. atlantica* cannot be excluded from the identification. The identification as *P. terebinthus*, however, is probably correct, as *P. atlantica* is not a component of the flora of northern Greece. The distinction between the two species, in any case, is unimportant for the evaluation of the uses of the plant as both species are closely related to each other in their properties and uses (Nielsen 1986).

P. terebinthus nuts are encountered at several Neolithic sites located primarily in northern Greece, albeit in limited numbers. They have been found at Neolithic Thermi B, Giannitsa B, Makri, all sites with a coastal setting. The plant is absent, on present evidence, from most inland sites that have been studied, such as Mavropigi-Fillotsairi, Toumba Kremastis Koiladas, Kleitos, Apsalos, and Promachon-Topolnitsa, with the exception of Dikili Tash, which, although inland, was not far from the coast (Valamoti, unpublished archaeobotanical evidence).

The richest concentration encountered so far comes from the Late Neolithic site of Makriyalos, in central Macedonia, a flat extended site southwest of the modern city of Thessaloniki in northern Greece, inhabited during the Late Neolithic (middle of the sixth millennium BC and the middle of the fifth millennium BC) (Pappa and Bessios 1999; Pappa 2008). The nuts originate from the interior of a pit, in close spatial association to the southern end of a rectangular, bipartite building that contained, besides flax and terebinth, other plant remains. The complete archaeobotanical assemblage from the pit used during the later phase of habitation at the site seems to underscore its individuality (Pappa 2008; Valamoti 2011c:Figure 3, 2013). About 70 whole nuts were found there in a sample dominated by this plant (Figure 4.2). It appears that the pit contained exceptional burned remains, given that most of the samples from Makriyalos are chaff dominated. They were most certainly harvested from the wild, probably in the vicinity of the settlement, and somehow ended up getting burned and deposited in the pit. Their use is certain, but their precise intended use can only be speculated on based on our knowledge of the plant's properties and on ethnographic information (Valamoti 2013b). They may have been intended for spicing up some dish, for oil extraction, or for the aromatic smoke that would result from their burning. A special, healing use for the nuts found at Makriyalos is also likely.

Terebinth nuts are edible and were heavily consumed by Persians; a special meal consisting of terebinth nuts, a figcake, and a bowl of sour milk

Figure 4.2. Terebinth *Pistacia terebinthus* nuts found charred in a Late Neolithic pit at Makriyalos II. (Photograph by Stavros Dimakopoulos. Courtesy of PlantCult.)

formed part of the Persian royal initiation rite, according to Plutarch (Alcock et al. 2001; Briant 2000; Dalby 1997). These nuts are rich in oil; the resin included in the plant's woody parts has medicinal and aromatic properties, known already in antiquity (Gennadios 1997; Özcan 2004; Facciola 1990; Kunkel 1984; Marcopoulos 1965; Merousis 2016; Rosengarten 1984).

The culinary uses of the nuts and oil continue nowadays in parts of the Mediterranean and the Near East. The nuts are used for oil extraction on a small scale in the modern Aegean (e.g., Chios and the Cyclades, Merousis 2016; Delatola-Foskolou 2006) as well as for spicing up breads in parts of Greece (Delatola-Foskolou 2006; Psilakis and Psilakis 2001). In recent times, at the village of Vassileoniko on the island of Chios, the "chickpea looking" terebinth nut (τσίκουδο ρεβυθάτο/*tsikoudo revithato*) occurs in a verse from an old song and corresponds to a variety best suited to be eaten as a nut, while two more varieties, an early one (ψηλάντικο/*psilantiko*) and a high-yielding oil one (χατζημηνάτο/*chatziminato*), were used for special Christmas sweets (Abanoudis 2007). Terebinth oil is also reported from Anatolia, in Turkey, where the steps involved in its production have been observed at the village of Seklik in the Yund mountains with an estimate of 25 kilos (55.12 pounds) of oil extracted from 100 kilos (220.5 pounds) of terebinth nuts (Doğer and Sezgin 2015). The difference between descriptions of terebinth oil extraction from Chios island and Seklik is that, in the latter case, heating of the crushed nuts is involved, a step not used on Chios island, the heat probably altering the quality of the oil extracted from the nuts. A reexamination of the Linear B texts has led Merousis to propose the use of terebinth oil among Minoan and Mycenaean societies in the Aegean (Merousis 2016).

Linseed

Flax (*Linum usitatissimum* L.) yields seeds rich in oil (Figure 4.3). It is not only an oil plant though. Nearly all parts of the plant have many uses for food, medicine, fiber, and oil production, hence its Latin denomination *usitatissimum*, meaning the "most useful one." Linseed remains from prehistoric Greece are charred and their finds range from a few seeds to dense, rich concentrations. Because of their high flammability, however, even these few remains could be an indication for flax cultivation (Kroll 1993), and their low visibility in the archaeobotanical record should be unsurprising. At present

Figure 4.3. Charred seeds of flax (*Linum usitatissimum*) from Late Neolithic Makriyalos (*top*) and end of third millennium BC Archondiko (*bottom*). Note the seed size difference between the two sites. (Soultana Maria Valamoti)

no stems or capsules have been reported from Greek sites. Chances for preservation of these components of linseed are highest among waterlogged remains (e.g., Hornstaad-Hörnle IA, Maier 2001, 2004; Herbig 2009).

Remains of flax from the Early Neolithic of Greece have been reported from the Thessalian sites Sesklo and Otzaki Magoula, where flax is present in three out of five samples (Kroll 1981). Besides a few Middle Neolithic (5800–5500 BC) records from Thessaly (no quantitative data is provided, Kroll 1983), flax is encountered in a few samples from Middle Neolithic western Macedonia, at the site of Apsalos, in very small numbers (only four seeds from four samples out of a total of 101 samples sorted for plant remains) (Valamoti 2006a). Thus, overall, linseed finds are rather scarce for the Early and Middle Neolithic (approximately 6600–5500 BC). Although no exact figures are provided for most of the early Thessalian and northern Greek finds, given the limited chances of preservation of this plant, they probably indicate that this crop was already cultivated by the first farmers that appeared in Greece in the seventh millennium BC, confirming the contribution of flax to the "Neolithic package" reaching southeast Europe from Anatolia and the Middle and Near East (e.g., Zohary and Hopf 2000).

Linseed remains become more prominent among Late Neolithic assemblages. Most of the sites dated to this period fall in the region of Macedonia in northern Greece. This geographical distribution is most likely the result of a sampling bias, with a more intensive retrieval of plant remains being recorded for the north of the country over the last two decades, a picture likely to change in the near future (see, e.g., Livarda 2014 and Livarda and Kotzamani 2013). Small numbers of flax seeds have been identified at Late Neolithic Toumba Kremastis Koiladas, Makri, and Pefkakia (Kroll 1981; Karathanou and Valamoti 2011; Valamoti 2004). Besides these sporadic finds at a number of sites, flax seeds have been identified as pure, relatively rich concentrations at Makriyalos, Arkadikos, Mandalo, and Dikili Tash (Valamoti 2004; Valamoti and Jones 2003; Valamoti 2011c). The flax concentration of Makriyalos, where approximately 2,100 linseeds were found in a pit, had quite a distinct composition, unlike most samples from the site, which were dominated by cereal chaff.

Linseed is prominent both in the north and south during the Bronze Age at several sites in Thessaly, Macedonia, and the Peloponnese. Early archaeobotanical explorations in the 1960s by Maria Hopf revealed an impressive find, among the first to be reported from Greece, from the site of Lerna in the Argolid, dated to the Middle Bronze Age (Hopf 1961): linseed was found fused in a lump, most likely due to charring. Further rich concentrations and/or conglomerations of fused flax seeds are reported from the north of the country, from the end of third millennium BC Archondiko (Valamoti 2011c) (Figure 4.4) and Late Bronze Age Kastanas (Kroll 1983) and Assiros (Jones and Valamoti 2005).

These finds clearly imply that during the Neolithic and the Bronze Age, flax was a cultivated crop at least at these sites where abundant evidence is available and possibly at the other sites where only a few seeds have been identified. The rich, pure caches of flax seeds at Archondiko have been encountered at various locations in the excavated area, corresponding to different

Figure 4.4. Agglomeration of flax seeds from Archondiko, end of third millennium BC. (Photograph by Chryssa Petridou. Courtesy of PlantCult.)

postframed habitation units (Figure 4.5). This suggests that flax was a common crop, regularly cultivated and used, grown perhaps by different households. At Mandalo, a large number of flax seeds were found in one sample, associated with what appears to be an early, specialized flax weed flora with a number of *Lolium* sp., *Lolium remotum*, and *Silene* sp. seeds, probably representing weeds of flax fields (Valamoti and Jones 2003). Rich flax concentrations have also been identified at Arkadikos and Dikili Tash in northern Greece, representing clean grain, stored for consumption or sowing.

The cultivation of flax is highly demanding in terms of soil and moisture requirements as well as weed elimination (e.g., Dewilde 1999), something

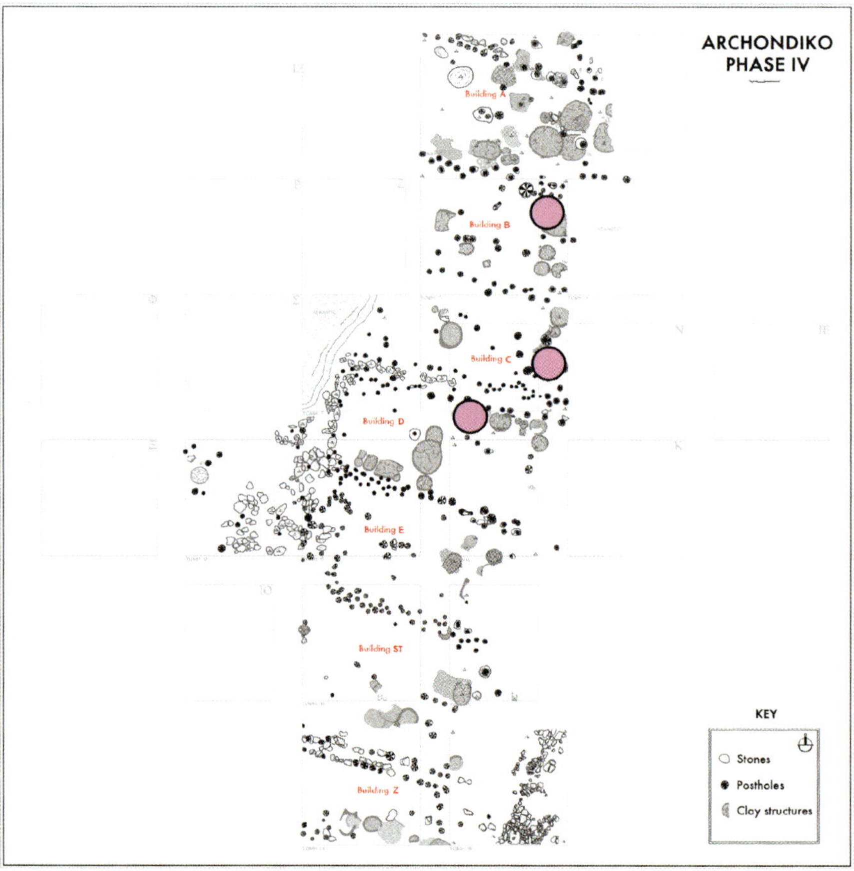

Figure 4.5. Site plan with circles suggesting separate concentrations of linseed among the structures excavated at Archondiko, phase IV, end of third millennium BC. (Courtesy of Aikaterini Papanthimou.)

already well known from Classical and Roman Antiquity (e.g., Kroll 1993; Tzahili 1997, referring to ancient writers such as Pliny and Collumela; contra Megaloudi 2006). Thus, the Neolithic and Bronze Age flax fields that can be inferred from the presence of linseeds in archaeobotanical assemblages from Greece would have required skilled agricultural knowledge, necessary for maintaining high nutrient levels in the flax fields through fallowing, manuring, and/or crop rotation with pulses. Flax fields would have probably required manuring, a practice indicated by very high nitrogen values obtained from isotopic analyses of linseed from Dikili Tash in northern Greece (Bogaard, personal communication).

Although no measurements for individual seeds are as yet available, it is interesting to note that the flax seeds from Archondiko appear overall to be of a smaller size than those encountered at the Late Neolithic sites of Makriyalos, Arkadikos, and Dikili Tash (Valamoti 2011c). Varieties of flax are usually small seeded when used for fiber and large seeded when used for oil, while varieties intended for both uses are also reported (Zohary and Hopf 2000). This size difference (see Figure 4.3) could be related to different uses of the plant, for example, the presence of an oil or fiber variety developed during the Early Bronze Age, or a landrace with specific uses or characteristics (see Herbig and Maier 2011 for flax varieties in fourth/third millennia BC Germany). Whether the small size of flax seeds at Archondiko is related to specific uses of the plant, in this case for fiber, or to harvesting of plants not fully ripe, a practice adopted when good fiber quality is desired (e.g., Dillman 1928; Gennadios 1997; Wilson 1979), is difficult to tell. Varieties intended for different uses could have either been developed locally or introduced through contact networks with other regions, evidenced through lines of artifactual evidence, including plant species (Valamoti 2007b, 2017), something worth investigating in the future through morphometric analysis. Even if the small-sized flax seeds found at Archondiko could be related to a variety intended for fiber extraction, the seeds, once ripened in the capsules, could have been used for the extraction of their oil.

Flax has usually been left out of discussions of economic organization and daily activities, yet, the regular occurrence of flax at several prehistoric sites from Greece, despite the admittedly limited chances of preservation, as well as a distinct weed flora and indications for manuring and storing in several household units, indicates that it was a very important crop. Late Bronze Age Linear B texts highlight its significance in providing flax yarn for the numerous weavers (the "flax ladies") working for the Mycenaean palaces (Killen 1984; Robkin 1979; Foster 1981; Tzahili 1997). The flax seed that survived within the archaeological deposits could represent seed kept for sowing the next crop intended for oil or fiber or both, seed kept for extracting oil, or seed kept for medicinal uses (see, for example, Chauhan et al. 2009 for a recent review of the uses and properties of *L. usitatissimum* products, seeds in particular).

Linseed oil is rich in omega-3 fatty acids, and it is edible when extracted using a cold-water method (Serpico and White 2000; Breslow 2006; Hu et al. 1999; Kapoor and Huang 2006). Linseed oil was produced in parts of Anatolia (Ertuğ 2000) and in Kappadokia in particular as olives do not grow in the region (Bozi 1997; Ertuğ 2010). As indicated by ethnographic evidence from Anatolia, the ideal timing for extracting linseed oil could have been in winter, when animals needed high-quality fodder such as pressed linseed (Ertuğ 2010).

The Olive

Among the plants that could potentially be used for oil extraction in prehistoric Greece, the olive stands out as the oil plant par excellence, associated with the Mediterranean and Greece in particular, a plant that holds an eminent role in the archaeological literature despite the fact that the archaeobotanical remains of the olive tree suggest a low visibility in the archaeological record at least for the Neolithic period (Hansen 1988; Hamilakis 1996; Valamoti et al. 2018). As early as 1986, Runnels and Hansen and later Hansen (1988) called for caution regarding the status of the olive as a systematically exploited crop for the Neolithic and much of the Bronze Age. The issue of olive exploitation by prehistoric communities of Greece has been investigated over the years, bringing together a wealth of archaeobotanical information related to this plant (Hamilakis 1996; Sarpaki 2000; Bottema and Sarpaki 2003; Megaloudi 2006; Valamoti 2009; Valamoti et al. 2018). Despite the great role of the olive for Bronze Age and later economies of the east Mediterranean and the Aegean, the origins of its cultivation/exploitation in the Aegean and mainland Greece remain obscure and largely speculative (see, for example, Blitzer 2014; Valamoti et al. 2018).

The presence of the olive in Greece as part of prehistoric natural vegetation is confirmed by the remains of wood and pollen of the plant. Fossils of olive wood and leaves have been found in Pleistocene layers from Santorini, dated to 50,000 and 37,000 years ago (Friedrich 1980; Velitzelos 1990) (Figure 4.6). Based on the pollen diagrams from Greece, olive pollen has been found in diagrams from the last 10,000 years, in low percentages and with a discontinuous presence (e.g., Athanasiadis et al. 1996). The available

Figure 4.6. Olive leaves fossil from Santorini dated to Palaeolithic times. (Photograph courtesy of Museum of Palaeontology, Aristotle University of Thessaloniki.)

palynological data indicate that the earliest appearance of olive pollen during the Holocene comes from Crete and is dated to the Late Neolithic (end of fifth/fourth millennia BC), either the result of introducing the tree on the island during that period or due to air transportation (Bottema and Sarpaki 2003). Such a timing for olive's appearance in the Aegean has now been confirmed by high-resolution pollen spectra from the Kouremenos marshland in eastern Crete showing the presence of olive trees, in a landscape dominated by open, shrubby, phryganic vegetation around 4300 BC, that is, the Late Neolithic, while during the Final Neolithic and throughout the Bronze Age, despite some fluctuations, there is a significant increase in olive pollen, especially toward later periods (Cañellas-Boltà et al. 2018), something also observed in the macrobotanical record from the island (Valamoti et al. 2018). The palynological record suggests that the olive may not have been a constant element of the natural vegetation of the Aegean, at least for a long time during the Holocene, and also during the Pleistocene, as the plant cannot survive in cold climates and cannot stand constant low temperatures or temperatures below -13°C (8.6°F). During glacial periods, *Olea europaea* survived in *refugia*, which, thanks to their orientation and altitude, provided conditions suitable for the plant, for example, as indicated by the Navarino pollen record in Messenia (Bottema and Sarpaki 2003 quoting Kraft et al. 1975); when climatic conditions improved, the plant spread to other parts of the Mediterranean basin (Neef 1990; Zohary and Hopf 2000). A core from Navarino in Messenia shows an early peak for the olive, dated to 8900 BC (Kraft et al. 1980). Olive pollen in the Peloponnese is prominently present in the Neolithic in some locations (e.g., Zakynthos), and it becomes more visible from the Bronze Age onward, with varying peaks in the Early and Middle Bronze Age periods, depending on location (Weiberg et al. 2016; Wright 1972; Athanasiadis et al. 1996).

In western mainland Greece, olive pollen is discontinuously present in extremely low numbers and appears at somehow higher percentages at about 3500 BC based on pollen analyses from Lake Voulkaria (Jahns 2005, 2009). A discontinuous, low presence of *Olea* pollen is observed farther to the north as is indicated by diagrams from central mainland Greece, dated to the fourth millennium BC (e.g., Athanasiadis et al. 1996; Greig and Turner 1974; Wright 1972; Valamoti et al. 2018). The appearance of olive pollen in northern Greece comes from the Tenagi Philippon diagram near Kavala, much later than in the south, dated between 1900 and 1350 BC based on pollen zone correlations (Greig and Turner 1974). These northern finds on their own, are rather problematic not only because of the insecure dating but also due to the fact that olive pollen travels long distances. Recent palynological investigations in the peninsula of Chalkidiki in northern Greece, however, seem to lend support to the earlier evidence: olive pollen appears at the site of Tristinika in Chalkidiki in the Late Bronze Age (second half of second millennium BC) (Panagiotidis and Papadopoulou 2016) lending further support that the tree was growing in the north Aegean in the second millennium BC.

The presence of olive wood in Greece during the Neolithic is extremely limited, and, based on archaeobotanical micro- and macroremains including charcoal, it can be argued that the plant was nearly absent from most archaeobotanical assemblages of the first two millennia of Neolithic farming in Greece and was probably not used during most part of that period. This picture begins to change toward the end of the Neolithic in the south of Greece (Valamoti 2009; Valamoti et al. 2018). Olive stones are present on the island of Keros (Margaritis 2013a) and on eastern Crete (Livarda and Kotzamani 2013; Cañellas-Boltà et al. 2018). Material currently under study has also yielded olive stones from the Early Bronze Age site of Skarkos on the island of Ios, in the Cyclades and from the Late Bronze Age palatial complex of Dimini/Iolkos (Valamoti, observations on unpublished archaeobotanical material) (Figure 4.7).

The remains of olive stones and wood make their appearance in a more or less systematic way during the Bronze Age (Appendix A:Table 4.2). The interesting feature of these remains is their geographical distribution, as they are found in southern Greece and mainly in the southern Aegean, Crete, and the Peloponnese (Valamoti et al. 2018:Table 1, Figure 1). In light of the available evidence, it seems that the olive tree is exploited toward the end of the Neolithic in southern Greece and that it is absent both from the Neolithic and most of the Bronze Age of the northern part of the country, except for the two pollen diagrams from Philippoi and Tristinika, aforementioned. On present evidence, the northernmost occurrence of olive stones falls around the area of the Pagasitic gulf and Sporades islands, on the coastal and insular regions to the east of Thessaly and the northeast Aegean islands (Valamoti et al. 2018). This differentiation between north and south should

Figure 4.7. Olive stones from Dimini-Iolkos, Late Bronze Age. (Soultana Maria Valamoti)

be unsurprising as the environmental conditions in the regions of Macedonia and Thrace are a limiting factor for a long-term survival of the olive tree. Ongoing research in the context of project PlantCult is expected to offer a more complex and nuanced picture of the presence and use of the olive in the north Aegean. For example, olive charcoal has been recently detected in Middle Bronze Age layers at Skala Sotiros on the island of Thassos (Maria Ntinou, personal communication, Theodosaki 2021).

Despite modern attempts to distinguish between wild and domesticated olives on the basis of olive stones (Newton et al. 2014), it may prove harder than expected to attempt this distinction when very small stoned varieties, like the *koroneiki* of southern Greece, are brought into the analysis. Moreover, wild olive trees growing together with cultivated ones may have been exploited in the same way as the domesticated/cultivated ones, thus blurring the distinction between wild and domesticated in terms of treatment of the plants and uses of its fruit. Once the wild olive is pruned, of course, it falls in the realm of "cultivation" due to tending. Morphometrics of prehistoric and later olive stones from Greece fall within the range of the wild looking olive, demonstrating the characteristic round shape of wild olives despite the definitive cultivation of the olive in historic periods on the basis of artifactual and textual evidence (Pagnoux 2016). The difficulty in distinguishing between wild and domesticated plants is also present in olive wood and pollen. Thus, the archaeobotanical distinction between the two subspecies remains problematic and impossible at present to determine whether the stones found in prehistoric Greece belong to domesticated or wild plants (Pagnoux 2016).

The available archaeobotanical evidence from Greece raises several questions in relation to the origins of olive cultivation in the region. What triggered the increasing appearance of archaeobotanical remains of olive in the south of Greece? A close connection between southeastern Crete and the olive tree is observed by Blitzer (2014), Valamoti et al. (2018), and Cañellas-Boltà et al. (2018). Blitzer (2014) argued that olive cultivation must have been introduced in the Aegean through the Near East, also emphasizing the role of olive oil in elite rituals. Was therefore incipient tending and cultivation a local process (Margaritis 2013b; Cañellas-Boltà et al. 2018) or subject to a wider need for olive oil among certain communities of the east Mediterranean (Blitzer 2014; Valamoti et al. 2018)?

Local wild olive trees were probably part of the natural vegetation of Greece, while the idea of managing them for their oily fruit might have been introduced, together with knowledge on managing the tree to increase fruit production. The idea to cultivate the olive tree in Greece may have been introduced via contacts with the east Mediterranean, where olive oil production appears somewhat earlier (see Valamoti et al. 2018). It is possible that already from the Late Neolithic the olive was cultivated for its oil-yielding fruit in eastern Crete, initiated through contacts with the Levant. From eastern Crete, olive cultivation quickly spread to the south Aegean and northward to the islands of Sporades, the northeast Aegean, and Thessalian coast, during the course of the Bronze Age, culminating in the Late Bronze Age via the maritime trade networks of the Mycenaean world with elites of the east Mediterranean.

The increasing popularity of the olive evidenced through plant macroremains corresponds with increasing evidence for the circulation and use of oil(s) and unguents, a trend possibly associated with ritual uses (Hamilakis 1996, 1999; Fappas 2008; Valamoti et al. 2018). It is very likely that olive cultivation and olive oil production increased during the course of the Bronze Age to meet the demands in oil for ritual pursposes, especially in rituals underlining the role of elites (see Hamilakis 1996).

Some archaeologists view olive oil as an essential element of prehistoric Aegean cuisine (e.g., Sarpaki 2009a). Hamilakis in his comprehensive overview in the 1990s (Hamilakis 1996) placed the emphasis on the luxury, ritual uses of oil by Bronze Age elites rather than a culinary use, a view shared by Boulotis who examined the uses of olive oil as suggested by Linear B texts (Boulotis 1996). Foxhall (2007), in her thorough review of the available evidence related to the olive, considers the use of its oil as a luxury practice, associated with the wealthier parts of society in ancient Greece.

Although the exploitation of the olive is attested in the archaeobotanical record of prehistoric Greece, the culinary uses of olive oil remain largely a field of speculation. Many scholars have expressed skepticism regarding the contribution of olive oil to the prehistoric diet of the region based on the limited presence of its remains (cf. Hansen 1988; Runnels and Hansen 1986). Indeed, the remains of the olive from prehistoric Greece are not as impressive when compared with later deposits of historical times where the finds of olive stones are abundant, despite a lack of systematic archaeobotanical recovery at Classical and Roman sites (e.g., Adam-Veleni and Mangafa 1996; Margaritis and Jones 2008; Megaloudi 2005; Megaloudi et al. 2007; Pagnoux 2016). If we consider that the percentage with which a species appears at archaeological deposits corresponds to a certain extent (after eliminating taphonomic factors) to its contribution toward fulfilling different needs of people, the cautious approach of Hamilakis (1999), Hansen

(1988), and Runnels and Hansen (1986) regarding a widespread use of olive oil for culinary purposes in Bronze Age Greece is still valid in light of the recently retrieved achaeobotanical data. Further support for the use of olive oil in special contexts is corroborated by artifactual and textual evidence, as will be discussed in the last section of this chapter. The bowl of olives retrieved from a Minoan well at Zakros on Crete, powerfully marked in the memory of Yannis Sakelarakis who brought it to light from its murky, waterlogged depositional environment (Sakelarakis 2005), points to other uses of the olive tree, for its fruit. The consumption of olives in a culinary or ritual context (see Margaritis 2014) can by hypothesized, yet further evidence is much needed to this end.

Bronze Age Novelties: Tiny, "Exotic," and More Than Just "Oil-Rich"

A wide range of potentially oil-yielding plants have been identified in the archaeobotanical record of Bronze Age Greece. Besides linseed and terebinth, known already from the Neolithic, new oil plants, some of exotic origin, make their appearance during the Early Bronze Age: opium poppy (*Papaver somniferum* L.), mustard (*Brassica/Sinapis*), gold of pleasure (*Camelina sativa*), and *Lallemantia*. These plants could have been used for their oil-rich seeds as well as for other properties.

Brassicaceae: Seeds of the Mustard Family

Seeds belonging to three taxa of the mustard family, *Brassica/Sinapis* and gold of pleasure (*Camelina sativa*), have been found in Bronze Age contexts of northern Greece, the former at Assiros (Jones and Valamoti 2005) and the latter at Kastanas (Kroll 1983). Both species yield edible, oil-rich seeds while this oil is edible. *Brassica/Sinapis* is widely cultivated in modern Europe for the production of edible vegetable oil and is promoted for the production of biofuels (Hedrick 1972; Woods et al. 1991; Lust 1983; Hill 1952; Brown 2005). Similar uses are known for *Camelina sativa*, whose oil-rich seeds contain linolenic acid, like linseed (Pilgeram et al. 2007). As is the case with many other plants, both species are also used in traditional healing remedies.

Lallemantia sp.

Lallemantia, a genus native to Transcaucasia and central Asia but not Europe, has seeds that are rich in oil (Jones and Valamoti 2005). *Lallemantia* oil can be used for lamps (Hedrick 1972), for making paints (Helmut Kroll, personal communication), and as a dietary element as it has similar properties as linseed, that is, it is rich in omega-3 fatty acids and linolenic acid. Seeds of a *Lallemantia* species that were impossible to identify further to *L. iberica* Fisch. and Mey, *L. canescens* Fisch. and Mey, or *L. peltata* Fisch. and Mey, whose seeds resemble each other, have been found at several Early and Late Bronze Age sites in northern Greece (Figure 4.8). A single seed reported from Late Neolithic Kremastis Koiladas by Karathanou (2009) is most likely a Bronze Age intrusion. *Lallemantia* has been found at Mandalo and Archondiko Giannitson in deposits dated to the third millennium BC, 2950–2200 BC and 2100–1900 BC, respectively (Jones and Valamoti 2005). These are the earliest finds from northern Greece and southeastern Europe, and later finds from northern Greece come from Assiros Toumba and are dated to the second half of the second millennium BC (Jones and Valamoti 2005). Farther north than Greece, *Lallemantia* has been identified

Figure 4.8. Seeds of *Lallemantia* sp. fused in agglomeration, Archondiko, phase IV, end of third millennium BC. (Photograph by Chryssa Petridou. Courtesy of PlantCult.)

at deposits of the first half of the third millennium BC at Dabene Sarovka in Bulgaria (Marinova and Valamoti 2014) and later, in the second millennium BC, at Feudvar (Kroll 1998) and Zidovar (Medović 2003). To the east of Greece, seeds of this genus have been found as a weed of cereal crops at Hattousa, dated to the second millennium BC (Neef 2001). Recently, a single seed of *Lallemantia* has been identified in the archaeobotanical assemblage from Adji Kui 1 in Murgab, Turkmenistan, dated to 2270–1960 cal. BC (Spengler et al. 2018).

The plant's natural distribution spreads from central Asia to Anatolia and the Near East and nowadays is used in Iran as an oil-yielding and medicinal plant (Hedrick 1972; Benedikt 2007 [1895]; Samadi et al. 2007; Ghannadi et al. 2015). The Murgab find is interesting in this context, as the area where it has been found falls in a zone that might have acted as a corridor through which different plant species might have spread eastward and westward. Our knowledge about the plant's geographical distribution in prehistoric times is very limited and patchy, the biggest gap existing precisely in the area of its natural distribution. Thus, the routes via which *Lallemantia* reached northern Greece remain obscure and speculative for now. Metal trade routes, of tin and tin bronzes in particular, could have been one of alternative reasons for which *Lallemantia* may have spread westward and reached northern Greece and the Balkans (Valamoti and Jones 2010). This has been suggested based on *Lallemantia*'s modern distribution and its overlap with ancient sources of tin. Indeed, some of the ancient tin sources can be detected in Transcaucasia, Iran, and Afghanistan, where the plant grows wild and where nowadays it is still cultivated and used for its seeds.

Were *Lallemantia* seeds imported to Greece on a regular basis, or was it a crop introduced and cultivated in gardens for its seeds and bee-attractive flowers? Were these used for their oil content or for their medicinal properties or for both? Given that linseed was already known since the Neolithic and its cultivation continued into the Bronze Age, what may have led to the introduction of yet another crop with similar properties? As flax is a demanding crop both in terms of nutrients and moisture, *Lallemantia* might have presented an alternative oil source from a plant less demanding in its growth requirements as it performs well under low soil moisture (Abdolahi and Maleki Farahani 2019; Omidi et al. 2018). *Lallemantia* thus presents an additional possible source of plant oil, an exotic plant, intriguing in the way it was introduced to the Aegean and later farther north in the Balkans.

Opium Poppy (Papaver somniferum L.)

Another "exotic" plant, introduced to prehistoric Greece from the west (Iberian Peninsula) or the north (central Europe), is opium poppy. Opium poppy is almost absent in Neolithic Greece, unlike the Neolithic of the Iberian Peninsula and Italy and of the Linearbandkeramik cultures of Europe (Valamoti 2009). Sporadic finds of the plant are known from Late Neolithic Makriyalos (four individual seeds identified), whereas at Final Neolithic Mandalo *Papaver somniferum* occurs at a few samples in low numbers (five definite and fourteen probable seeds reported in the fractions studied (Valamoti and Jones 2003; Valamoti 2001a). Given the very limited chances of preservation of opium poppy seeds as has been recently shown by Märkle and Rösch (2008), it is possible that these scarce Late Neolithic finds might indicate that the plant was known and used already from Neolithic times in Greece. Nevertheless, the possibility that these seeds might be intrusive from later levels cannot be ruled out. Recent archaeobotanical research at Archondiko (near Giannitsa) in central Macedonia has provided evidence for the presence of

Figure 4.9. Seeds of opium poppy (*Papaver somniferum*) from Archondiko Giannitson, phase IV, end of third millennium BC. (Soultana Maria Valamoti)

opium poppy in the end of the Early Bronze Age in northern Greece, around 2100 BC (Figure 4.9). The Late Bronze Age finds of opium poppy, by contrast, are dense and rich, such as the poppy seeds retrieved from Kastanas and Assiros Toumba in central Macedonia, suggesting that opium poppy was a crop in its own right, stored and therefore used by the inhabitants of Greece. In the absence of solid evidence for a Neolithic presence of opium poppy in the study area, the view holds that opium poppy was a crop of the Bronze Age (Valamoti 2003, 2009) a picture that may be valid also for the wider region of southeastern Europe (Marinova and Valamoti 2014).

Opium poppy seeds, like those of flax, have many uses, including oil extraction from the crushing of the seeds. Culinary uses include the consumption of the seed, whole or crushed, as is the case, for example, with various bread and cake types some known from the ancient Greek world (Valamoti et al. 2022a). There are recipes in ancient Greek texts that clearly mention the culinary use of opium poppy seed. An example is the poppy seed bread (ἄρτος μακωνίδης/*artos makonides*) as described in ancient Greek poetry and recipe texts (Psilakis and Psilakis 2001:127; Micha-Lampaki 1984). Opium poppy seed is probably free of the narcotic substances contained in the plant's head from which opium is extracted (Azcan et al. 2004; Özcan and Atalay 2006), although other sources report a limited content of alkaloids, including morphine, in poppy seed (e.g., Bjerver et al. 1982; Grove et al. 1976; Buddha et al. 1996). The seeds are a rich source of edible plant oil, with high nutritive value (Özcan and Atalay 2006; Rahimi et al. 2011), rich in linolenic, linoleic, and other fatty acids, with other uses in soap and cosmetic manufacture as well (e.g., Frick and Hebeisen 2005). Opium poppy seed oil is edible and contains stearic, palmitic, oleic, linoleic, and linolenic fatty acids.

Today the production of opium poppy seed oil is carried out on a domestic level, in small presses, in countries such as India and Turkey (Alagesaboopathi 2012; Özcan and Atalay 2006). The mass of the pressed seeds after oil extraction can be used as human food, while poppy seeds pressed into a "cake," using a specialized mill, form an ingredient of various recipes in different parts of central and northeastern Europe (e.g., the Polish sweet *makowiec*) (Sheen 2012). As with *Lallemantia* and other species with small-sized, oil-rich seeds, the intended uses of the stored caches of poppy seeds remain unknown, with oil production being one among several possible culinary uses as well as medicinal ones. Of course, opium poppy is more famous, both nowadays and in the past, for its medicinal and hallucinogenic properties, contained in the sap of the unripe poppy head.

Prehistoric Plant Oils in Context: A North-South Divide?

Plant oils required an investment of labor for their extraction and thus acquired a special value that contributed toward a status as luxury commodities centrally produced and distributed in the Bronze Age Aegean (Sherratt and Sherrat 1991). Therefore, uses other than food may have been more likely, in the context of healing and/or for cosmetic purposes (Andreou et al. 2013).

The plants presented in this chapter could have been used for oil extraction and for several other uses. Olive oil, stereotypically associated with the Aegean since prehistoric times, would have been limited in space and time, following the distribution of this plant's use established on the basis of the archaeobotanical record as presented here. The olive, a source of raw material for the production of oil and perfumes in the palaces of the south, was probably not growing in the north Aegean before the second millennium BC. Alternatives, such as terebinth nuts or the tiny seeds of poppy, linseed, mustard, *Lallemantia*, and gold of pleasure, could have been used as sources for local oil production in the north where the olive could not successfully grow or yield fruit (Jones and Valamoti 2005; Andreou et al. 2013). Ethnographic evidence documents the use of plant oils other than olive oil in regions unsuited for its cultivation: *Brassica* in northern Europe, poppy in France, linseed in Kappadokia, and *Lallemantia* in Iran (Hedrick 1972; Benedikt 2007; Bozi 1997; Azcan et al. 2004; Özcan and Atalay 2006). Where olive trees could flourish, however, other oil-yielding plants could have also been grown for the same purpose: oil extraction. Modern oil production from terebinth nuts offers an interesting example, both for prehistoric times and for today, as is the case on the island of Chios, where both trees, terebinth and the olive, are used for oil production.

Processing the oil-rich seeds or fruit of the plant species discussed in this chapter would have required some basic equipment for breaking and crushing of the plant material. Oil could have been extracted using grinders and grinding stones as are commonly encountered at many prehistoric sites in Greece (see Figure 4.1). Pounding equipment, stone pestles and mortars, or a combination of stone and wooden elements could also have been used to this end. In this respect, pestles and mortars, which were rare during the Neolithic in Greece, increase in number during the Bronze Age and appear almost exclusively in the south (Bekiaris et al. 2020). There is a strong geographical gradient for the Bronze Age finds as they are encountered in areas of mainland Greece and the Aegean where the use of the olive has been detected archaeobotanically. This observed increase in pounding tools correlates with archaeobotanical indications for olive cultivation in the south during the Bronze Age, especially in Minoan and Mycenaean times, suggesting perhaps a close relationship of olive oil extraction with pounding equipment in the south Aegean, Peloponnese, and Crete. Crushed olive stones have been linked to olive oil production (Sarpaki 1999a), yet no straightforward criteria have been offered to present to distinguish postdepositional fracture of olive stones from that resulting from processing by pressing (Neef 1990; Marinova et al. 2011; Braadbaart et al. 2016).

The increased archaeobotanical visibility of olive stones in Late Bronze Age southern Greece, culminating in Mycenaean times, correlates with another line of evidence that leaves no doubt that olives were grown for oil: the Linear B tablets. Olive oil production is mentioned during the Late Bronze Age within the context of perfume production in the Mycenaean palaces of southern Greece (e.g., Shelmerdine 1984). It appears that oil from wild olives was preferred for perfume production (Melena 1983), the suitability and superiority of which are confirmed by recent research (Gulfraz et al. 2006).

The Mycenaean palaces were involved in the production and distribution of sophisticated perfumed substances, demonstrating a variety in terms of ingredients and consistency (Fappas 2008). Certain Mycenaean vase types have been interpreted as containers for perfume oils and ointments, often occurring in mortuary and other deposits (see Mountjoy 1993:127–128). A variety in the contents may be reflected in the variety of pots associated with perfumes such as alabastra, piriform jars, stirrup jars, perhaps each shape corresponding to a different content, combining different ingredients or consistencies (Figure 4.10 and Figure 4.11) (Melena 1983; Shelmerdine 1985). These appear to have been consumed in the Aegean and the wider eastern Mediterranean (Haskell et al. 2011; Cavanagh and Mee 1998:72, 74, 119). Their popularity and circulation across the Mediterranean, including the west, reach a peak in the Late Bronze Age, with Crete being the center of production (e.g., Ben-Shlomo et al. 2011; Day et al. 2016). Even if the idea of oil production from olives might have originated in the Levant, the Minoans and Mycenaeans managed to generate a "brand name" for their perfumed oils across the Mediterranean. This "branding" may have operated in a much more nuanced way, with regional centers of production or specific contents in stirrup jars being indicated by variations in their decoration and shape as has been suggested for postpalatial contexts of the end of the second millennium BC (see Koh and Birney 2017 and their reference to Rutter's "regional markers," pp. 26 and 33).

Figure 4.10. Cylindrical alabaster perfume container found in the Mycenaean chamber tomb K-3 at Englianos, Messinia. (From Malapani 2015, 604, Figure 490. Photograph courtesy of Evangelia Malapani.)

Figure 4.11. Stirrup jar found in the Mycenaean chamber tomb K-3 at Englianos, Messinia. (From Malapani 2015, 596, Figure 468. Photograph courtesy of Evangelia Malapani.)

As northern Greece is an area where the olive is at the limit of its cultivation, alternative oil plants could have been used for the preparation of luxury substances analogous to those produced in the Late Bronze Age palaces of southern Greece. The impressive variety of oil plants, found in rich concentrations at a large number of sites systematically investigated for their plant remains, could point toward their use for oil extraction, given the lack of olives. In the north, the plant oils may have been associated with a possible local oil/ointment production, making use of locally produced, special small vessels intended as oil/fat containers, a use supported by recent residue analyses (Figure 4.12) (Andreou et al. 2013). The local kantharoid pots were widely distributed over a large area extending from central Europe to northern Greece and most likely contained oils and fats that perhaps were perfumed (Andreou et al. 2013).

Olive oil consumption in the Minoan and Mycenaean worlds was, according to several authorities, an element of rituals associated with special occasions, a view based on contextual evidence, Linear B textual information, and ritual practices involving oil consumption taking place in Assyrian cities of the Middle East as described in cuneiform texts (e.g., Hamilakis

Figure 4.12. Comb-incised and paste-filled pottery, Assiros Toumba, without datable context (*left*) and from destruction level of Phase 9 granary (Room 9), mended with lead rivets, approx. 1350–1300 BC (*right*). Not to scale. These incised vessels may have been associated with plant oils and potentially with unguents or perfumes. (Photograph courtesy of Ken and Diana Wardle.)

1996; Blitzer 2014; Fappas 2008, 2011). Stirrup jars, the specialized liquid containers closely linked with the Mycenaean perfume industry and trade, were predominantly manufactured in Crete. Their very wide circulation in the Eastern Mediterranean emphasizes the prominent position of Mycenaean oils and/or perfumed oils among Late Bronze Age societies with a long tradition in the ritual use of such products, either locally produced or imported (Leonard 1994; Hankey 1993; Mee 2008:375, 377; Smith 1987:128). The continuing association of Crete with the production of stirrup jars throughout the periods of their use could reflect what clearly emerges from the archaeobotanical record, that is, the first appearance of olive cultivation and oil production on the island of Crete in the Aegean. It cannot be coincidence that both the ingredient (olive oil used for perfume/ointment production) and the stirrup jar (its container par excellence) appear to have been first produced on the island of Crete (Valamoti et al. 2018; Haskell 1985). Hamilakis notes that "Crete with its large variety of native spices and other plants with aromatic and even psychoactive properties . . . must have become renowned for its perfumed oils, the production knowledge of which might have been controlled by the palatial authorities" (Hamilakis 1999:49). What Hamilakis (1999) observes for the end of middle Bronze Age in Crete (Neopalatial) probably originated in the eastern part of the island and the Late Neolithic, when the intensive management/cultivation of the olive began, while the emergence of a specialized vessel, the stirrup jar in the Middle Minoan III (1750–1490 BC) period, coincides with an increase in the archaeobotanical evidence for the olive on the island that culminates in the Late Bronze Age (Valamoti et al. 2018). Special oils might have been produced and consumed as early as the Late Neolithic, where the olive was managed, yet their systematic production, circulation, and "marketing" through a special container (the stirrup jar) were rooted in the Middle Bronze Age, culminating during the Late Bronze Age.

The association of vessels used as oil/ointment containers with funerary contexts (though not exclusively with such contexts), both in the south and in the north of Greece, as well as within the wider eastern and central Mediterranean, further underscores the symbolic, ritual nature of the use of oils and unguents in prehistory (Haskell 1985; Koh and Birney 2017; Andreou et al. 2013; Leonard and Cline 1998; Spagnoli 2016). In the north Aegean the number of imported Mycenaean perfume containers is negligible (Andreou and Psaraki 2007), and local pottery variants (alabastra, small piriform jars, and amphoriskoi) seem to have this function (Andreou et al. 2013). Thus, olive oil, if at all known in the north during the Bronze Age, would have reached the shores of the north Aegean in these rare Mycenaean containers. Olive systematic cultivation and olive oil production may have arrived there late, probably toward the end of the Bronze Age, perhaps the outcome of Greek colonization and the introduction of lifestyles requiring the use of olive oil (Valamoti et al. 2018).

Assuming our interpretation of the archaeobotanical evidence is correct, oil seems to acquire a special position in the Bronze Age, both in the north and the south. The oil derived from olives or other plants available in prehistoric Greece was probably a luxury product as it required special ingredients, a set of techniques and know-how, as well as specialized equipment (Koh and Birney 2017; cf. Sherratt and Sherratt 1991). The nonculinary use of olive oil in prehistoric Greece was very convincingly argued by Hamilakis, who took into consideration all available evidence for prehistoric times as well as later periods (1999). The consumption of oil and aromatic oils seems to have held a special position in the Mycenaean world: it is mentioned in Linear B tablets as one of the commodities supplied to those in charge of organizing feasts within the context of religious festivals (Bendall 1998–1999, 2007; Shelmerdine 1984, 1998; Bunimovitz 1996; Boulotis 1996; Fappas 2008). The ritual use of an elite, luxury produce like oil in special places and contexts, related to ancestors (e.g., cemeteries) and deities (e.g., sanctuaries), would have probably been associated with power relations and the negotiation and/or reinforcement of social position and authority (Sherratt 1991b; Hamilakis 1999; Fappas 2009).

The special position of oil plants in Bronze Age Greece, as has been highlighted in this chapter, is probably connected to rituals and lifestyles requiring a luxury substance, oil. Plant oils could have been used in healing rituals, through their consumption or external use. Special uses of oil are indicated by residue analyses that associate plant oils with particular types of pottery, often found in funerary contexts. Oils in special containers might have been involved in social networks of bonding and/or status display through access to these networks and to ingredients of possibly limited availability and accessibility (see Hamilakis 1996). In light of the archaeobotanical and ceramic containers associated with plant oils, different networks of oil circulation may have existed during the Bronze Age, one connecting northern Greece with central Europe (pots with incised decoration) and another connecting southern Greece with the east Mediterranean world (the stirrup jar). Although in the north of Greece, no prominent remains suggestive of elites, as is the case with the palaces in the south, have been found, local elites may have existed, using luxury items such as plant oils as a means of social distinction.

5

Beyond Prehistoric Fields

The Harvest of Fruits and Nuts

With the onset of the Holocene, forest vegetation gradually expanded from *refugia* existing during the preceding glacial and interglacial periods. Neolithic landscapes of Greece were wooded, perhaps more so in the north, where a dense mixed oak woodland existed well into the Bronze Age (Marinova and Ntinou 2018). Fruits and nuts of some of the wild shrubs and trees growing in these forests were collected by the prehistoric inhabitants of the region, and some of them were probably being tended or even brought into cultivation. The opening up of land for fields and pastures would have encouraged fruit production as more sunlight would have accessed fruit-bearing trees, leading the way toward domestication for some. Moreover, access to those trees and bushes would have been easier at the margins of fields and pastures and the borders of grazing paths.

Unlike the stereotypical division between "wild" and "domesticated" often encountered in discussions of prehistoric farming, wild plant food resources continued to be exploited by the prehistoric inhabitants of Greece from the Early Neolithic onward. Crop fields may have corresponded to a part of the landscape more closely controlled by human populations. Wild trees, however, would have probably been the thread connecting farmers to ancestral gatherer-hunters (see Valamoti 2015a). Neolithic farmers would have constantly struggled against forest regeneration and weed regrowth, and some weeds could well have been brambles that give tasty fruit but are difficult to eradicate when a weed-free soil is required for a cereal or pulse crop.

We have no way of knowing how the continuum from wild vegetation to managed crop fields might have been perceived on a symbolic level by Neolithic and Bronze Age populations inhabiting the region. Such a dichotomy between what ancient Greeks thought of as civilized life, that is, cereal cultivation, on one hand, and wild plants, raw food, and humanity before civilization on the other (see, for example, Vidal-Naquet 1983:44–61), may not have existed in prehistoric Greece. Judging, however, from a people-plant perspective, archaeobotanical remains offer ample evidence for the presence of morphologically wild grape pips even in historic times and for a constant presence of wild vegetation in prehistoric lives in the region. People would have interacted with wild plants when removing weeds from crop fields or when poisonous or intoxicating wild plants such as the grapevine were needed by the community. That part of the landscape where wild plants grew might have constituted a potentially dangerous realm as crop fields expanded during the course of the Neolithic and the Bronze Age. Fruits like figs and grapes that were cultivated in later periods could have corresponded to the wild part of nature, connecting farmers to their distant gatherer-hunter past (Valamoti 2015a). Several millennia later, in the historic periods, these two worlds of vegetation correspond to the domains ruled by two ancient Greek gods, Demeter the goddess of cultivation and Dionysos the god of the grapevine and wild vegetation. The fruits and nuts encountered in the archaeobotanical samples of prehistoric Greece could have corresponded to margins between fields and the woodland proper, a transitional zone between a tamed landscape and wilderness.

Although fruits and nuts in most cases may not represent staple foods of Neolithic and Bronze Age communities of Greece, they must have constituted an important dietary contribution of vitamins, minerals, and trace elements. Some among them may have even constituted staples or sources of special, mind-altering preparations, as will be discussed further on. A long list of fruit and nut species emerges through the archaeobotanical record of Neolithic and Bronze Age Greece (Appendix A:Tables 5.1 and 5.2). These included grapes, Cornelian cherries, blackberries, elderberries, figs, wild pears, acorns, almonds, and terebinth nuts, although not at all sites and all periods. Occasionally a stone of a Rosaceae fruit (such as plums and cherries) is reported, although such single finds are not discussed here due to their low representation in the archaeobotanical record. Fruits and nuts would have been seasonally available, in abundance in some years, scarce in others.

Where rich concentrations of fruits and nuts are encountered in the archaeobotanical record, the most likely interpretation is that they were harvested and stored for later use. Such concentrations are often encountered inside burned houses or refuse pits. Sometimes they are found in small numbers, one or two seeds among rich concentrations of cereal chaff. In this case, the occasional stones, fruit seeds, or nut cotyledons may represent a few fruits consumed by an individual, a human or even a goat, seeds that after their journey through the digestive system ended in dung and ashes resulting from dung burning.

Two of the species that could have been included here, terebinth (*Pistacia terebinthus* L.) and the olive (*Olea europaea* L.), as they yield fruits that could have been used as food, have already been discussed in Chapter 4, devoted to oil plants.

Grapes and Figs: From the Wild Forests of the Neolithic to Mycenaean Orchards

Grapes

The fruit that has been widely associated with prehistoric Greece is the grape. The main product associated with viticulture, wine, is discussed in the next chapter. The importance of the grapevine and wine has been emphasized by many researchers of the Aegean, especially in discussions about Bronze Age societies. Colin Renfrew considered the cultivation of the grapevine of great importance for the emergence of Bronze Age elites: together with the olive, the two plants allowed for the exploitation of marginal lands (not ideal for cereal and pulse cultivation), leading to a specialization of production, which in turn called for redistributive centers (C. Renfrew 1972). Likewise, Antonio Gilman (1981) considered the cultivation of the grapevine of key importance for the emergence of the hierarchical communities of the Bronze Age; he argued that the high demands of grapevine cultivation tied people to their land, rendering them more vulnerable to being controlled by an authority. The process through which viticulture emerged and its relationship to hierarchical societies and elites have been the subject of ongoing research and debate (see, for example, Hamilakis 1996, 1999; articles in Wright 2004b).

Greece falls within the broader zone where the grapevine (*Vitis vinifera* L.) is native, according to Vasileios Logothetis, a Greek agronomist who studied wild grapevine populations in Greece (Logothetis 1962, 1974). The grapevine is present in prehistoric Greece as is indicated by the pollen finds of the plant, identified in pollen diagrams at many locations (Athanasiadis et al. 1998; Bottema 1982). Pollen evidence from northern Greece suggests that the grapevine was present in *refugia* during the last ice age and perhaps in other cold periods during the Pleistocene (Mommersteeg et al. 1995; van der Wiel and Wijmstra 1987; Wijmstra 1969). The *Vitis* pollen record, however, is not continuous, neither geographically nor temporally during the Holocene. Palynological analyses provide information on the presence of the plant in the vegetation of a region and periods of *Vitis* pollen increase could suggest the appearance or expansion of vineyards (see Valamoti et al. 2020). An increase in *Vitis* pollen is observed around 3,500 years ago, considered in the literature as indicative of cultivation of the plant (Bottema and Woldring 1990; Bottema 1982). Archaeobotanical macroremains allow a better, high-resolution understanding of the relationship between prehistoric communities of the region and the grapevine.

The grapevine is probably the most commonly occurring fruit in prehistoric Greece, as a large number of the sites investigated archaeobotanically have yielded remains of these fruit plants, some among them in rich concentrations. The archaeobotanical macroremains of the grapevine, encountered at archaeological sites in Greece, consist mainly of charred pips and to a lesser extent stems and skins, the latter occasionally attached to pips (Figure 5.1). In a few cases, grape pips are found mineralized (e.g., Valamoti 2004). Besides parts of the fruit, *Vitis* wood has been identified at a few prehistoric sites (Valamoti et al. 2020). Comprehensive accounts of *Vitis*

Figure 5.1. Grape (*Vitis vinifera*) pips, charred during a conflagration episode, from House 1 at Dikili Tash, Late Neolithic, approx. 4300 BC. (Soultana Maria Valamoti)

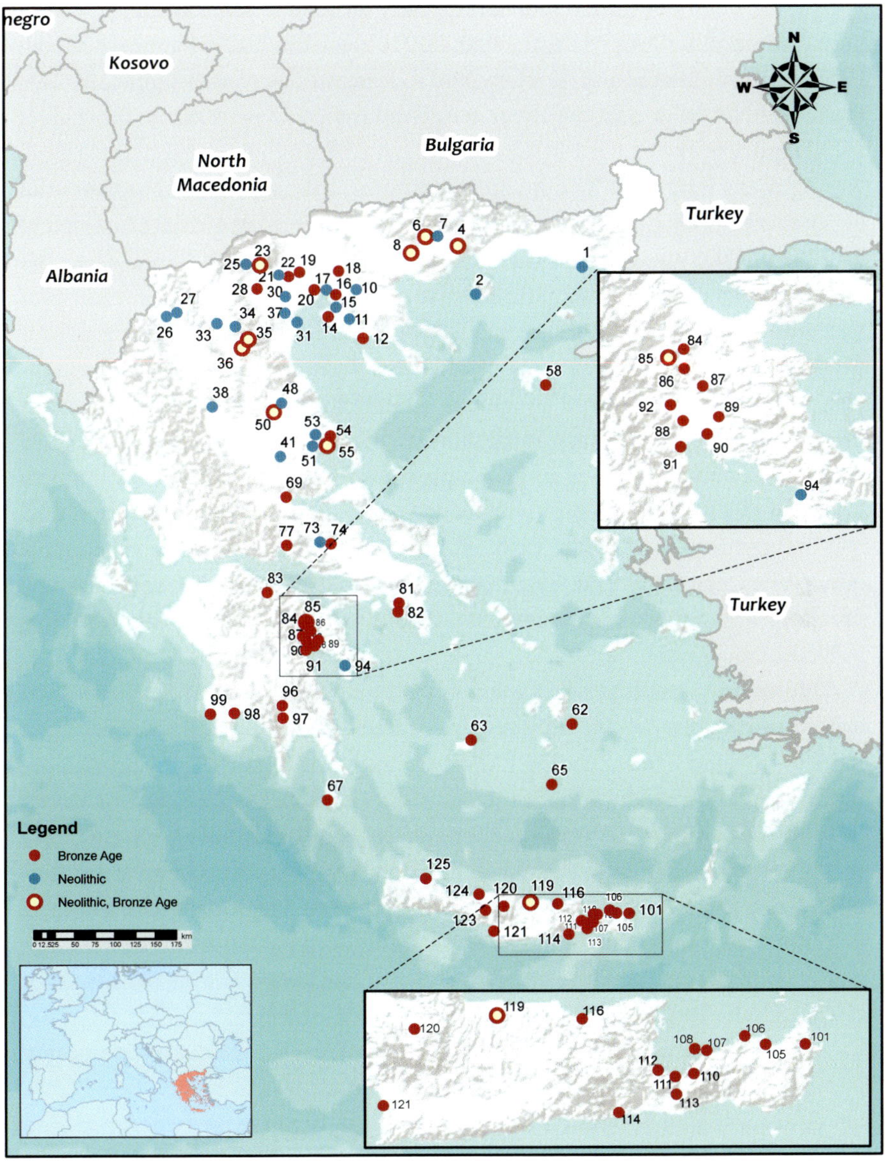

Figure 5.2. Map of Greece showing the presence of grapevine in the Neolithic and the Bronze Age. (Base map sources: Esri, US Geological Survey, and National Oceanic and Atmospheric Administration. Map by Themis Roustanis. Courtesy of PlantCult.)

finds have been compiled and have greatly aided the discussion presented here (Hamilakis 1996; Mangafa et al. 2001; Valamoti 2009; Pagnoux 2016). Grapes are present at nearly every site investigated archaeobotanically dating to the Neolithic and the Bronze Age in Greece (Appendix A:Tables 5.1, 5.2) (Figure 5.2).

At many Early and Middle Neolithic sites, the grapevine is represented by a few pips only, dispersed among the various archaeological features excavated. At Makriyalos, in an area with several clay structures, mainly hearths and/or ovens, dated to the first half of the fifth millennium BC, several pips have been identified. Their number, however, does not exceed fifty pips in a single context, a low percentage of the total plant items present in the samples, which are dominated by glume wheat chaff (Valamoti 2004). From the second half of the fifth millennium BC onward, the number of sites where rich concentrations of grape pips are present increases. At Dimitra in northern Greece, 234 pips were found, and 2,400 grape pips were found at a burned house at Dikili Tash (Valamoti et al. 2007). Many more have now come to light at Dikili Tash, where besides the grape pips, several grape pressings and a few stalks were also found (Valamoti 2015a; Valamoti, unpublished material).

Numerous remains of charred grape pips from the Bronze Age have been found. In the north, hundreds have been identified at Kastanas (Kroll 1983), Toumba Thessalonikis (Andreou and Kotsakis 1997), and Assiros Toumba in central Macedonia (Jones 1983). In southern Greece rich concentrations of pips were found at Lerna in the Argolid and at Myrtos in Crete (Hopf 1962a; Renfrew 1972). Recently, rich concentrations have been identified from P.O.T.A. Romanou (an area of economic regional growth) in the Peloponnese, suggesting intensive exploitation during the Early Bronze Age and possibly an early form of cultivation of the grapevine in the south of Greece (Valamoti et al. 2020).

Samples with rich concentrations of grape pips occur at an increasing number of sites as one moves from the Neolithic to the Bronze Age. Thus, for Greece, the raw material from which wine could have been produced was available to the prehistoric inhabitants of Greece. Of the thirty-four sites investigated archaeobotanically in the north, only Arkadikos Dramas did not yield grape pips. Comparing Neolithic with Bronze Age finds from Greece, it becomes obvious that the number of sites where grape pressings and/or grape pips in large concentrations increases, especially in the south of Greece. In the Neolithic, the grapevine is encountered in 37 out of 61 sites; among those, two bear indications for intensive use due to their large numbers, both located in the north, while in the Bronze Age the grapevine is encountered in 76 out of 80 sites with seven cases corresponding to rich concentrations, four of them located in the south (Peloponnese and Crete), suggesting a more intensive use of the grapevine during the Bronze Age (Appendix A:Tables 5.3 and 5.4) (see also Pagnoux et al. 2021).

One of the research questions addressed when grape pips are found among archaeological remains from Neolithic and Bronze Age sites in Greece concerns the form of the plant, that is, whether the pips originated from wild (*Vitis vinifera sylvestris*) or domesticated grapevines (*Vitis vinifera vinifera*). This is a particularly interesting issue: the distinction between these two forms would provide a reliable criterion for the recognition of a close relationship between human communities and the grapevine, that of cultivation. Moreover, the ability to identify domestication in prehistoric times would allow us to detect the existence of a larger scale of plant use and labor investment related to soil preparation, pruning, fertilizing, and planting in specific areas, unlike the small labor investment when plants are harvested from the wild.

For the Aegean, Jane Renfrew (1995), in a comprehensive review of the grape finds from the region, suggested that by the fifth millennium BC the cultivation of the grapevine is indicated with both wild and domesticated forms being present in archaeobotanical assemblages. Any investigation of early grapevine exploitation needs to take into account the types of grape pips represented in the archaeological deposits. The wild subspecies (*Vitis vinifera sylvestris*) produces small and roundish pips with short stalks, while domesticated ones (*Vitis vinifera vinifera*) display bigger and more elongated pips with longer stalks. The only way available to archaeobotanists to distinguish the two subspecies is based on various measurements of features of the grape pip and their relative proportions in combinations, starting from the basic breadth/length index of Stummer (1911) and continuing with various attempts to resolve the issue with more refined formulas, taking charring also into consideration (e.g., Smith and Jones 1991; Mangafa and Kotsakis 1996; Bouby et al. 2013; Terral et al. 2010). Morphometrics have formed the basis for an identification of several morphotypes of domesticated grapevine, which in turn could shed light on the spatial and temporal dynamics of cultivated grapevine diversity (cf. Pagnoux et al. 2015; Pagnoux 2016). Yet, recent work suggests that the attribution of archaeological grape pips to the wild or domesticated grapevine can be affected by the method selected to achieve this (Pagnoux et al. 2022). Despite morphometric approaches that achieve a high proportion of accurate classification of pips into the wild and domesticated groups (Bouby et al. 2018), distinguishing between wild and cultivated grapevines in the archaeobotanical record is not a straightforward process. Most significantly, morphologically wild pips do not directly reflect the nature of the relationship between the grapevine and human societies that used it in prehistoric times.

A basic problem in using the various ways of classifying archaeological pips for the identification of domestication and thus cultivation of the grapevine lies in the fact that we know hardly anything about the time that might have been required for the development of the morphological changes of the grape pip that serve to distinguish the domesticated subspecies from the wild one: for an unknown period of time people might have cultivated the plant without any visible changes in its morphological characteristics (Valamoti 2009:97). By selecting a wild grapevine that would produce sweet grapes, as has been suggested by Miller (2008), or big bunches of grapes of a certain color and taste, people would have reproduced the plant vegetatively through cultivation, without this necessarily corresponding to pips looking domesticated. There is nothing to confirm, at least for the time being, that the grapevines selected for an incipient cultivation had fruit with morphologically domesticated pips. Morphometric studies indicate that this was a process (e.g., Pagnoux et al. 2021; Valamoti et al. 2020). Thus, one needs to exercise caution in using grape pip morphology, something also underscored by the grape pip characteristics of the modern variety of Pinot noir, used in France for wine making, which according to its grape pip measurements fall in the wild grape category (Rivera-Nuñez and Walker 1989). In fact, recent research has shown that even in historic periods for which grapevine cultivation is well known from ancient sources, wild-looking grape pips continue to be present in the assemblages in significant proportions (Bouby et al. 2013; Pagnoux 2016). Similar observations performed on material from Petra in Jordan, dated between the first century BC and the first century AD, show that the grape pips fall within the wild grape category (Jacquat and Martinoli 1999), an unexpected outcome, given the late dates for the finds, that is, in a period when the grapevine was widely cultivated.

Both Renfrew (1995) and Logothetis (1970) point out that grape pips in the early phases of cultivation would most likely have closely resembled wild grape pips. Terral and his team (Terral et al. 2010) have argued that grape pip measurements change toward a more elongated shape, a trait of domesticated grapevines, developed as a result of vegetative propagation through cultivation, as the pip gradually lost one trait associated to germination efficiency. The time required for wild-looking grape pips to change their shape as a result of cultivation remains unknown, however, as wild-looking pips need not necessarily correspond to grapes harvested from grapevines growing in the wild but from plants under some form of management (Valamoti 1998; Valamoti 2015a). As wine could have been produced from wild fruits, however, cultivation was not a prerequisite for wine making per se but for large-scale wine production (Valamoti 1998; Valamoti et al. 2007).

The means therefore to detect grapevine domestication in the archaeobotanical record are problematic for two main reasons: (1) a considerable overlap in pip measurements between wild and domesticated grapes, aggravated by distortion caused by charring (Logothetis 1970; Stummer 2011; Smith and Jones 1991) and (2) early cultivated/domesticated grapevines need not have necessarily developed seeds looking morphologically domesticated in shape and dimensions. The wide variation and overlap of pip morphology depend on how well developed the domestication traits (pip morphology, the length of the pip, and the stalk in particular) are in the material examined (e.g., Rivera-Nuñez and Walker 1989; Smith and Jones 1991; Stummer 1911; Terrall et al. 2010). The morphometric traits of pips remain open to alternative interpretations when considering the boundaries between domestication and cultivation. Thus, cultivation cannot be excluded as a possibility when morphologically wild grape pips dominate an assemblage (Valamoti 1998, 2015a; Valamoti et al. 2020).

The implications of this for researchers of prehistoric viticulture and wine is that, for an unknown period of time, the prehistoric grape pips that we describe as morphologically wild might have in fact belonged to cultivated plants. On a more optimistic note, once pips that yield indexes that attribute them to the morphologically domesticated subspecies are found, one can be certain for the cultivation of the plant. The rich Dikili Tash finds, including pips and grape pressings, seem to have originated from morphologically wild plants as the formulas of Mangafa and Kotsakis (1996) suggest. The same results have been obtained by a recent morphometric study by Pagnoux (2016). Yet, the large quantities of grape pips and/or the presence of *Vitis* charcoal could suggest incipient cultivation (Valamoti 1998, 2015a; Valamoti et al. 2020). *Vitis* charcoal identified at the site is an outcome of pruning within a framework of cultivation of the grapevine (Miller 2008) and could correspond to a cultivated grapevine with morphologically wild pips. The Neolithic find from Dikili Tash in northern Greece may also be an indication that wine making, viticulture, and domestication could have started independently in different parts of the broader zone of wild *Vitis* distribution, northern Greece being one of them. This view seems to be strengthened by DNA analysis of *Vitis* plants (Arroyo-Garcia et al. 2006), which suggest at least two different regions where the grapevine was domesticated. In other words, it is not necessary that viticulture and wine making originated in solely a single region such as the Near East (Zohary 1995), Anatolia, or the Caspian (e.g., Olmo 1995; McGovern et al. 2017). In any case, wine making and early viticulture should not necessarily be linked to morphologically domesticated grape pips in the archaeobotanical record as the remains from prehistoric Greece clearly suggest. Moreover, the shift from morphologically wild pips to increasingly more morphologically domesticated ones over time, as supported by morphometric studies spanning the Neolithic through to the Late Bronze Age from Greece (Pagnoux et al. 2021), would have required a considerable time under cultivation to become visible in the archaeobotanical record (Valamoti et al. 2007). This recent investigation of several assemblages of archaeological grape pips (Pagnoux et al. 2021) provides clear evidence for a process of systematic management of originally wild plants that were locally available, both in the north and the south of Greece, leading to domesticated-looking pips during the course of time, in particular the end of the Early Bronze Age. This is further confirmed by recent morphometric studies on grape pips from Agia Paraskevi in central mainland Greece (Ntinou et al. 2022) and Tsoungiza in the Peloponnese (Pagnoux et al. 2022). This would have been achieved initially by encouraging wild plants, pruning them so as to be able to reach the fruit and increase yield and at perhaps soon after by employing vegetative propagation of certain individuals with desirable properties such as sweet or large fruit (see Zohary and Spiegel-Roy 1975 for a detailed discussion of Old World fruit domestication processes). Although there is no direct evidence for vegetative propagation in the archaeobotanical record, we may assume that this may well have been practiced as early as the end of fifth millennium in the north of Greece, a practice that would have led to the large number of grapes harvested to produce wine at Dikili Tash.

Based on the archaeobotanical remains from Greece in the Neolithic, it appears that since the beginning of the Neolithic, people used the fruits of the grapevine. Moreover, grape juice was extracted in northern Greece during the fifth millennium BC and throughout the country during the Bronze Age, rendering wine making a strong possibility. The numerous remains of grape pressing at northern Greek sites may be due to a more intensive archaeobotanical investigation in the region. Recent research in the Peloponnese has doubled the available evidence for the south (Valamoti et al. 2020). Among prehistoric sites in Greece, Neolithic Makri, and Dikili Tash, where numerous wild-looking grape pips have been identified (Mangafa and Kotsakis 1996; Pagnoux 2016), *Vitis* charcoal is also present (Makri, Maria Ntinou, personal communication; Dikili Tash, Valamoti et al. 2015). According to Miller's suggestion, these charcoal finds would indicate pruning and therefore cultivation, as pruning would not be needed or even possible to carry out on wild-growing plants (Miller 2008). The rich northern Greek finds may suggest a close association of the region with the grapevine and its products, in which case the mythical association of ancient Thrace (partly corresponding to the regions of Macedonia and Thrace in modern Greece) to one of the birthplaces of Dionysos, god of wild vegetation, wine, and intoxication, might reflect some very early wine-drinking episodes in the region.

The grapevine seems to be absent from regions farther north than Greece with few exceptions of a few grape pips found at fifth millennium BC Vinča in Serbia (Dragana Filipović, personal communication) and Mursalevo in Bulgaria (Marinova et al. 2016). It also appears in Stillfried in Austria in the early first millennium BC (Kohler-Schneider 2001). Thus, compared with the Greek finds that are abundant, macrobotanical finds from the Balkans and central Europe suggest that the grapevine was probably not part of the fruits exploited by the prehistoric inhabitants of those regions. The *Vitis* pollen detected in pollen profiles from Lake Constance in southern Germany probably shows that wild grapevines were initially present in the Early Holocene, climbing on high trees. Later on, however, these were reduced, perhaps due to human activity disturbing the natural vegetation of those parts of the landscape where the wild grapevines grew (Rösch 2016). Turning to the west Mediterranean, grape pips appear at several sites of the Iberian Peninsula as sporadic finds, and it is after the colonization by Greeks and Phoenicians that viticulture becomes established in the west (Buxó 2008; Pérez-Jordà et al. 2017; Iriarte-Chiapusso et al. 2017), and the same applies to the south coasts of France (Bouby 2014). *Vitis* remains are more numerous in Italy during the Neolithic and especially in the Late Bronze Age, something attributed to contacts with the Mycenaean world (Marvelli et al. 2013), contacts attested via ceramic finds (e.g., Blake 2008). Such contacts and especially Greek colonization in the early first millennium BC have been considered as the driving force for the establishment of viticulture in Italy (Lentjes and Saltini-Semerari 2016).

Viticulture and wine making probably spread gradually westward from the east Mediterranean and the Aegean, through maritime contacts, and this is visible in the archaeological record during the Late Bronze Age and the Mycenaean period (see Soares 2019 for a recent review of the evidence).

Such contacts, well attested for the Late Bronze Age, might have begun earlier. In Sardinia, an island with contacts with the Mycenaean world (Usai 2017; Recchia and Cazzella 2017) the third millennium BC rich concentrations of pips suggest the systematic use of the grapevine (Ucchesu et al. 2015) the outcome perhaps of influences from cultures in the east already familiar with wine-drinking combined with local processes involving the management of wild-growing grapevines. Thus, around the third millennium BC a greater connectivity via maritime routes might have contributed to the spread of wine and viticulture, a "bloody" process if ancient Greek myths reflect a collective memory of the roots of viticulture and wine.

Figs

Fig trees (*Ficus carica* L.) constitute an important element of Mediterranean vegetation and would have attracted prehistoric people with their sweet fruits. It has been argued that fig was probably the first plant to be domesticated, based on observations on finds from Gilgal I in Israel, dated to the tenth millennium BC (Kislev et al. 2006; but see Denham 2007). Domesticated figs produce both male and female fruits while in the wild form, the fig is a dioecus, that is, with male and female plants. The distinction between wild and cultivated figs in archaeobotanical material is problematic. The fig tree producing edible fruit (*Ficus carica* var. *domestica*) is in reality the female plant, while the male (*Ficus carica*, var. *caprificus*) produces pollen and hosts the insect *Blastofaga psenes* that fertilizes the female plants (Gennadios 1997; Zohary and Hopf 2000; Kislev et al. 2006). As the edible fruit corresponds to a female fruit, there is at present no way to distinguish archaeobotanically the type of tree from which a fig has been harvested (wild-growing or cultivated). The domestication of the fig tree corresponds to the selection of shoots for reproduction from plants that produced sweet fruits (Zohary and Hopf 2000), and there is no way of distinguishing the seeds from a wild female plant and those from cultivated female figs.

Whole figs that had been dried, most likely, have been unearthed at many sites such as Dikili Tash and Makri (Valamoti 2004), Rachmani in Thessaly (Renfrew 1973a:Figure 36), and Sternes at Chania in Crete (Sarpaki 1999b). Fig fruit fragments have been found in ditch A at Makriyalos, probably the remains of food for people, as their possible consumption by animals as fodder would have led to the preservation of the seeds only and not of the whole fruit. This is clearly shown in an experiment carried out where figs and einkorn were fed to a goat and the subsequent dung pellets were studied for their contents of identifiable plant remains (Valamoti and

Figure 5.3. Charred figs (*Ficus carica*) from a stored concentration of more than 800 whole figs, excavated at Kyparissi, northern Greece, Late Neolithic. (Photograph courtesy of Maria Pappa.)

Figure 5.4. Charred figs (*Ficus carica*) from a stored concentration of more than 800 fragmented figs, excavated at Kyparissi, northern Greece, Late Neolithic. (Photograph courtesy of Maria Pappa.)

Charles 2005). Recently a rich concentration of figs has been unearthed from a house floor at a Late Neolithic house at the site of Vasilika Kyparissi (Pappa and Nanoglou 2019; Lathiras 2020), showing that figs, a rich source of energy and nutrients, were systematically harvested and stored, dry, for later consumption (Figure 5.3 and Figure 5.4).

Berries and Other Fruit

Woodland clearings and riverbanks and those margins where the forest would have given way to the managed landscapes of prehistoric Greece would have been a rich source of colorful and perfumed berries and other fruit (Appendix A:Tables 5.1 and 5.2). Blackberries (*Rubus fruticosus* L. agg.) are frequently present in samples from the Neolithic and the Bronze Age, while raspberries (*R. idaeus* L.) are very rarely encountered, which might reflect their availability close to the settlements: blackberries thrive alongside fields and in woods while raspberries are more difficult to find and are usually encountered in forests in the northernmost parts of the country. Wild strawberries (*Fragaria vesca* L.) are only reported from Kastanas in the Early Geometric period in the beginning of the first millennium BC (119 seeds, Kroll 1983) and so far have not been found at Neolithic and Bronze Age sites.

Plant remains from northern Greece testify to the presence of Cornelian cherry (*Cornus mas* L.) during the Neolithic (Figure 5.5 and Figure 5.6); it has been found in the Early Neolithic deposits of Mavropigi-Fillotsairi in western Macedonia (Valamoti, unpublished material) and continues to occur at several sites in northern Greece. The only rich charred concentration of Cornelian cherries comes from Sitagroi, where forty charred stones, dated to the Early Bronze Age, were identified. Perhaps the regular but low numbers of Cornelian cherry at most sites of northern Greece is due to taphonomic procedures related to the preservation of the stones in the archaeological deposits (see, for example, Jacomet and Brombacher 2005). If Cornelian cherries were not in contact with fire, it is unlikely that their stones would have been preserved to be recovered during excavation and flotation. This scenario is supported by the fact that at sites where waterlogging is the means of preservation, Cornelian cherries have been found in large quantities, as is the case at the site of Limnochori II, in the four lakes area of western Macedonia in northern Greece (see Figure 5.6) (Valamoti, unpublished material). No Cornelian cherries have been reported from the south of Greece, although it is difficult to tell whether this is due to differences in natural vegetation between the north and the south or to differential food preferences or intensity of archaeobotanical research.

Figure 5.5. Charred Cornelian cherry (*Cornus mas*) stones from Early Neolithic Mavropigi-Fyllotsairi. (Photograph by Stavros Dimakopoulos. Courtesy of PlantCult.)

Figure 5.6. Waterlogged Cornelian cherry stones from Limnochori II in the four lakes region of northwestern Macedonia, northern Greece. (From Valamoti 2009, Figure 8.10.c. Photograph by Katerina Moniaki. Courtesy of Panagiotis Chrysostomou.)

Sambucus appears in the archaeological deposits either as *Sambucus nigra* L. (elderberry) or as *Sambucus ebulus* L. (dwarf elder). Its presence is constant, that is, it occurs at many sites but in low numbers that do not allow an evaluation of its contribution to the diets of people at the time. It is possible that its fruits were used as food, in the case of *Sambucus nigra,* as the fruits of *S. ebulus* are poisonous (Gennadios 1997). The few charred seeds of these two species could correspond to fruits consumed by animals of course. A recent exception is the case of Kleitos Kozanis (Stylianakou 2013; Valamoti and Stylianakou 2015), where 70 seeds of elderberry (*S. nigra*) have been

Figure 5.7. Concentration of elderberry (*Sambucus* cf *nigra*) seeds from Kleitos, Kozani, Late Neolithic. (Photograph by Stavros Dimakopoulos. Courtesy of PlantCult.)

found in a, more or less, pure concentration (Figure 5.7). Elderberries can be used as food and in a wide range of traditional medicinal preparations.

Wild pears (*Pyrus amygdaliformis* Vill.) were found in Mesolithic deposits at Franchthi Cave (Hansen and Renfrew 1978). They are not such a frequent find among archaeobotanical remains from prehistoric Greece. Despite this, it is possible that they were seasonally harvested from trees as is indicated by a very impressive concentration of over 1,000 whole wild pears found charred within House 1 of Dikili Tash, destroyed by fire during the second half of the fifth millennium BC (Figure 5.8 and Figure 5.9). These pears correspond to two concentrations, one found stored inside a jar with a hole at its bottom, broken during the destruction of the house, the other concentration being associated probably with a mat, the impression of which was found on a piece of clay (Valamoti 2015a). The pears found in the jar still had some of their stems attached. Besides these impressive concentrations, archaeobotanical evidence from other houses from the same phase at Dikili Tash show that it was probably widely used at the site, having been found harvested and stored in other houses, too.

Figure 5.8. Prickly pear (*Pyrus amygdaliformis*) found during excavation in a smashed pot inside House 1 at Dikili Tash, Late Neolithic, approx. 4300 BC. (Photograph by Pascal Darcque. © Dikili Tash Research Project.)

Figure 5.9. Prickly pears (*Pyrus amygdaliformis*) found in a smashed pot inside House 1 at Dikili Tash, Late Neolithic, approx. 4300 BC. (Soultana Maria Valamoti)

"Nutty" Tastes: Almonds and Acorns

The almond (*Prunus dulcis* [Miller] DA Webb and *Amygdalus communis* L.), a plant very well known to the modern inhabitants of Greece, its nuts (seeds in reality) widely used in cuisine, especially confectionery and the preparation of a sweet syrup called *soumada* (like the *orgeat* of France), has a rather limited presence in the archaeobotanical record of prehistoric Greece when regarding rich concentrations, whereas it occurs at Neolithic and Bronze Age sites primarily of southern Greece. Remains of almond wood identified among anthracological remains as *Amygdalus* sp. are reported from the wider Aegean region since the upper Paleolithic and Mesolithic from sites on the western part of Turkey and south mainland Greece (e.g., Asouti et al. 2018), while almond shells have been reported since the Epipaleolithic/Mesolithic at sites such as Öküzini in Turkey (Martinoli and Jacomet 2004) and Franchthi Cave in the Peloponnese in Greece (Hansen 1991). The Öküzini almond shell fragments are abundant, and their high number (1,280) and rounded fracture surfaces have been considered as indicative of their systematic collection and use as part of the Epipaleolithic diet of the cave inhabitants (Martinoli and Jacomet 2004).

In northern Greece, the almond does not appear as part of the Neolithic landscape according to anthracological studies (Marinova and Ntinou 2018), yet, a shell fragment was identified at Late Neolithic Makri (Valamoti 2004). The almond occurs in the form of charred shells sporadically at some Neolithic sites in the south and appears more frequently during the Bronze Age in Crete. Among the Neolithic finds, those from Franchthi Cave and Knossos consist of a significant number of shell fragments: 66 in the Early Neolithic and 95 in the Middle Neolithic (Sarpaki 2013). Those found at the Early Neolithic II levels at Knossos are interpreted as an indication for the cultivation of the almond, as they are found in larger numbers than those of the earlier layers (Sarpaki 2013:80). At Knossos, however, the largest concentration consists of 15 fragments of almond in a single sample, thus they could correspond to collection from the wild, perhaps from selected trees that yielded less bitter fruits or wild ones rendered edible/palatable through processing. Either way, almond cultivation is a strong possibility, as suggested by Sarpaki; these trees, as wild pears, already used in the Mesolithic, as well as figs, would have benefited from the anthropogenic intervention to natural vegetation through forest clearance, allowing access to more light that would encourage fruit production (Valamoti 2015a). On present evidence, almond shells, although present and increasingly so during the Bronze Age, do not seem to make a significant contribution to the archaeobotanical record of prehistoric Greece, yet, ongoing research, especially in the south of Greece, might alter this picture in the future (Appendix A:Table 5.2). Almond (*sensu lato*) wood charcoal has been reported from sites investigated in southern Greece since the Paleolithic (Maria Ntinou, personal communication) and is common at Akrotiri on the island of Santorini during the Bronze Age (Mavromati 2017). This would point to the use not only of almond wood but also of its fruits in a more widespread and/or systematic way than the limited macrobotanical remains of the fruits of the tree allow us to guess.

Hazelnuts (*Corylus avelana* L.), although found in pollen (e.g., Willis 1994) and charcoal spectra (e.g., Marinova and Ntinou 2018) from southeastern Europe, are very rare as plant macroremains among charred assemblages from Greece. Hazelnut charcoal has been found at a few cites in central and western Macedonia (e.g., Revenia, Kleitos, Avgi, Marinova and Ntinou 2018) and appears more frequently in Bulgaria (Marinova and

Ntinou 2018). As for natural and cultivated chestnuts and walnuts, there are no archaeobotanical macroremains from prehistoric sites that would testify to their use. Indirect indications for their presence in southeastern Europe

Figure 5.10. Charred acorns (*Quercus* spp.): Final Neolithic Mandalo (*top*); Dikili Tash, Late Neolithic, approx. 4300 BC (*bottom*). (Photographs by Stavros Dimakopoulos. Courtesy of PlantCult.)

are provided through their patchy occurrence in the palynological record of the Late Pleistocene and their identification in pollen diagrams in deposits dating to the end of the Bronze Age in the area (e.g., Bottema and Woldring 1990; Willis 1994).

Despite the fact that at Greek prehistoric sites almonds are rare, while chestnuts and walnuts probably absent, in prehistoric Greece we see a dominance of another "nut": the acorn (*Quercus* spp.). Acorns are found at many Neolithic and Bronze Age sites and in particularly rich concentrations that suggest their intentional harvest. They usually occur as cotyledons, that is, without their cups, from which they become detached when ripe. Cotyledons of acorns have been found in large quantities at many sites mainly of prehistoric northern Greece, with rich concentrations encountered at Dikili Tash, Mandalo, and Archondiko (Figure 5.10) presumably corresponding to stored harvests perhaps from selected trees.

Preparing Fruits and Nuts for Consumption

For people in the recent and prehistoric past, in a context when sugar did not exist and honey was probably a rare luxury, the anticipation of seasonal sweet delicacies must have been great, and equally great the excitement when fruit was converted into a dish, its taste hitting the palate (Figure 5.11). The seasonal availability of fruits and vulnerability would have been counteracted by boiling into syrups, drying on mats under the summer and autumn sun, and then storing for consumption until the next harvest. In this way the sweet or sweet-sour taste of figs, plums, blackberries, elderberries, wild pears, Cornelian cherries, and perhaps other fruit that occurs only sporadically such as plums/cherries could be enjoyed in the form of a syrup, soup, or thick cream, whereby the juice of the boiled fruit was thickened with cereal flour. On the island of Samothrace in recent years, inhabitants would harvest all types of fruit when fully ripe, dry them well under the sun, then slice them in half, place them in a basket that would then be immersed in a vessel with boiling water mixed with various herbs such as oregano and bay leaves, and then let them completely dry before storing; this acted as a preservative, preventing worm attack and provided fruit that was converted into a sweet soup by boiling piecemeal over the winter months (Maltezou 2004:215). A very similar recipe for drying pears has been described to me by Sophia Papageorgiou at the village of Kosmati in northern Greece (2017) and Chryssi Trakaki at the village of Kakodiki, on Crete (Valamoti 2009).

Figure 5.11. Wild harvest in the autumn; top row: acorns (*left*) and *Sorbus* sp. (*right*); bottom row: prickly pears (*left*) and blackberries (*right*). (Soultana Maria Valamoti)

The fruits of the grapevine could have been consumed fresh or dried, or they could have been pressed for the conversion into fruit pulp or the extraction of juice. Fruit pulp could be used for the preparation of dried fruit paste, while grape juice could be converted into a syrup through boiling (*petimezi/pekmez* in Turkish), or for the preparation of fermented products such as wine and/or vinegar. One could argue that there is a strong bias in favor of wine when the products of the grapevine used in prehistoric times are mentioned in the literature, and this is a justifiable interpretation especially in cases as the Dikili Tash grape pips and grape pressings corroborated through residue analyses and artifactual evidence. Grapes could have been sun dried for later use as is the case with a wide range of wild fruit dried and consumed during periods when no fruit is available in nature (Wiltshire 1995). The archaeobotanical distinction between raisins, grape pressings, and whole fresh grapes is possible, as has been shown by experimental research carried out by Mangafa and colleagues (1998): grape pressings do not change after charring, fresh grapes are converted into a fragile, porous mass, while raisins swell, resembling fresh grapes. These experimental observations were later confirmed by further experimentation by other scholars (Margaritis and Jones 2006; Cartwright 2003). No finds that could correspond to prehistoric raisins are reported from Greece, with the exception of a single find from Late Neolithic Makriyalos (Valamoti 2004).

The preparation of raisins would have been simple and straightforward as described in later ancient Greek sources. Hesiod in his *Works and Days* (609–614) provides the recipe: bunches would be harvested and brought home, exposed to the sun for ten days, then placed in a shady place for five more days, and then on the sixth day they were placed in a jar. We do not know what the jars for storing raisins looked like; however, the chemical residue mark they would leave on the jar walls would have been tartaric acid, a cautionary note on tracking the history of ancient wine on the basis of residue analyses alone. Grape juice fermentation past the stage of wine would naturally lead to the making of vinegar, which could have been used as a preservative of other foods and a condiment. The preparation of *petimezi* would have been possible with the prehistoric cooking facilities available. It would have required equipment for crushing the grapes and a strainer for removing the pips, pressings, and stems. The liquid would then need to be simmered over fire until it was reduced considerably. It would then be mixed with a special kind of soil that clarifies the boiled juice. After the mixture was left to settle overnight, it would be decanted into another container and further reduced into a thick grape syrup, regularly removing the scum (Figure 5.12). The mixture is ready when it has the consistency of a thick syrup, and it would last long stored in a clay jar (personal observations, Kosmati village, Grevena, northern Greece, November 2017). The exclusive use of ash for clarifying grape juice is mentioned for *petimezi* production on the island of Lefkada, in the Ionian Sea (Voutsina 2008). *Petimezi* can be prepared from fruit other than grapes, such as figs (e.g., Kiziridou 2002; Gialousi-Chatzichristou 2009) and wild pears. *Petimezi* is a natural sweetener and could act as a fruit preservative, used for the preparation of spoon sweets. Its significance as a sweetener can be better appreciated if one imagines a world without sugar (Kochilas 2003:431). Fruit such as figs and *petimezi* were thus boiled together (e.g., Voutsina 2008:24) and placed in clay pots. In the region of Kozani in northern Greece, slightly dried bunches of black grapes were preserved in *petimezi* in the following way: the fruit was separated from the stalks and placed in a clay, narrow-necked pot, then covered by *petimezi*, and four tablespoons of *Brassica* seeds tightly wrapped in a cloth were placed together. In this way the grapes were preserved and served to visiting neighbors. This sweet is called *staphylarmia* (Tsikritzi-Momtsiou and Ftaka-Tsikritzi 2006:191). Similar syrups like *petimezi* can be prepared from other wild and cultivated fruit such as wild pears and figs.

The preparation of a syrup from grape juice called σίραιον/*siraion* or ἕψημα/*epsema* was known in ancient times. It was a natural sweetener, known for its medicinal properties (Hippocrates, *De mulierum affectibus* 66). Grape juice could have been used to prepare sourdough by mixing it with flour and letting it ferment over a series of days with regular admixtures of juice and flour (Karagiannis 2015), while the use of a diluted form of grape juice or of the by-product of alcoholic fermentation of grapes for breadmaking is mentioned in *Geoponika*, dated to the tenth century AD (Valamoti et al. 2022a). The preparation of a bread called οἰνοῦττα/*oinouta*, where the dough included wine, is mentioned in ancient texts (Aristophanes, *Plutus* 1121), yet

Figure 5.12. Preparation of grape juice *petimezi* at the village of Kosmati, Grevena, northern Greece, November 2017. After clarification of grape juice with the use of a special soil, the liquid is boiled until it reaches the desired thickness. (Soultana Maria Valamoti)

this later, textual information provides only possible ways in which grape products might have been used and is no proof that they were indeed used in this way in prehistoric times. Their value, however, lies in that they reveal the multitude of ways past fruit products might have been consumed.

In the archaeobotanical record, the drying of figs for consumption could be indicated by the lack of fruit distortion (Kislev et al. 2006), something that is confirmed by my unpublished experimental charring of figs. The archaeological finds of figs suggest that they were stored in a dry form. The dried figs could have been stored whole or kneaded (Figure 5.13 and Figure 5.14), in a form of "bread," like the so-called συκομαγίδα/*sykomagida* ("precisely kneaded figs"). *Sykomagida* is known from Greece, Cyprus, and other parts of the Mediterranean nowadays as well as in antiquity. Figs, due to their

Figure 5.13. Pressed figs in a figcake (*sykomagida*), wrapped in fig leaves, purchased in Corfu town market, 2019. (Soultana Maria Valamoti)

Figure 5.14. Pressed figs in a figcake (*sykomagida*), wrapped in fig leaves, split-open interior, purchased in Corfu town market, 2019. (Soultana Maria Valamoti)

high sugar content, could have also been used as a sweetener, added to other foodstuffs, or used to prepare a syrup as reported for the islands of Chios (Gialousi-Chatzichristou 2009) and Samothrace (Maltezou 2004:101). The medicinal properties in folk medicine are widely acknowledged for treatment of various conditions related to the endocrine, reproductive, and respiratory systems as well as against infectious diseases (e.g., Badgujar et al. 2014).

Cornelian cherries are very popular nowadays in Greece and are mainly used for the production of jams and homemade liquor. They are widely used in many forms in the region of Caucasus, both as sweet and savory dishes, mixed with various other ingredients, and they are appreciated for a wide range of medicinal uses (Batsatsashvili et al. 2016). The preparation of a mixture of Cornelian cherry fruit juice boiled with starch, in the form of a *kissel* (boiled fruit juice with starch), is reported for the region of Azerbaijan, and it is a traditional sweet prepared by the Pontic Greeks under the same name (Kiziridou 2002), the name according to Davidson (2002) being of Russian origin and the food preparation similar to the German *rote Grütze* and a Scandinavian equivalent known as *røtgrøt* made with red fruit and thickened with semolina. Their frequent occurrence in Neolithic Italian sites has led to the suggestion that it may have been used for the production of some kind of fermented beverage (see Bouby 2014:159 for a review of the evidence from Italy and France), and similar use is mentioned for the wider Caucasus region (Batsatsashvili et al. 2016). The species identified as Cornelian cherry by ancient Greeks was used, according to Pausanias, for the construction of the Trojan horse, and the fruits of this tree were given as food, together with acorns, to Ulysses and his colleagues by the witch Circe (Gennadios 1997).

Blackberries, seasonally available, could have been dried and stored for later use, and the same would have been true for Cornelian cherries and elderberries. The dried fruits could then later have been soaked in water and cooked, upon demand (Wiltshire 1995). Elderberry fruit and flowers are consumed by humans, and elderberry is used for the preparation of a drink (Bown 1995). Preparations such as infusions could have used elderberries. Its medicinal properties are concentrated mainly in the flowers (Gennadios 1997).

Wild pears, commonly known as *gortzia* in Greek, are bitter in taste until they become fully ripe in November, when they are sweet and acquire a strong and pleasant smell. They were known also in antiquity for their late maturity (Theophrastus, *Historiae Plantarum* 3.4.4) and for being very aromatic (Theophrastus, *De Causis Plantarum* 6.16.1), continuing to ripen even after they had been harvested, although not in the same way as if they had been left on the tree (*De Causis Plantarum* 2.8.2). As they are subject to worm attack, even when they are unripe (*Historia plantarum* 3.12.8) according to Theophrastus, we can assume that when placed in storage, as the finds from Dikili Tash, they would have required a special treatment to avoid this attack.

The drying of wild pears usually involves cutting them in half and processing with oil, water, and herbs (Renfrew 1973a; Valamoti 2009). Their drying and storage are mentioned for Greece (Gennadios 1997). In parts of Crete, they have been traditionally used as an ingredient of a special bread prepared by dried and ground wild pears mixed with flour (wheaten or barley) (Psilakis and Psilaki 2001:148). A similar practice whereby ripe pears are sun dried, ground into flour, and stored for piecemeal use by mixing with boiling hot water is included among the recipes of the Pontic Greeks (referred to as *chavourta*) (Kiziridou 2002). Until recently they were harvested by collecting mainly the fallen fruit from the trees or directly from the tree, then stored in straw (Chaido Koukouli-Chrysanthaki, personal communication 2020, village of Tsotyli, western Macedonia). At villages in the northern Peloponnese (Corinth) and central Greece (Agrafa mountain region) they were harvested unripe and kept in straw where they were left to ripen; for longer-term storage the wild pears would be covered by some form of fat, olive oil or animal fat (Elias Sverkos, personal communication, spring 2021). Wild pears could have been consumed in various ways, as fruit soup, dried and ground into flour, or in the form of a syrup (see Krikou 2019). In parts of northern Greece they have been used for the production of a sweet syrup, like *petimezi*, following a long sequence of boiling, squeezing, and mixing with a particular kind of soil (see Figure 5.12), as observed in the village of Kosmati in November 2015 (Figures 5.15, 5.16, and 5.17) (Valamoti and Mimi 2016). A very similar procedure is described in a book on the cuisine of the Greeks of the south Black Sea coast, known as *Pontioi*: when ripe, the wild pears were boiled, placed in a sack and left to drain into a bowl hanging underneath the sack. The juice and fruit pulp extracted from the sack were then boiled (Kiziridou 2002:359).

The wild pears encountered inside the Dikili Tash houses were probably stored in a dry state for later use, especially those found associated with the mat impression, spreading over a large area of the room. They might have been contained in a basket or a bag hanging from the ceiling, the fruit spreading during destruction by fire, or they might have been contained in a flat

Figure 5.15. Boiling wild pears into a syrup in metal cauldron at the village of Kosmati, November 2015. (Photo courtesy of Georgios-Vily Kapetanakis.)

Figure 5.16. Wild pears after having boiled in water in order to prepare a syrup at the village of Kosmati, November 2015. (From Valamoti and Mimi 2016. Photograph courtesy of Georgios Vily Kapetanakis.)

Figure 5.17. Preparing wild pear syrup at the village of Kosmati, November 2015: Ioanna Mimi and her grandmother Sophia extract soil used for the clarification of the pear juice. (Photograph courtesy of Georgios-Vily Kapetanakis.)

container of perishable materials, covered with clay and thus sealed. Those in the big jar with a hole near its base (see Figure 5.8) could either have been stored dry, the hole perhaps serving for ventilation. Alternatively they could have been stored as pickled fruit, the hole of the jar serving to easily remove the brine. The preservation of *Pyrus amygdaliformis* as pickled fruit is reported from islands of the northeast Aegean (Bozcaada/Tenedos) and the area of Balikesir in the northwestern part of Turkey where they are consumed as food as well as for medicinal purposes as they are known to reduce blood sugar levels (Bulut and Tuzlacı 2008; Arıtuluk and Ezer 2012). Wild pears (*diva krusha, masleva krusha*) pickled in brine are also known from traditional Bulgarian cuisine (Nedelcheva 2013). Although a wild plant, it will only yield fruit when adequate light is available; therefore, some form of tending to increase fruit yield can be envisaged since at least the fifth millennium BC (Valamoti 2015a).

In a traditional Greek context, grape juice or *petimezi*, mixed with flour, would be converted to a delicious dessert known as *moustalevria* from *moustos*, the Greek word for grape juice. In a prehistoric context, where no refrigerators or canning technology were available, grape juice would only be seasonally available for a short time, whereas *petimezi* would have lasted for at least one year. Such natural sweeteners would have been greatly anticipated in prehistoric times when sugar was unavailable and honey was probably much rarer than it is today.

The sporadic almond finds from prehistoric Greece, if corresponding to wild harvests, would have required processing to detoxify them. Martinoli and Jacomet (2004), despite suggestions in the literature that wild almonds could not have constituted food due to their bitterness, argue convincingly that their consumption was feasible if consumed by slowly chewing them and consuming them over a long period of time, while their processing by mere grinding and exposing to the air would suffice for removing the poisonous volatile hydrogen cyanide. Moreover, treatment with heat would have also been effective toward removing the bitter taste (Davidson 2002). As has been the case with the grapevine, however, the almond could have been cultivated during the Neolithic or later in the Bronze Age, as orchards began to form in the landscape, consisting of grapevines and fig trees. Almonds would have been a valuable food resource, rich in calcium, proteins, and fatty acids. Almonds mixed with honey were a common sweet in ancient Greece (Dalby 1997:81), yet in a prehistoric context such a sweet, if prepared, would have been most likely a luxury food for special occasions as, honey at least, was most likely a rare ingredient.

Despite a widespread perception of acorns as pig food, or at least fodder, acorns are suitable for consumption by people and have sustained populations both in the recent past and antiquity. There are relevant quotes in ancient Greek literature (Theophrastus, *De pietate* 2) as well as in modern records for the consumption of acorns in different forms (see, e.g., Mason and Nesbitt 2009 for Turkey; Ayerdi et al. 2016 for the Pyrenees). Acorns could have been consumed as food by humans after special processing for the removal of bitter tannins: roasting, soaking, and discarding of water (see, for example, Mason 1995; Renfrew 1973a; Hosoya 2011). The experimental removal of their bitter tannins was realized in two ways by Jørgensen (1977): (1) after removing the shell surrounding the cotyledons, they were boiled for one hour, and (2) acorns were left to soak in water for a day and then were roasted for 15 minutes. In both cases the bitterness was removed. Ethnographic work and further experimentation have proposed a range of ways suitable for removing the bitterness of acorns for human consumption including grinding, leaching, and adding red clay with several variations (Johns and Duquette 1991) as well as soaking in water with red clay prior to boiling (Driver 1953) and roasting, followed by leaching (Ayerdi et al. 2016). Archaeobotanically, the various steps in processing acorns may be difficult to preserve, yet, an integrated approach using ethnographic and experimental evidence, archaeological finds of processed food remains, different types of processing/cooking facilities, and associated plant micro- and macroremains could potentially shed light in the use of acorns in prehistoric diets (see, for example, Primavera and Fiorentino 2013).

The acorns found stored in Neolithic and Bronze Age contexts from Greece might have been harvested from individual trees with sweet-tasting acorns, as has been the case in Crete in recent times (Psilakis and Psilaki 2001). Harvest time would have usually been around November, as is the case in recent times in parts of northern Greece, and October in the south, as is the case on the island of Kea (personal observations). Acorns, besides being consumed roasted or raw, as a snack, as recorded for Spain and Turkey, respectively (Parsons 1962; Chamber's Encyclopedia 10, 1959; Mason and Nesbitt 2009), could have been ground into flour, consumed as a bread or a porridge, as is recorded for parts of Turkey and Iran (Mason and Nesbitt 2009), Sardinia (Pignone and Laghetti 2010), as well as California and other parts of the United States, where sometimes the acorn cakes would be dried for later consumption (Johns and Duquette 1991; Driver 1953; Bettinger et al. 1997, Šálková et al. 2012). Acorns could have also been mixed with cereal flour and used in breadmaking (see Psilakis and Psilakis 2001). Nowadays, bread made of acorn flour is mentioned for certain areas of Crete and a special mention is provided for the techniques involved in the removal of the bitter tannins, depending on the type of acorns used (Psilakis and Psilaki 2001:148): certain acorns, classified as sweet (γλυκοβέλανα/*glykovelana*), need only a simple roasting after removal of the shell, while the bitter ones were boiled, dried in the oven, and then ground into flour. Perhaps grinding for making bread was what the einkorn, barley grain, and acorn cotyledons were initially intended for, as they were lying on a grinding stone in prehistoric Raskopanitza in neighboring Bulgaria (Renfrew 1973a). No indications for acorn processing have as yet been reported from Greek sites; further experimentation could contribute comparative material for understanding prehistoric acorn processing.

From Wilderness to Orchards

The remains of wild fruit that appear in the Neolithic settlements and more prolifically in the Bronze Age were likely harvested for the provision of the needs of the household. Harvesting the wild in the Early and Middle Neolithic probably gave way to management of fruit trees by increased light exposure through pruning and planting of selected plants in specific areas. At some point during this management of fruit-bearing wild trees, vegetative propagation could have been invented, which would have led to the domestication of certain wild tree plants that can be vegetatively propagated, as is the case with the olive, grapevine, and fig (Zohary and Spiegel-Roy 1975). Vegetative propagation most likely focused on stock from individual trees with desired properties such as sweet (see Miller 2008) or large-sized fruit. By the historic period we know from ancient texts that a wide variety in fruit trees in terms of their fruit properties existed already, especially for grapes and olives, reflecting this long-term interaction with taming the wild vegetation and selecting fruit trees for cultivation (Dalby 1997).

The wild trees and shrubs growing in the first millennia of the Holocene were valuable to gatherer-hunters and later to farmers. They continued to be in use for their colorful fruits throughout prehistory as the archaeobotanical evidence we have explored in this chapter clearly demonstrates. They would have provided not only an additional source of calories but also of vitamins and trace elements as well as a buffer wild food resource, in case of crop failure (see Halstead 1981, 1994; Zohary and Spiegel-Roy 1975). The ethnographic record provides ample evidence for the use of wild fruit and nuts in cases of food shortage (e.g., Łuczaj 2016). Many among these plants possess medicinal properties, and it is possible that they would have also been valued for this reason, too. Over time, some among these wild plants became domesticates that gradually occupied the landscape in orchards, vineyards, and olive groves. Although narratives about fruit domestication turn to the Near East for the origins of fruit cultivation (e.g., Fuller and Stevens 2019), the evidence examined in this chapter clearly shows that, for some species at least, local domestication could have started in the Aegean, too, in agreement with earlier views on the matter (Zohary and Spiegel-Roy 1975). This is especially true in the case of the grapevine (see Zohary and Spiegel-Roy 1975:22–23) and may also be the case for the olive in the south Aegean and Crete in particular (Margaritis 2013b; Canellas-Bolta et al. 2018; Langgut et al. 2019), although a Levantine influence initiating olive cultivation cannot be excluded (Zohary and Spiegel-Roy 1975; Blitzer 2014; Valamoti et al. 2018; Langgut et al. 2019; see also Oflaz et al. 2019 for a recent overview on the issue).

Tree cultivation would have increased the areas in the landscape dedicated to provisioning for prehistoric societies of the area and would have contributed to an increased diversity in the plant food resources under human control. This would have gone hand in hand with a greater commitment to the land as fruit trees require a long-term investment of labor and resources (Gilman 1981), thus their association with emerging Bronze Age hierarchies in the Aegean is fully justified, while their potential to make use of more marginal land, compared with land dedicated to cereals and pulses, would have contributed to centralized control of land production and redistribution (Renfrew 1972). Of the cultivated fruit species, the grapevine and the olive are also the sources of substances that might have been used in special contexts due to the mind-altering properties of wine and the luxury nature of olive oil, thus providing means of social distinction and power manipulation in the context of rituals controlled by emerging Bronze Age elites (e.g., Hamilakis 1996). As mentioned in Chapter 6, however, it would be far-fetched to consider their cultivation as the trigger for the emergence of elites, rather, their cultivation lent itself to such manipulation in specific social contexts.

Figs and grapevines appear in Linear B tablets either grown separately or in close association to one another, suggesting that grapevines may have been trained on fig trees in some cases (Palmer 1995:276). Orchards specializing in grapevines called *wo nasi* are also mentioned, although it seems that a variety of fruit trees grew on Mycenaean orchards (Palmer 1995). More fruit trees than those visible in the archaeobotanical record may have been grown in the orchards of Bronze Age Greece under cultivation, such as the almond, for which limited evidence exists, and the pomegranate, for which only charcoal evidence testifies to its cultivation (Asouti 2003; Mavromati 2017). Although the almond was locally available since the Upper Paleolithic in the Aegean (Asouti et al. 2018) and could have locally been brought to domestication, the pomegranate would have been an "exotic" introduction as it is native to Iran, Afghanistan, and surrounding regions (Levin 2006; Preece and Moersfelder 2016), the south Caspian belt, and northeastern Turkey (Zohary et al. 2012). It has been found in the form of actual archaeobotanical remains or representations of the fruit and flowers in different contexts in the Near East, and it seems that it corresponded to a luxury item that circulated in the eastern Mediterranean (Zohary et al. 2012). It is very likely that the Bronze Age contact networks that were operating in the wider area of the eastern Mediterranean, Anatolia, and southeastern Europe introduced plant species in our study area such as millet and *Lallemantia* sp. (see Chapter 9 for a discussion of contact networks); likewise, the pomegranate, a fruit tree with an overlapping geographic origin as the other "foreign" crops, could have been introduced through such contact networks. The fruit of the pomegranate had a strong symbolic meaning in the later historic periods, especially in relation to the worship of the cereal goddess Demeter, her daughter Persephone, and the underworld, associated with *hierophagy*, the consumption of food that transcends the world of the living and the dead (e.g., Warren 2017). The charcoal of pomegranate is clearly indicative of pomegranate trees growing in the Bronze Age Aegean, corresponding to stock brought in the form of living plants from areas to the east where it was presumably grown earlier than in Greece. The circulation of plant stock in the Bronze Age Aegean, contributing to the development of new varieties of cultivated fruit trees of the species discussed in this chapter, is very plausible (see Valamoti et al. 2020 for grapevine stock in Greece and wider region; Kourakou-Dragona 2000 for grapevine stock in antiquity; Zohary and Spiegel-Roy 1975 for the main fruit trees of the Near East and the Aegean) and would have accounted for the large variety of local landraces in antiquity, some famous for their taste and aroma as were for example figs from Rhodes, Caria, and Smyrna, and various local grapevine varieties used for dessert grapes (e.g., Dalby 1997).

Beyond the fields where the seed and grain crops were grown, Neolithic and Bronze Age farmers of Greece used a wide range of wild fruit and nuts. We can assume that the trees and shrubs that yielded these additional sources of nutrients, vitamins, and minerals grew in the extensive woodlands of prehistory that existed in the Neolithic, regionally varying as regards the composition of forest canopy and its density. The fruit and nut trees selected for their edible yields started out from patches in the wilderness and became tamed and tended in orchards. In a way they form a thread connecting the gatherer-hunter past with the farming communities of later prehistory. This primeval wilderness that harbored some of the components of the orchards of the Late Bronze Age may be reflected in the much later sacred forests of ancient Greece, patches of biodiversity maintained through religious prohibition containing wild but also cultivated tree species (e.g., Bowe 2009). A connection between fruit trees (fig or olive) and Minoan-Mycenaean religion has been suggested on the basis of artistic representations of trees (Kourou 2001), yet, at the same time plantations of large numbers of fig trees are mentioned in the Linear B tablets (Bottino 2014), clearly suggesting extensive orchards and large-scale production of figs. Figs were distributed as rations, possibly in the context of religious festivals (Killen 2004). Interestingly, the ancient tradition of associating trees and religion seems to reach modern times as is evident for example in the sacred forests of Epirus, in the landscape surrounding Konitsa and Zagori where rituals and taboos are associated with access and use of the wild plant resources in these areas, trees in particular (Marini Govigli et al. 2020; Stara et al. 2016). Such sacred forests are also found across Europe, often situated close to settlements or agricultural landscapes. Their sacred status influences human interference with vegetation in these landscapes, protecting wild vegetation and therefore affecting tree composition (see Covigli et al. 2020 for a recent review of the evidence and a detailed case study in Epirus).

The majority of the trees that appear as cultivated in the orchards of famous kings of the Bronze Age, attested in Linear B texts (fig, olive, grapevine) and in greater detail in the Homeric poems (e.g., Dalby 1997), probably started out as elements of these prehistoric forests. The first farmers of Greece cleared patches of this woodland to create fields and pasture. The knowledge of the landscape and of its wild vegetation, well rooted in the Paleolithic past and the regular tending of crops and flocks of animals, provided the opportunity to harvest wild fruit and nuts. Some among them, through the systematic use, were at some point selected for propagation, management through planting, pruning and replanting, and clearing surrounding vegetation to increase fruit yield. Thus, gradually, the fruit-bearing wilderness became tamed, providing a wide range of exciting foodstuffs. Perceptions about wild vegetation are of course likely to have changed over time, yet, already in Linear B texts the concept of wilderness and land boundaries have been recognized (Palaima 2014), acquiring variable meanings and boundaries in the historic periods (see Vidal-Naquet 1983). It is useful to consider the role of plants themselves in this close interaction between prehistoric people and their surrounding vegetation. In this process of taming a selection of wild-growing fruit trees, both humans and plants played their role. As Oliver Rackham notes, "plants and animals are not a generalized nature, not the passive recipients of whatever mankind chooses to inflict on them: they are thousands of individual species, each with its own behaviour which has to be understood" (Rackham 1996:17). Indeed, the selection and vegetative propagation of certain wild plants with desired properties clearly reflect this complex interaction between people and fruit plants, leading to the cultivation and domestication of some of them as well as to creating perceptions and connections on a symbolic level about this relationship.

6

Special Fermented Brews

Exploring the Alcoholic Drinks of Prehistoric Greece

A Very Old Taste for Alcohol

Alcohol consumption has been considered a universal trait of human societies, characterized as an "evolutionary hang-over" (Dudley 2002). Primates that were able to ingest ethanol in ripe fruit on the forest floor (Carrigan et al. 2015) probably contributed genes responsible for a new metabolic capacity to modern humans, a capacity also shared by certain primates who enjoy consuming fruit that has matured to a state of incipient fermentation (McGrew 2011; Hockings et al. 2015). What is particular to humans is their ability to develop technologies to ferment and thus control the amount of alcohol available to them, as well as to produce higher alcohol levels in food, compared with those encountered in nature (Carrigan et al. 2015). Over time, this biological adaptation to ingest alcohol was transformed into a significant cultural element among prehistoric societies who were able to control alcohol production, moving beyond the mere consumption of rotting fruit lying on the forest floor. Indeed, archaeological indications for past alcohol production by human societies go back in time to prehistory (e.g., Moore 1989; McGovern and Michel 1995; Wang et al. 2016). It has been argued that alcohol production would have been possible even in Paleolithic times (McGovern 2003:7–11): harvesting wild fruits and letting them ferment with the natural yeasts on their skin would have been a rather straightforward process for Pleistocene gatherer-hunters. Alcohol, due to its mind-altering effects, has also been associated with special contexts of consumption and ritual practices of past societies (see Guerra Doce 2014).

Beer has featured in discussions related to the onset of early farming in the Near East (Braidwood 1953; Hayden et al. 2013; Hayden 2003). In contexts of communal feasting that brought different gatherer-hunter groups together, the need for larger quantities of food (Bender 1978) including alcohol (Hayden 1995) to provide for the participants has been considered a possible trigger toward increasing the available cereal grain that was required for holding these feasts and thus for cereal domestication. Yet, the available archaeological evidence in support of an early association of alcohol production, feasting, and the emergence of agriculture (e.g., Liu et al. 2018; Dietrich et al. 2012) is rather slim and ambiguous.

Beer making, requiring fermentation of cereal starches, unlike the straightforward process of alcoholic fermentation that takes place when fruit is crushed and juice is in contact with fruit skin yeasts, is a much more complex recipe: sprouting of grains would have been necessary to render the sugars in the kernel susceptible to yeasts and fermentation when mixed with water (McGovern 2003). Thus, it is unlikely that beer preceded wine among the alcoholic drinks potentially prepared in prehistoric Greece.

Direct evidence for early beer making, although featuring in publications, is problematic: calcium oxalates, used to identify beer in the archaeological record, constitute a controversial line of evidence among specialists (cf. McGovern et al. 1996; Mukherjee et al. 2007; Evershed 2008; Pecci et al. 2013; Garnier and Valamoti 2016; Mazow et al. 2014). Detecting wine in ceramic vessels appears less problematic, it is not, however, a straightforward process (see Drieu et al. 2020 for a recent review of the various methods). Even when acids potentially related to wine such as tartaric, succinic, and syringic acid have been detected (see, for example, McGovern et al. 1995; Garnier and Valamoti 2016), it is impossible to tell apart wine from alternative plant ingredients (e.g., pomegranate, *Crataegus*, *Viburnum*; see Barnard et al. 2011; Altmeyer 2018; Drieu et al. 2020) and/or the different grape products (wine, grape juice, grapes/raisins) that may have been contained in the analyzed vessels (Valamoti et al. 2007; Valamoti and Stika 2019 for a recent review). Residue analyses and ceramic types suggest wine production as early as the sixth millennium BC at the site of Hadji Firuz Tepe in Iran (McGovern et al. 1996). Transcaucasia has been considered as a possible center of the earliest wine in the world as well as the area of grapevine's systematic exploitation and domestication in prehistory (e.g., Terral et al. 2010; McGovern 2003; McGovern et al. 2017). Nevertheless, all claims to the earliest wine in western Eurasia are based on residue analysis and pollen diagrams indicating the growth of grapevines in the vicinity of settlements such as Areni and Early Neolithic sites of the Shulaveri-Shomutepe culture in Transcaucasia (Barnard et al. 2011; McGovern et al. 2017).

Archaeobotanical macroremains, however, can provide direct indications for grape juice extraction through the presence of pressed grapes in the form of grape skins and grape pressings. Compared with residue analyses, archaeobotanical remains offer an undisputed indication for juice extraction, the main ingredient for wine making.

Grapes for Wine: The Earliest Wine in Europe?

While the first ingredient for beer making, cereals, were cultivated and domesticated very early on in Holocene human history, spreading from the Near East to Europe, the harvest of grapes, grapevine cultivation, and wine making may have followed different paths from those of the "founder crops" of the Early Neolithic. Wild grapevines have a different distribution from those of cereals and pulses, with the Caucasus and the Mediterranean being areas of grape's natural distribution, along rivers and water (Terral et al. 2010) (Figure 6.1). In regions where wild grapevines grew, Paleolithic gatherer-hunters could have intoxicated themselves by a rich wild harvest, even perhaps by eating very ripe fruit, like our nonhuman primate relatives. The production of some early form of wine did not require viticulture as a prerequisite, as grape juice from wild grapes can also be converted to alcohol (Gorny 1995). The process through which wine can be produced from grape juice is rather straightforward: alcoholic fermentation starts once the fruit has been crushed and contact between its juice and the yeasts present on the skin is under way (e.g., Singleton 1995; McGovern 2003). Wine from wild grapes therefore may well have been the first alcoholic beverage

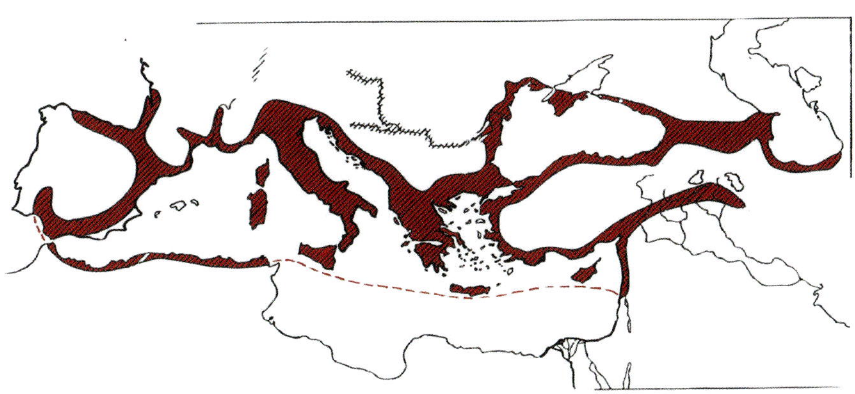

Figure 6.1. Natural distribution of the grapevine. (From Zohary et al. 2012, Map 16. Drawing by Danai Chondrou. Courtesy of PlantCult.)

produced in the Old World, in regions and/or years where wild grape harvests were abundant.

The archaeobotanical remains that form the backbone of this book can contribute significantly toward the archaeological investigation of prehistoric wine and beer making in Greece. Grape pips are numerous in the archaeobotanical record, sometimes occurring together with grape pressings (Appendix A:Tables 5.3 and 5.4). Large concentrations of grape pips indicate deliberate harvests and the production of some special produce. There is a range of products, however, deriving from grape juice, such as the juice itself, wine, vinegar, or a syrup produced from a rather long and complex procedure (see *petimezi* discussion in Chapter 5).

The earliest finds of pressed grapes suggesting juice extraction are recorded from the late fifth millennium BC site of Dikili Tash in the north Aegean. Grape juice could have been transformed into a range of food products. Alone, pressed grapes only provide indications for wine making as one

Figure 6.2. Grape pip in grape skin (*left*) and pressed grape skin (*right*) from Late Neolithic Dikili Tash, 4300 BC. (Soultana Maria Valamoti)

Figure 6.3. Coarse ware jar, Dikili Tash, House 1, 4300 BC. Its sherd fragments together with grape pips and grape pressings were found among the destruction floor debris. Preserved height: approx. 62 cm (24.4 in.). (Photograph by Philippe Collet. © Dikili Tash Research Project and French School at Athens.)

cannot exclude that they were harvested and pressed for other types of food products. In the case of Dikili Tash, however, by using different lines of evidence, it was possible to show that grape juice was being fermented into wine; wine making was under way or wine had been stored with the pips and skins inside a clay vessel in the interior of a house destroyed by fire. The grape pressings (Figure 6.2) and numerous pips (see Figure 5.1) were found together and around a simple pot of a capacity of approximately 130 liters (ca. 34.34 gallons) (Figure 6.3) (Zoï Tsirtsoni, Dimitra Malamidou, personal communication). The grape pips and pressings extended to a relatively large area of approximately 2 m^2, perhaps the result of the flow of the liquid content of the pot as it became smashed presumably during the fire. The pot itself was impregnated by a liquid that contained tartaric and syringic acid, which are both characteristic of grape products and fermentation (Garnier and Valamoti 2016).

The three lines of evidence (archaeobotany, pottery, residue analysis) suggest that the pot contained some liquid from grapes and grape pressings. The coexistence of juice and pips provides the necessary conditions for fermentation of the juice and therefore wine production. Moreover, residue analysis showed the co-occurrence of tartaric, syringic acids, which indicate grape, and succinic and pyruvic acids, which are markers of fermentation. This analysis leaves no doubt that wine was being produced. In addition to this exceptional find, the house contained vessels for serving and consuming some form of drink, in small quantities (Figure 6.4). Thus, it is from the Neolithic period and the late fifth millennium BC in northern Greece that the earliest evidence for wine making in Europe has come to light (Valamoti et al. 2007; Valamoti 2015a), based on direct, dated, multiproxie evidence. In this respect, the Dikili Tash finds correspond to the earliest grape wine identified in the Old World. This is based on dated, solid archaeobotanical remains, ceramic finds, and residue analyses, while evidence such as prehistoric grape pips or grape pressings is still missing from the Transcaucasian sites from which early dates for wine making and presumably viticulture have been obtained by residue analyses.

Palynological evidence and archaeobotanical macroremains as well as the grapevine's natural distribution suggest that northern Greece may have

Figure 6.4. Ceramic drinking vessels from Dikili Tash, House 1, 4300 BC. *Left*: height 10.4 cm (4.09 in.). (Photograph by Stephanos Stournaras.) *Right*: height 6.9 cm (2.7 in.). (Photograph by Stavros Zachariadis.) (© Dikili Tash Research Project.)

been one of the first places in Europe where wine was first made. This places Greece and the Aegean region in a wider area within which viticulture and wine making could have originated, a large area that spreads from Transcaucasia to southeastern Europe and the wider Mediterranean basin (see Figure 6.1). In fact, the beginning of viticulture may have been triggered by the need for more wine. Wine need not necessarily have been a fermented beverage that was invented once in one location; it is very probable that this occurred in locations within this wide area, where wild grapevines grew in great enough numbers to yield a decent harvest. Although often Transcaucasia, or the Levant, is considered the cradle of viticulture (McGovern 2003; Olmo 1995; Terrall et al. 2010; Zohary et al. 2012), the tending and domestication of the grapevine could have occurred at different places and times rather than once in one place (see McGovern 2003:14).

In the south of Greece, archaeobotanical finds in support of pressing grapes for juice extraction and wine making are dated to the third millennium BC. Grape pressings in large numbers have been found on Crete at Myrtos, dated to the Early Bronze Age, and Monastiraki, dated to the Middle Bronze Age (Appendix A:Table 5.4). Rich concentrations of grape pips without pressings have been retrieved at several Bronze Age sites: Lerna and P.O.T.A. (area of tourist development) Romanou in the Peloponnese (Early Bronze Age) and Kastanas in the north (Late Bronze Age). In all three cases, pedicels were either absent or too few. A similar absence of pedicels is observed at the much earlier Dikili Tash grape pip and grape pressings assemblage in Neolithic northern Greece, suggesting that the grapes had been removed from the grape bunches for some purpose.

Finds of charred grape pressings show clearly that juice was being extracted from pressed fresh grapes. This juice could subsequently ferment into an alcoholic beverage. Juice extraction is probably also indicated by rich concentrations of grape pips without skins encountered at several Bronze Age sites from Greece. The absence of skins may be due to taphonomic processes affecting the preservation of grape skins currently under investigation by the PlantCult project. Thus, archaeobotanical indications for wine making occur already in the Neolithic record of Greece while they become far more abundant during the Bronze Age both in the north (Kastanas, Assiros Toumba) and the south (P.O.T.A. Romanou, Lerna, Myrtos, Monastiraki) (Appendix A:Tables 5.3 and 5.4). The recent finds from P.O.T.A. Romanou suggest that, well before the Mycenaean Palaces, wine was produced in the Peloponnese and grapevines probably carefully tended (Valamoti et al. 2020).

In light of the available evidence, wine making had long been established in prehistoric Greece since the Neolithic, continuing to be produced in the Bronze Age. Given the timing of the appearance of the grapevine and indications for grape cultivation or grape juice extraction in Anatolia, the Near East, and Egypt (Miller 2008; Zohary et al. 2012), the available archaeobotanical evidence is probably pointing to an additional, alternative trajectory to that of an *ex oriente* introduction of wine making into Europe, one moving from the Aegean eastward and westward. The only prehistoric finds of grape pressings and/or large grape pip concentrations so far from Anatolia have been reported from Kurban Höyük in southeastern Turkey, dated to the third millennium BC (Miller 2008). On the basis of the earliest date for wine making in fifth millennium BC Dikili Tash, one might also be tempted to interpret the archaeobotanical evidence as indicating the spread of wine making from the north to the south. Those who search for a center from which a plant or a recipe later spread to other regions might argue that the knowledge for making wine reached the south of Greece and the coasts of the eastern Mediterranean from the north. As noted, however, wine making probably originated in more than one place, where/when grapevines yielded rich harvests. The finds from the Peloponnese indicate wine making sometime around the third millennium BC (Valamoti et al. 2020) from grapevines bearing wild-looking pips. Thus, the process of bringing wild grapevines to cultivation was not specific to the north of Greece, reflected in later myth in the many birthplaces of Dionysos, the ancient Greek wine god. Both myth and archaeobotanical evidence from Greece suggest that wine making had multilocal origins, both in the north and the south.

Wine: The Ambivalent Gift of a Tamed Nature

Unlike crops from carefully tended fields, like wheat and lentils, grapes and wine refer to another part of the landscape where wild trees grew, which is where the grapevine would have initially grown, along riverbanks, for example. Once the desire and need to produce wine were established, the next step toward domestication was to take special care of these wild plants, helping them to have more light and thus yield more fruit. Managing wild grapevines would have been a first step toward viticulture, and wild grapevines might have been encouraged to have access to more light by the felling of trees around them and/or pruning. Thus, the first steps toward viticulture may lie in the desire to obtain a rich harvest from originally wild grapevines to produce an intoxicating drink, wine. The intention to increase the quantity of an alcoholic beverage available for consumption and thus also the number of people consuming it, would have inevitably led to the management of wild grapevines, thus generating a zone of interaction between wild and cultivated (see Valamoti 2015a).

The Neolithic grapevine finds from the north of Greece and the Bronze Age ones from the south suggest that the long-established relationship between the grapevine and the prehistoric inhabitants of Greece led to such a management of the plant that would have allowed very rich harvests. Thus, despite the morphologically wild grape pips that have been identified in the Neolithic and Early Bronze Age, the relationship between the plant and people might have been that of an early form of cultivation (Valamoti 2015a; Valamoti et al. 2020).

For Mycenaean Greece, viticulture and wine making are well documented in the Linear B archives, especially those of Pylos. In vineyards as well as orchards, a mixture of trees, including grapevines, was kept. Wine, a rather rare, luxury commodity, was probably produced in different parts of the countryside while the palace was in charge of collecting and distributing it on special occasions and to high-status individuals (Palmer 1995). There is also ample artifactual and pictorial evidence for Late Bronze Age drinking vessels, drinking in pairs, and feasting, where wine presumably was the main drink consumed and/or offered (Wright 2004a).

The time necessary for wild grape pips to change from morphologically wild to cultivated is unknown (Valamoti 1998), and it is certain that some of the cultivated grapevines of today do not have morphologically domesticated looking pips. So, wine making could have started early among farming communities that had accessible wild grapevines growing in the vicinity of their settlements and that appreciated the intoxicating effects of its consumption. Clearing the forest, especially along riverbanks, where the land would have been moist and fertile, would have given light to wild grapevines, thus increasing yield, which early farmers surely would have noted.

The gradual process of morphological change of grape pips from wild to domesticated is clearly visible in measurements of pips from Greek sites from the Neolithic to the historic periods. Some, like those from Kastanas, bear the morphometric features of domestication while others dated to the historic periods have a large proportion of wild-looking pips (Pagnoux 2016; Pagnoux et al. 2021). Domesticated grapevines evidenced through morphometrics or/and large numbers of pips are strong indications for wine making even in the absence of grape pressings.

Wine, as suggested by archaeobotanical data, emerges as the most likely produce of the grapevine, especially if one takes into consideration the large number of cups that appear in Late Neolithic times and become abundant from the third millennium BC onward (Rutter 2017). In regions where early indications for wine making are available, one can assume that wine was the intoxicating produce of that part of the landscape lying beyond crop fields, the forest. Of course, these plants constituted food, but at the same time they were probably involved in classifications and perceptions related to different parts of the landscape, to different relationships between plants and people, seasonal modes and tasks, food taboos, and gender relations. A grape harvest from wild or seminatural vegetation around a settlement, on a symbolic level, may have somehow corresponded to perceptions of nature and culture ambivalent, contradictory but ultimately reconciled once housed under the same roof as stored crops and wild harvests. To the extent that morphometric analyses of the prehistoric grape pips reflect to some extent the relationship between people and the grapevine, harvesting of grapes in Neolithic and Bronze Age Greece had the characteristics of tending, managing the wild plants, and incipient cultivation (see Valamoti 2015a; Valamoti et al. 2020). The lack of a clear-cut distinction regarding the domesticated features of grape pips well into historic times, observed through morphometrics (Pagnoux 2016), matches elements of Dionysos, the Greek wine god. Dionysos, already mentioned in the Late Bronze Age Linear B tablets (di-wo-nu-so), was the god of wild vegetation, eating raw food, and ecstatic behavior, his worship including acts transcending human societal norms. At the same time, wine was a focal element of symposia and therefore of civilized life (Schlesier 2011).

The need for wine consumption to be used in feasting contexts might have triggered the cultivation of the grapevine in parts of Greece (and elsewhere), where it was growing naturally. The ambivalence observed through morphometrics regarding the domesticated features of grape pips shows a lack of a straightforward, linear pattern from wild grapevines to domesticated ones. This may be quite telling about the processes of domestication of the grapevine in Greece and southeastern Europe where wild grapevines formed part of the natural vegetation: the emergence of domesticated grapevines, viticulture and wine making were probably not uniform, entirely introduced, or locally shaped. Different regional processes might have been under way, occurring at different times: people bringing wild stock into cultivation and/or "importing" (even "smuggling") stock from other areas where domesticated varieties had already been developed. In such a scenario, the spread of grapevine cultivation and wine making may have not been a straightforward process, an inevitable outcome of the availability of wild grapevines and knowledge of fermenting wild grapes into wine. Given the mind-altering effects of wine consumption, the production and consumption of such a potent drink were most likely laden with strong culture-specific regulations, variable across time and space. The association of wine production and consumption with funerary rites and rituals on the island of Crete in the Early Bronze Age at sites such as the Phourni cemetery at Archanes and tombs at Mesara containing wine press models (Platon and Kopaka 1993; see also Pratt 2021:64–69 for a recent synthesis of the evidence) provides hints for the strong symbolism and connections with the underworld associated with wine in the context of Minoan civilization. This may be reflected in the later Dionysos myths about the difficulties encountered in the spread of his intoxicating gift, wine, to different regions of Greece: ancient Greek myths referring to the spread of viticulture and wine involve violation of social rules and murder, with the extreme case of the king of Thebes, Pentheus, who, in his desire to ban the worship of Dionysos (which involved wine consumption), led to the furry of the god, leading to the murder of the king by his own mother, Agave, by dismemberment (*diamelismos*, part of maenad rituals of Dionysos worship), in her ecstatic frenzy, helped by the other women celebrating the god.

The much later image of Dionysos sailing in a boat with a grapevine growing on its mast evokes images of grapevine stock moving along sea routes to other regions, thus spreading viticulture and wine making. Of course, stock could have circulated already in prehistoric times, from other regions where early indications for grapevine exploitation are available spreading different varieties across space (Kourakou-Dragona 2000; Miller 2008). Westward, other places in the Mediterranean such as Sardinia seem to have also enjoyed wine making since the Bronze Age, but only much later does wine reach the shores of southern France and the Iberian Peninsula. By the Late Bronze Age, wine seems to be produced in many different parts of the Mediterranean, apparently spreading westward from the east Mediterranean and the Aegean traveling along the sea routes of the Mediterranean (e.g., Knapp 1993; Cline 1994; Blake 2008). Mycenaean contacts with the Adriatic coast and farther to the west are well known (see Soares 2019), and they may have contributed toward the spread of wine consumption westward. In terms of drinking sets, styles of consumption must have also circulated along presumably the same routes (e.g., Rutter 2012; Rutter 2017). In light of this discussion, it is possible that some of the ancient Greek wines of appellation (Logothetis 1970) mentioned by Homer were formed already in the Bronze

Age, despite the lack of any varietal uniformity of the type known today, in the shapes of the charred grape pips from archaeological sites.

Wine or Beer? Revisiting an Old Question

A large variety of drinking vessels has been observed in the archaeological record of Bronze Age Greece suggestive of changing traditions in etiquette of consumption regionally as well as over time (Andreou and Psaraki 2007; Rutter 2017). It is almost always assumed that the alcoholic drink consumed in prehistoric Greece was wine, and until recently, few inferences about beer consumption have featured in the archaeological literature (Jung 2006; Rutter 2008). Variability in the forms of drinking sets might have been related to different alcoholic drinks requiring special vessels, as is traditionally the case nowadays with wine, beer, or cider. Beer has featured in connection to a special vessel from Lerna on the basis of similarities to one from Anatolia for which it has been hypothesized that beer was drunk from straws fixed inside the openings at the rim of the vessel (Rutter 2008). These interpretations were partly based on pictorial evidence from the Near East of collective drinking from a large, open-mouthed vessel (Rutter 2008). Such vessels are extremely rare in the archaeological record and were probably exceptional, reserved for special occasions and liquids. Could beer have been made then in prehistoric Greece despite the deeply rooted assumption that it was always wine that was drunk? At Early Bronze Age Myrtos Pyrgos, beer was considered a possible drink, based on scarce archaeobotanical remains of barley and residue analyses (McGovern 2003). Yet, in both Lerna and Myrtos, beer consumption lies in the sphere of speculation rather than solid evidence. Thus, until recently, beer was not mentioned for prehistoric contexts from Greece, despite its widespread use by prehistoric communities of the eastern Mediterranean in the Neolithic and the Bronze Age with which contacts existed at least since the Neolithic.

As with grape wine, archaeobotany recently contributed convincing evidence that some form of fermented beverage based on cereals might have been produced in certain parts of prehistoric Greece, on the basis of archaeobotanical remains from central and northern Greece, dated to the end of the third millennium BC. The finds come from excavations carried out in the 1950s in Thessaly and between 1995 and 2000 in the region of central Macedonia in northern Greece, at the sites of Argissa and Archondiko, respectively. Finds of sprouted cereal grains provided the first "hint" that beer might have been brewed in Bronze Age Greece.

The sprouted grains from Archondiko (Figure 6.5) come from one among a series of rooms, where many clay facilities have been unearthed, related to various types of food preparation given their wide variability in shape, dimensions, and function (Papadopoulou 2010). At Archondiko, the sprouted grains are relatively few and poorly preserved. Thus, at first sight, they could correspond to just a few cereal grains that happened to sprout accidentally. The same archaeological context, however, had already yielded some unusual ground cereal food preparations (see Figure 2.29), initially interpreted as some form of precooked cereal food of the type of *xinochondros* or *trachanas*, that is, ground cereal grain boiled in milk and dried. In addition to these finds, loose cereal fragments, generated prior to charring, were also present in the site's archaeobotanical assemblage. Further study of the samples from Archondiko revealed the presence of different types of cereal fragments, demonstrating different features, especially in relation to their fracture surface (Petridou and Valamoti, work in progress). Some of these fragments show a glassy, shiny surface of the type experimentally recognized in cereal grain fragments that had been boiled prior to charring, before or after grinding (Valamoti 2002a; Valamoti et al. 2019; Valamoti et al. 2021). Other fragments present in the samples are bulging but show no signs of this shiny surface. The material consisting of ground cereal grain at Archondiko could correspond to ground malt. Recent experimental charred malt shows specific features that will help interpret the finds (work in progress, project PlantCult; Valamoti et al. 2017, 2021). It is possible that, at Archondiko, ground cereals were mixed with water and malt, toward the preparation of a fermented drink (Valamoti and Petridou 2022; Petridou and Valamoti, work in progress), in a way similar to those described for Egypt and Mesopotamia (Bottéro 2004; Samuel 2006; Jennings et al. 2005). The lumps of cereal fragments could have been used as starters for beer as is the case at the site of Tall Bazi based on archaeological, archaeobotanical, and experimental evidence (Zarnkow et al. 2011) (Figure 6.6 and Figure 6.7). The lumps found at Archondiko might have been generated accidentally during charring (see Valamoti 2002a). Alternatively, they could have been deliberately prepared and left to dry to be later used as starters for brewing or making bread (Valamoti 2018).

There is further, corroborative evidence, however, pointing toward brewing at Archondiko, provided by a complex clay structure found at the site (Figure 6.8 and Figure 6.9): experimental investigations of the temperatures that can be reached in such a structure have shown that the back chamber does not exceed a temperature of 60°C (140°F) (Papadopoulou 2010). This

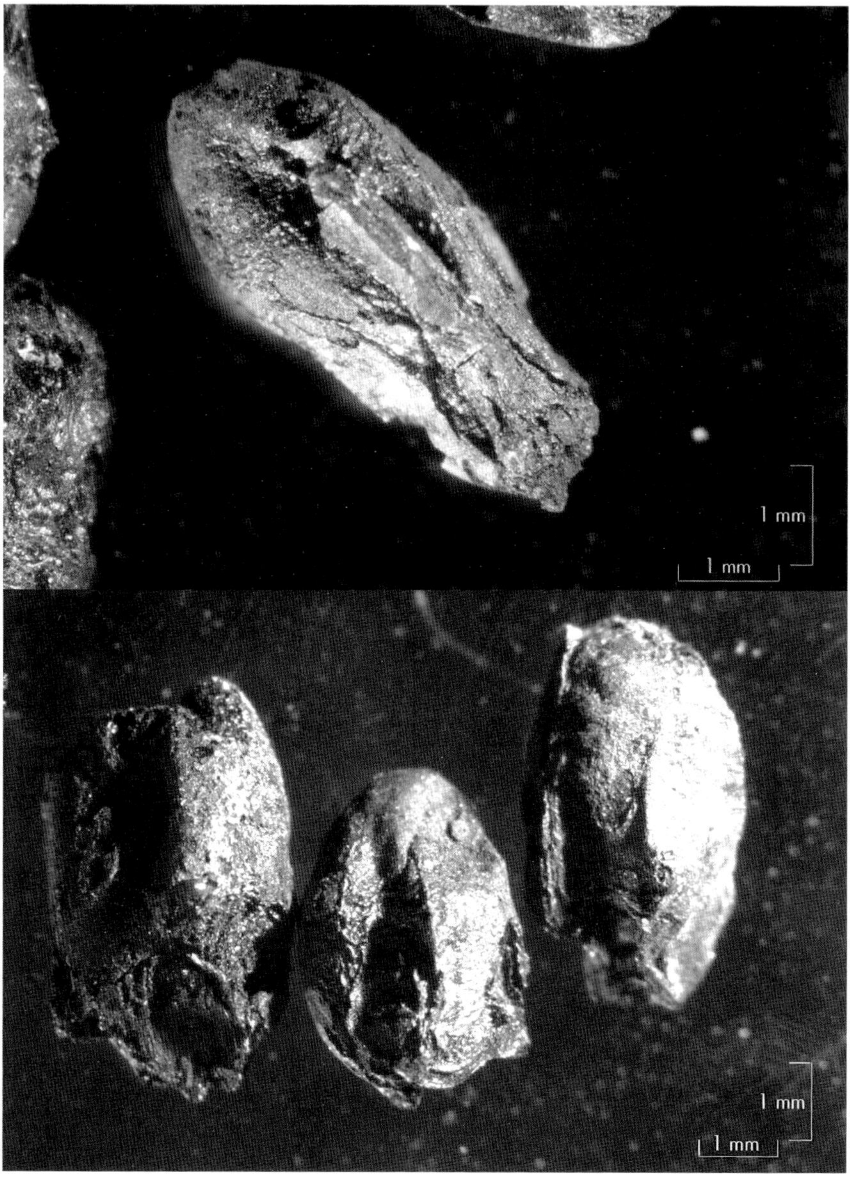

Figure 6.5. Charred, sprouted grains from Archondiko Giannitson, end of third millennium BC. (Photographs by Chryssa Petridou. Courtesy of PlantCult.)

Figure 6.6. Instant beer produced with a mixture of water, ground malt, ground cereal grain, and *bapir.* (Photograph courtesy of Martin Zarnkow.)

Figure 6.7. (*below*) *Bapir* is a partly baked bread that can provide "instant beer" within twenty-four hours. (Photograph courtesy of Martin Zarnkow.)

is indeed the maximum temperature for creating the mash for making beer, after which the liquid is allowed to ferment. At Archondiko, one of the main vessels found in the houses is of an open type, tapering at its base, leaving a relatively wide surface exposed to the air (see Figure 6.8). It is possible that some of these vessels could have been used for brewing a liquid mixture of water, ground malt, ground bulgur, and lumps of ground grain. These open-mouthed vessels are very similar to those found in Celtic Europe and modern Ethiopia as well as ancient Egypt that are associated with beer making. Yet, no evidence for internal vessel-wall erosion, similar to that observed in modern ethnographic examples from the Gamo in Ethiopia (Arthur 2014), has so far been reported for these vessels from Archondiko.

Another site of a similar date, Argissa in Thessaly, has yielded even clearer indications for the intentional preparation of malt. More than 3,000 charred

Figure 6.8. Two-chamber thermal structure from Archondiko with open clay vessel. (Photograph courtesy of Aikaterini Papanthimou.)

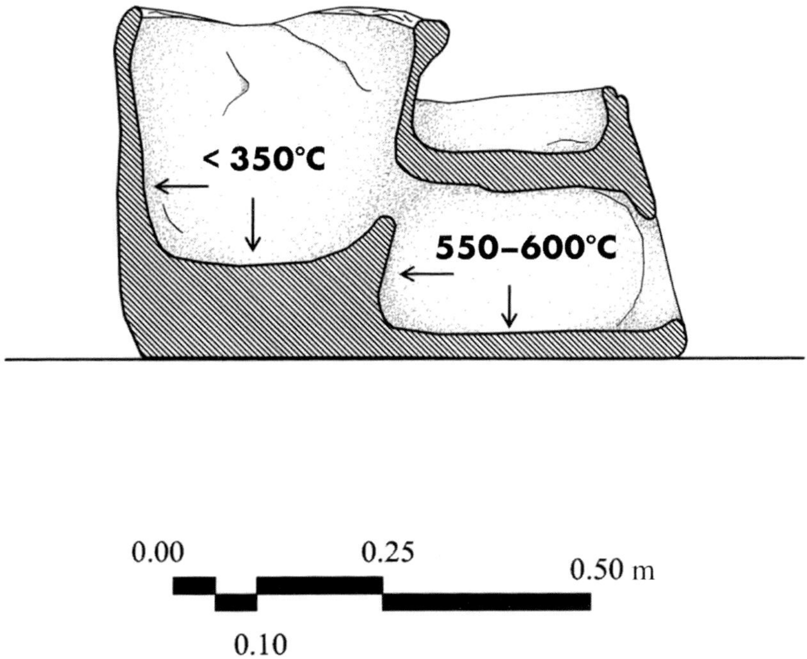

Figure 6.9. Section of a two-chamber structure showing highest temperatures reached in each chamber. Experimental replication by Papadopoulou 2010 showed that temperatures up to 60°C (140°F) were maintained in the back chamber. (From Papadopoulou and Maniatis 2013, Figure 5. Courtesy of Evanthia Papadopoulou and Aikaterini Papanthimou.)

cereal grains, predominantly of wheat, were uniformly sprouted, suggesting a controlled process (Figure 6.10). The house where the rich concentration of malt has been found could well qualify for a space closely linked to brewing. It contained a hearth as well as a particular structure with parallel stone walls. Similar structures have been identified in all houses of the Early Bronze Age phase at Argissa (Figure 6.11) and could have been used for preparing the malt, in other words for germinating the grains in a controlled way, the stone walls serving to support reed mats for spreading the grain for sprouting and keeping it moist and well aerated until germination of the grain was complete and had reached the point of terminating the process. This interpretation for the Argissa remains is inspired by similar-looking constructions from Egypt where the remains of mats have been preserved resting on parallel walls at the site of Tell el-Fara'în-Buto (Hartung 2021; see also Lehner 2009 for a thorough discussion of malting and related installations in

Figure 6.10. Sprouted cereal grains from Argissa, 2100–1700 BC. (Photograph by Felix Bittmann. Courtesy of Niedersächsisches Institut für historische Küstenforschung.)

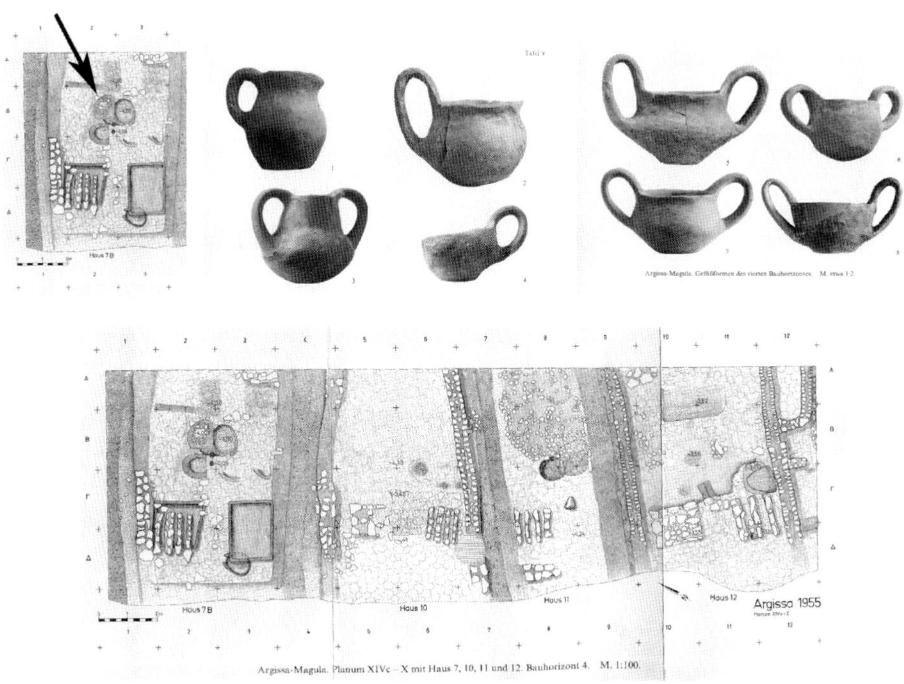

Figure 6.11 Middle Bronze Age Argissa house plan where charred malt, drinking cups, hearths, and a special feature of parallel stone walls have been identified. These walls might have supported mats for the preparation of malt. (Courtesy of Joseph Maran, Heidelberg University.)

Figure 6.12. Bronze Age drinking vessels from Archondiko found in the houses of the phase IV destruction debris. (Photograph courtesy of Evangelia Voulgari.)

ancient Egypt). The hearths identified in this room could have provided the heat for different stages of beer making: drying of the malt and preparing the wort, the liquid of ground malt and water that required a controlled temperature not higher than 60°C (140°F). The hearth remains excavated at Argissa house 7B, where the concentration of sprouted grain was found, are inadequately preserved for determining whether they could correspond to the complex structure with two chambers as that identified at Archondiko.

Overall, the available archaeobotanical and artifactual finds provide adequate indications for brewing at Archondiko and Argissa. At Argissa the evidence is much stronger in terms of numbers of sprouted grains and special installations that could be related to malt production. At Archondiko malt remains are few, yet brewing appears as a very likely possibility on the basis of the complex hearth finds and of the large quantities of ground cereal grain and ceramic vessels that could have been used for brewing. If indeed beer was made at the two sites, was there a special type of container for consuming it? A large number of drinking vessels (see Figure 6.11) (Figure 6.12) has been identified from both sites, which could have been used for special drinking events where a special drink, potentially beer, might have been consumed.

Given that parts of mainland Greece and the Aegean islands in the Bronze Age were in contact with societies of the eastern Mediterranean where beer was produced at places such as Hierakonpolis (Attia et al. 2018), Tell Farkha (Kubiak-Martens and Langer 2008), Mesopotamia (Bottéro 2004), and Cyprus (Crewe and Hill 2012), it is surprising that beer has not featured more in discussions of feasting and alcohol consumption in prehistoric Greece. This may probably be a projection to prehistoric times of ancient Greek habits and the concept of the Mediterranean triad, whereby wine holds a prominent role, with beer corresponding to a drink of foreign people (see, for example, Nelson 2005).

The recent archaeobotanical evidence presented here provides clear indications that during the Bronze Age, in some parts of Greece, beer was being brewed. Whether this was prepared in addition to wine or as the sole alcoholic drink of certain communities is as yet impossible to tell. Archondiko has yielded only a few grape pips, while numerous pips are reported from Bronze Age levels at Lerna where the special "beer" drinking vessel was found. Yet, archaeobotanical remains are subject to various taphonomic processes, and absence of evidence should not be taken as evidence for absence. In light of current evidence, however, the existence of wine-drinking and beer-drinking groups in the third millennium BC in southeastern Europe and the Aegean cannot be excluded. This possibility is worth investigating further in the archaeological record, combining archaeobotany and ceramic vessels potentially used for alcohol production and consumption. What emerges with some certainty is that some groups of people made and consumed beer toward the end of the third millennium BC. Their relationship to wine drinkers, if such distinctions indeed existed, remains to be explored in the future.

Contexts of Alcohol Consumption

Neolithic wine at sites such as Dikili Tash or Dimitra in northern Greece was probably made by allowing a mixture of pressings, pips, and skins with the juice to ferment. It is unknown how long fermentation would have lasted. Once ready, the wine could have been consumed in a big, collective feast. There are no indications from Neolithic contexts for storing wine in large quantities. The Dikili Tash vessel with which the pips were associated and where residues of grape juice fermentation were detected (Garnier and Valamoti 2016) was quite large and would have contained a substantial quantity of wine, estimated approximately to 130 liters (ca. 34.34 gallons) (Zoï Tsirtsoni, personal communication). The cups that have been found in the same house were small, their capacity up to 250 ml (ca. 15.25 cubic in.) (see Figure 6.4); thus, the large jar would have contained servings to hold a feast for as many as 500 people, depending on the size and number of the servings per participant.

Wine, identified at Dikili Tash in the fifth millennium BC, takes us a step further in the exploration of contexts of alcohol consumption. In a

Neolithic house, a fruit harvest from beyond the crop fields, in the woodland, was transformed into an intoxicating substance and was probably served with specially designed decanters and consumed in small cups. Its consumption would have led to the experiencing of particular states of consciousness, and it might have prescribed specific rules of participation and etiquette of consumption. At Dikili Tash, in the cold days of winter, a celebration perhaps of communal character might have revolved around the village's new wine, served in the cups found inside House 1. The context of such an event, the opening of new wine and its consumption, for example, might have been marked by the use of specially decorated vessels for food and drink consumed for the occasion as well as special clothing and body decoration (see the importance of such paraphernalia in feasting contexts in Hayden 2001:40–41). This "Dikili Nouveau" might have been consumed in a feast where certain participants may have worn precious items such as the necklace made of numerous beads, including gold ones, found inside the house where a large quantity of wine was produced (Figure 6.13).

Figure 6.13. Beads found in a large concentration in House 1, Dikili Tash, 4300 BC. (Photograph by Pascal Darcque. © Dikili Tash Research Project.)

Yet, the remains of the feast were not found at Dikili Tash but at two other Late Neolithic sites of northern Greece, Makriyalos and Promachon-Topolnitsa. At Makriyalos thousands of pots had been smashed and huge quantities of meat consumed in one very large or several large-scale feasting episodes. The pots and animal bones were deposited in a pit shortly after the event(s) (Pappa et al. 2004; Pappa 2008). It is possible that the remains of the feasts were also deposited in smaller pits as well as the ditches, which were excavated. The spatial organization of feasting at Makriyalos might not have been merely confined in the large pit where the remains of the feast are abundant. Smaller pits at the site might have also served as the recipients of the remains of feasts, size and content depending on various factors such as occasion, number of participants, and their status within the community(ies) participating in the feast and their possible special function in a series of rituals (see Wiessner 2001:134, Figure 4.3). At Promachon-Topolnitsa, in a special-purpose subterranean building, gatherings took place, perhaps also spreading in the open-air space during the feast itself: large numbers of *boukrania* have been found in this building, suggesting special ritual practices (Koukouli-Chrysanthaki et al. 2007). Could these *boukrania* have been worn like masks in special rituals? Was alcohol consumed in this context? If so, was it wine? It is often the case that large quantities of alcohol or other psychoactive substances are produced specifically for different kinds of feasts in various parts of the world (e.g., among the Luo of Africa; Dietler 2001:89). These questions will remain unanswered at least for the time being. Neither Makriyalos nor Promachon-Topolnitsa has yielded evidence for wine, yet its consumption in the context of feasting remains a possibility, especially at Makriyalos, where some pips were found in a pit (Valamoti 2004). Insights to wine consumption in Neolithic contexts would require a detailed consideration of cups and serving vessels, found at different sites, especially those that yielded rich remains of grape pips.

From the Bronze Age on, indications for wine or alcohol consumption in general become more abundant (articles in Wright 2004b and Gauß et al. 2011). Alcohol consumption, wine in particular, has been linked to feasting in Mycenaean palaces, inferred not only from the impressive cups both in number and raw materials but also from the Linear B archives where the grapevine, wine, and Dionysos are mentioned. Feasting during the Bronze Age in the south would have taken place within the palatial complexes or in open spaces there or outdoors, such as at cemeteries (e.g., Hamilakis and Sherratt 2012). The Minoan and Mycenaean evidence clearly connects wine to the elites, used as a means of distinction and empowerment. In Bronze

Age Crete, specific contexts where smashed cups have been found deposited in quantity have been associated with drinking alcohol as part of funerary rituals, rituals that may have involved dancing and singing, according to Hamilakis (1998). For these strongly hierarchical societies of the south, wine consumption probably was subject to a more formal etiquette of consumption and sequence of rituals. Wine consumption took place in feasts organized by the palaces, and there is evidence to suggest that participation was subject to specific rules, including differential access to cups made of different materials and space(s) where the ritual events took place (e.g., Stocker and Davis 2004). It seems that the wine that was consumed in specific ritual contexts in Mycenaean times was a luxury item reserved for such occasions (Palmer 1995). The main ritual events took place in the central court where the central hearth was located. Drinking by the participants is probably depicted on wall paintings found inside the palace at Knossos (piano nobile), showing drinking in pairs (Evans 2013:384, 388–389, Figures 323, 324).

The produce of viticulture, wine, is often regarded as a luxury, usually related to rituals and feasting (e.g., Powell 1995). Such contexts of consumption might have rendered wine suitable for appropriation by elites (see Hamilakis 1996). Indeed, elites and alcohol seem to have had a close relationship in the archaeological literature whether the discussion revolves around Mycenaean wine drinking (Wright 2004a), Sumerian and Egyptian beer drinking (Damerow 2012), or Celtic beer (Dietler 1996; Arnold 1999). The origins of viticulture and wine making have been at the core of discussions concerning the emergence of social stratification among prehistoric societies of the Old World (e.g., McGovern et al. 1995; Hamilakis 1996). Evidence for prehistoric alcohol production on its own, however, cannot be used to indicate the existence of elites. In other words, hierarchical organization was not an inevitable outcome wherever in the world people produced and consumed alcohol. The way wine or other alcoholic drinks was appropriated determined whether or not they were used as a means of distinction. Regardless of whether wine was consumed by elites or the whole community, its power to transform the senses rendered each context of consumption special, generating feelings, memories, and a sense of belonging or exclusion.

Whether wine was consumed in a village in the north during the fifth millennium BC or in a palace in the south during the second millennium BC, it is beyond doubt that its consumption would have had mind-altering effects generating special levels of consciousness, emotions, and memories. For the Neolithic and Bronze Age inhabitants of prehistoric Greece (and elsewhere for that matter), consuming alcohol must have been a powerful experience by mere virtue of its mind-altering power, further enhanced by other elements that might have structured the rituals of each consumption event. To understand the special significance of its consumption, we need to distance ourselves from the stereotypical image we have of the present, with a glass of wine being an element of daily meals, at least along the coasts of Mediterranean Europe. Wine in prehistoric Greece could not have been produced on such a large scale and be readily available in the quantity a regular consumption as part of the diet would have required. As mentioned, even in Late Bronze Age contexts, it is known from Linear B archives that wine was considered a luxury liquid, reserved for special occasions (Palmer 1995).

Despite the long tradition of wine making in northern Greece in prehistoric times, spanning the Neolithic through to the Bronze Age, continuing through to the Iron Age, local elites remained inconspicuous even during the Late Bronze Age, when the Mycenaean palaces flourished in the south. The archaeobotanical evidence, combined with contextual evidence from the north and south of Greece, clearly shows that it took a lot more than mere wine making for elites to establish themselves and appropriate power in a way visible in the archaeological record. It is unlikely that wine making on its own led to the emergence of elites in prehistoric Greece. Already established elites, however, could have used wine for its mind-altering properties to exercise their power and possibly control access to the produce. In the south of Greece in the Bronze Age, there is insufficient research so far that would allow exploration of whether production of wine took place beyond palatial control and whether wine was widely available/accessible outside the palatial sector.

The archaeobotanical evidence available, as presented here, leaves little doubt that wine was produced, and it is reasonable to assume that this was the liquid that filled the numerous cups found at places of the living and the dead from the north to the south, in Neolithic and Bronze Age Greece. The Neolithic evidence for feasting and wine making as snapshots of practices caught in the debris of specific settlements suggests that wine consumption in special contexts might have been a plausible scenario. Other communities at least during the Bronze Age, but possibly earlier, seem to have been engaged in beer production as the evidence from Argissa and Archondiko suggests. There again, numerous cups leave no doubt that they were made to hold a special liquid. The distribution of cups at Archondiko, primarily in one house, suggests that the hosts may have been certain individuals, families, or lineages in charge of organizing the feast and providing for it.

On a more speculative note, feasts including rituals and alcohol consumption might have been related to the ancestors or at least included them, as well, as has been widely observed in the ethnographic record (e.g., Dietler 2001; Clarke 2001:149; Wiessner 2001:132–133; Kirch 2001:172; Dueppen and Gallagher 2021). At Makriyalos, the remains of the feast do not reflect a haphazard barbecue event: parts of the site comprise ancestral bones and some pits definitely contained the remains of feasts (Pappa 2008). Similar contexts associating wine consumption to funerary rituals are known from Bronze Age Crete. In this context, ritualized events of getting together, eating and remembering, eating with the ancestors, might have taken place. Yet, we still lack the site where both indications for ritual practice and evidence for wine or beer coincide on the same site. Although absence of evidence for alcohol production may be an artifact of taphonomic processes, we cannot suggest a widespread practice of drinking wine wherever there is evidence for feasting. Such a potent liquid might have been unwelcome in some places.

7

A Cuisine of Healing

The consumption of plants as food, spices, medicine, or psychotropic/hallucinogenic substances corresponds to a continuum from feeding the body to fulfill its basic needs to keep in good health, to healing illness as well as experiencing special states of consciousness. Perceptions about health and disease are culturally prescribed and generally reflect perceptions of each society about what constitutes regular and irregular, clean and unclean (Douglas 1966). Defining good health almost in all cases requires a juxtaposition to disease, as the latter is a divergence, some form of interruption of the normality of healthy life (Svenaeus 2001:59; Opler 1963). The definition of what signifies a healthy condition may also be related to appropriation of power as specific groups with authority may define what constitutes an accepted behavior/condition healthy or unhealthy and may also possess the right to provide medical treatment (Ólafsdóttir 2013; see Kaye 2018 for the Caribbean; Castle 1993 for an example from Mali; Popper-Gideon and Ventura 2008 for Palestinian female healers in Israel). Likewise, what constitutes a healing practice is culturally defined. For example, western medicine, rooted in ancient Greek Hippocratic medicine, is different from Ayurvedic or Unani medicine (e.g., Poulakou-Rebelakou et al. 2015; Mukherjee 2001) or from the health care provided by traditional healers as is the case in many parts of the world (see Kale 1995 for South Africa). There is a rich ethnographic, ethnopharmacological, anthropological, philological, and archaeological literature where one can observe the multiplicity in the perception of illness and disease and the ways in which these can be treated through the intake of plant ingredients and preparations (see, for example, Hsu 2010).

For prehistoric times, the notion of treatment through intake of an active substance isolated from its source would be an anachronism based on modern, western world pharmacopoeia and medical practice, not one usually encountered in healing practices performed among traditional communities. Plants may well have constituted important sources of active medicinal substances since the Paleolithic while foodstuffs have often been prescribed to heal specific conditions.

The role of plants in healing practices is found in different parts of the world and for different periods and cultural settings (e.g., Petrovska 2012; Weiner 1980 for the American Indians; Teall 2014 for the Sumerians). In Sumerian texts the apothecary (*simnu*) literally meant a grower of medicinal plants (Geller 2010). The rich ethnographic record is particularly instructive in this respect, showing the deep knowledge of indigenous people of their surrounding vegetation and the properties of plants (e.g., medicinal, narcotic). Lévi-Strauss (1962) offers many such examples of people like the Subanum of South Philipines, the Seminoles, and the Hopi and Navajo of the American Southwest. This knowledge is equally reflected in the terminology these peoples have devised to name the different parts of the plants and their corresponding properties in terms of nutrients and healing power. Moreover, plants form elements of a cosmological order, and they correspond to classification systems whereby connections between what may appear arbitrary to the western mind are meaningful and operate on many levels including the realm of the symbolic, as is recorded, for example, for the Tobas of Bermejo in Chaco, Argentina (Martínez 2010), for the Mixe in Oaxaca, Mexico (Heinrich 1994), for the Cherokee in the United States (Garrett 2003), and for the Talensi in Africa (Insoll 2011a). An example of such associations can be found in North America between various *Artemisia* species, the moon, the night, women, and the treatment of problems related to the female reproductive system (Lévi-Strauss 1977). Similar associations between specific plants, body parts, and their healing go as far back as the Bronze Age, as can be seen in ancient Mesopotamia (Teall 2014).

Such classifications are not as arbitrary as they may seem, however, and in the selection of plants for such associations, smell and taste seem to play important roles: a plant may be selected for the treatment of a specific illness on the basis of its smell and taste, something possibly related to the survival value resulting from such associations. The recognition of a strong-smelling plant often coincides with medicinal properties (Majno 1975). Traditional knowledge of medicinal plants relies on taste and smell not only for the selection of the plant for healing purposes but also for the transmission of the knowledge on such plants from generation to generation, as they both contribute to generating the memory of the plant and its properties (Molares and Ladio 2010; Heinrich 1998). At the same time, taste, odor, and the medicinal properties of plants are culturally defined and therefore variable, depending on the different social and ethnic groups people belong to, as is indicated by research conducted in various parts of the world (Johns and Keen 1985; Shepard 2004; Pieroni and Torry 2007; Molares and Ladio 2010).

The brief and selective exploration of the ethnopharmacological and anthropological record shows how different communities know the healing potential of their surrounding vegetation. For the Paleolithic gatherer-hunters, it is reasonable to assume that a similar knowledge was available, given the biological mechanisms that seem to be under way for securing treatment of mammals and primates. Recent investigations in animal behavior (e.g., Huffman 1997, 2007, 2016) as well as ethnobotanical studies (e.g., Pieroni et al. 2006) show how primates, animals as well as humans, know which specific plants to eat/graze when they are unwell. It is in the distant Paleolithic past of Greece that this interaction with the surrounding vegetation would have led to an acquired knowledge transmitted and enriched from generation to generation (Valamoti 2013b; see also Hardy and Kubiak-Martens 2016a for a general discussion) including the healing properties of plants (see Valamoti 2013b).

We certainly do not know how the prehistoric inhabitants of Greece perceived and classified their plants, whether they possessed distinct categories for food and medicine. The healing cuisine explored here focuses on plants identified in the archaeobotanical record that could have been used for modifying the condition of the human body. In this sense it covers the whole range from food as a means to recover from an illness to special substances used to alleviate pain or induce altered states of consciousness. Moving beyond the cautious interpretation of prehistoric plant resources as merely being used for feeding the physical body, the healing potential of some plants is discussed together with the psychoactive/hallucinogenic potential uses of some.

We tend to treat most of the plant remains we find in rich concentrations at archaeological sites as mainly used for securing subsistence, fulfilling basic needs for food, fuel, and building material. Wild species are more often than not considered as weeds of cultivation or components of dung, a trend that is gradually changing (see Antolín et al. 2016). Even some plant species that are famous worldwide for their "special" properties, such as opium poppy or the grapevine, are often interpreted as sources of food in the form of oil, fruit, or wine. It is only when plants with clear medicinal properties are present in archaeobotanical assemblages that a special section is dedicated to such plants in archaeobotanical publications (e.g., Kroll 1983:77–82).

The cautious interpretations of archaeobotanical data regarding their association with healing practices and rituals result probably in a huge underestimation of the experience and accumulated knowledge on the properties of the natural and managed vegetation that would have surrounded the built space of Neolithic and Bronze Age settlements. This chapter explores plants used in a "cuisine of healing" as potential ingredients of remedies that could have been used in healing practices, often grouped in publication lists as "oil crops," "other wild plants," or "weeds." It sheds some light on this marginalized aspect of prehistoric plant use in Greece, as it would have been an essential element of both daily life and special occasions such as rites of passage or the treatment of serious illnesses.

Good to Eat, Good to Heal

Sometimes a bowl of warm soup may have a soothing effect and help us feel better. This could also apply to a warm infusion at the end of the day with *Sideritis* flowers, also known as Greek mountain tea. Many traditional remedies, usually given as advice from grandparents, or found in traditional recipe books, involve some plant preparation that may ease stomach pain, cough, or fever. Preparing such recipes requires fire, containers, mixing, and serving. All these steps also are involved in cooking. The recipe for roasting linseed with mustard seed to prepare a poultice is found in modern herbal books (Bown 1995) as well as the works of Galen, a Greek physician of the second and third century AD (see, for example, Valamoti 2011c, 2013). These remedies might be the closest we can get to what may have been a healing cuisine of prehistoric times, a cuisine that has managed to survive in our rational, sterilized modern world (Figure 7.1). Many of the plants identified in the archaeobotanical record of Greece, presented in the preceding chapters, besides their potential use in healing also constitute at the same time foodstuffs. Drawing the line between food and medicine is probably difficult even in a modern context: physicians and nutritionists advocate specific dietary habits for the prevention of diseases such as cardiovascular disease, colon cancer, and diabetes that affect modern human populations.

The perception of food as medicine is not only encountered in medical texts, ancient and modern, but also in the ethnographic record. In Ethiopia emmer porridge called *ga'at* is given to sick people, children, and nursing mothers for being easier to digest and having medicinal properties (D'Andrea and Mitiku Haile 2002). A roasted barley porridge, *ptisane*, was prescribed for its medicinal properties by Galen (Dalby 1997; Valamoti et al.

Figure 7.1. Herbs collected from the wild and pomegranates for sale at Kalamata market, October 2018. (Photograph courtesy of Hans-Peter Stika.)

2022a). In the Indian Ayurvedic system, food and health are closely connected whereby food is not only intended for body nourishment but also connected to a cosmic moral cycle, the human body being the product of food. Different groups of food are associated with different qualities in people, for example, foods associated with wisdom (*satvika* foods) such as cooked vegetables, milk, and fresh fruits, or the lowest qualities in humans (*tamasika* foods) such as liquor, meat, and garlic (Sarkar et al. 2015:97).

In the works of ancient Greek physicians like Hippocrates and Galen we encounter a clear association between different types of plant food species and/or plant food recipes regarding their influence on health. It was believed, for example, that different cereal species affected health differently in terms of digestion, as did the consumption of plant foods cooked or raw (Krug 1997). These ancient sources correspond to medicine in its birth, based on ancient Greek philosophical ideas that were generated in the sixth century BC and on observations by professional physicians of antiquity. They searched carefully for the causes of illnesses and prescribed a wide range of remedies for which plants were an important element on the basis of a classification system of the four humors, which later dominated medicinal practices of the Middle Ages. The four main categories of the elements (earth, water, fire, air) corresponded to distinct powers and properties (e.g., Krug 1997). Foodstuffs are encountered in medicinal recipes of ancient Egypt (Nunn 2002) while bewitched food was believed to cause illness among the Sumerians (Geller 2010). Ancient Greeks placed great importance on diet as a means to maintain good health (Krug 1997). Thus, the connection between food and health is ancient and can be found in different cultural and spatiotemporal contexts from antiquity to the present.

Prehistoric inhabitants of Greece most likely did not theorize about how digestion transformed food into heat necessary for human survival and how cooking, that is, processing of food by fire, was necessary to avoid indigestion. Yet, it is fairly reasonable to think that they, as all prehistoric people who cooked food, might have observed how cooking raw ingredients facilitated digestion of many foodstuffs. Thus, a main tool for keeping good health in the first place might have been the pot and the hearth/fireplace: plants would have been cooked, fermented, infused, and processed to remove toxins and obtain beneficial substances necessary to keep in good health or to heal if such a need emerged. Indeed fire, pots, cooking, and healing plants have been considered an essential element of Early Neolithic societies of prehistoric Greece operating in "special" contexts of consumption. Individuals in charge of cooking plants were most likely those who also knew their healing properties, thus holding a special role in the first Neolithic farming communities in the area (Vitelli 1995).

Toward an "Herbal" of Prehistoric Greece

The Paleolithic knowledge of the healing properties of plants available to gatherer-hunters of prehistoric Greece was certainly enlarged and diversified with the onset of the Holocene. The new way of interacting with nature through farming offered a rich inventory of plants with medicinal properties that were out there, available for experimentation and use by Neolithic and Bronze Age farmers. Thanks to the Pleistocene *refugia* present in this mountainous tip of southeastern Europe, lying at the crossroads between Asia, Europe, and northern Africa, surrounded by sea and including many islands, a new, diversified flora emerged in the post-Pleistocene landscapes of Greece with the expansion of arboreal vegetation, the emergence of new habitats generated by humans such as fields, pastures, and open land. These habitats were home to numerous plants with aromatic and medicinal (including poisonous) properties. The opening up of fields for the cultivation of crops generated habitats for "weeds," some of which could have been harvested as food and medicine at the same time, a kind of "protonutraceuticals" (Leonti et al. 2006:139), their consumption in the form of greens being very common in Greece in the form of *horta* as well as in other parts of the Mediterranean (e.g., Forbes 1976; Lambraki 1997; Leonti et al. 2006).

The knowledge that accumulated over the millennia of the Neolithic and the Bronze Age led to the later pharmacopoeia of the ancient Greek physicians, corresponding in historic times to a parallel universe with roots in myth and the collective memory of a distant past. Myths inform us of goddesses-witches and half-human half-horse healers, possessing the knowledge of the powers of plants and of potent potions altering perceptions. Some healers from legend (e.g., Circe, Medea, centaurs) knew well how to use plants to evoke evil or good as well as how to communicate with gods, the underworld, spirits, and ancestors (Figure 7.2). Knowing plant ingredients, cooking plants, and preparing potions were probably closely linked (see Vitelli 1995, 2000). Those on the receiving end were at the mercy and power of these healers, as plants can be used to heal or kill, depending on the substances and quantities used as well as the specific circumstances of consumption (on empowerment through cooking, see Wrangham 2009:Chapter 7:174, in particular; Hastorf 2017:192–193).

Unlike the fantastical accounts in Greek myth, the archaeobotanical

Figure 7.2. Bowl depicting Odysseus pursuing Circe. In the air are Circe's magic wand and the deep drinking cup that contains the potion with which she transforms men into animals. Behind Odysseus, two of his men with features of a boar and a horse or mule gesticulate toward him. After the painting on a red-figure Attic calyx-krater (440 BC) attributed to the Persephone Painter. (Drawing by Danai Chondrou. Courtesy of PlantCult.)

record is less impressive, and the stories one can draw from it are always woven with great caution and uncertainty. It is very likely that certain plants would have been used to heal wounds, alleviate pain, and alter the senses and human consciousness. There is pictorial evidence from much later periods that shows treatment of wounds or women in their ecstatic dancing, while for prehistoric times the main indications for the treatment of illness come from the osteoarchaeological record where traces of pathologies and their healing are sometimes visible. The archaeobotanical identification of plants with healing properties offers a means to approach a world of prehistoric senses, smells, pleasure and pain, a world that is subjective and culture-specific (see Hamilakis 2002), and therefore difficult to approach.

How can the long species list of archaeobotanical publications be "translated" into a prehistoric "herbal" guide? A first concern is determining whether a plant was used for healing when identified in the archaeological samples. The association between the identification of a plant with medicinal properties and its use in a specific healing context is not straightforward. Some plants possess exclusively medicinal, aromatic, and/or hallucinogenic properties; others, however, constitute at the same time food that could be used to nourish, to heal, and to experience altered states of consciousness through one of its products, as is the case with wine and opium poppy. Medicinal plants harvested from the wild could correspond to field weeds or could have grown on pastures, meadows, riverbanks, and wastelands where animals grazed or where people wandered in search of raw materials necessary for their lives. Many of the plant parts used in healing practices such as roots, leaves, flowers, and mushrooms would have had limited chances of preservation during contact with fire. This of course drastically affects the survival of medicinal plants in archaeological contexts of Greece, especially if they had been stored as dry flowers, leaves, or buds (see Figure I.8 and Figure 7.1). In addition, use and disposal of such plants may have taken place in contexts away from household fires or even settlement space. Thus, the recent finds of garlic cloves from Late Bronze Age Akrotiri (Sarpaki 2021) and from Mycenaean levels at Tsoungiza in the Peloponnese (Allen and Forste 2021) stand out as rare exceptions at present.

A large number of species or genera with known medicinal properties present in various parts of a plant from the roots to the fruit has emerged from the archaeobotanical record of prehistoric Greece. Some among them have been found in dense, pure concentrations and correspond to crops or harvests from orchards or the wild. They fall in a large category of medicinal/food plants (Appendix A:Table 7.1) that could have been used by prehistoric societies of Greece in healing contexts and meals alike. Although plant ingredients such as barley, bitter vetch, lentil, mustard, gold of pleasure, linseed, and medic were field crops, they most likely were also known for their medicinal properties as one might infer based on ancient written sources, modern herbals, and modern pharmacological research (Valamoti 2001b, 2013a). Likewise, the wide range of fruits and nuts available, for example, oak, fig, olive, grape, blackberry, wild pear, wild rose, strawberry, wild plum, almond, terebinth, Cornelian cherry, *Sambucus* sp., could have had medicinal uses. Additional ingredients for healing remedies become available during the Bronze Age with the appearance of species that could have been used for their oily seeds, their scent, and their medicinal or hallucinogenic properties. These include *Lallemantia* sp., opium poppy, mustard, and gold of pleasure.

For the Late Bronze Age, the Linear B archives offer some insight about the range of aromatic plants used, although there are controversies about the plant species to which the Linear B words correspond (Appendix A:Table 7.2). The word "ki-ta-no," for example, has been interpreted as indicating terebinth resin (Knapp 1991) or terebinth nut oil (Merousis 2016); Sarpaki (2001a), however, does not agree with a terebinth interpretation and opts for its identification as labdanum. Sarpaki (2001a) and more recently Arnott (2014) have listed a range of species mentioned in the Linear B archives for their aromatic/medicinal properties (see Appendix A:Table 7.2). This clearly shows the problems inherent in deciphering texts on such plant taxa. Future investigations might resolve the disagreement.

In addition to potentially medicinal plants that are found in dense/pure concentrations, whether food staples, or, harvested from gardens and the wild, a long list of wild plants with medicinal properties can be compiled, based on the archaeobotanical record of prehistoric Greece. Their remains are often found mixed with cereal chaff or crop seeds (see Appendix A:Table 7.1). The list of such genera and species potentially used in prehistoric healing contexts is rather long, and in light of the rich flora of Greece it is likely that it might well have been longer. Archaeobotany offers solid evidence for a plant's intentional harvest and use; moreover, some of the plants that are potentially medicinal from prehistoric Greece have contextual associations that render their use in healing more likely than in other cases. Among those plants identified in Neolithic and Bronze Age Greece, some deserve special discussion for their possible prominent role in healing practices and in inducing altered states of consciousness. Oil-rich plants, grapevine, terebinth, and opium poppy have been discussed in previous chapters in terms of their distribution and potential uses as food.

Many of the oil-rich plants identified in the archaeobotanical remains possess medicinal properties and have been used in antiquity and recent times. Oils could have been used in healing practices either internally or externally. The medicinal properties of olive oil, for example, are many and have been well known since ancient times as well as nowadays (e.g., Caramia et al. 2012). Perfumed oils in particular have a long history of use, including their external application in a medicinal context (Brun 2000). Likewise, the oil from seeds of the mustard family is known to have healing properties. *Brassica/Sinapis* seeds have been found at Late Bronze Age Assiros in a rich concentration (Jones and Valamoti 2005) and could have constituted a very effective medicine against muscle aches and problems of the respiratory system. Its oil is also mentioned as an emetic in large concentrations (Bown 1995). Members of the Brassicaceae family are used in traditional healing remedies, as is the case of *B. napus* in Iran (Saeidnia and Gohari 2012). The medicinal uses of *Brassica/Sinapis* are also mentioned in ancient Greek texts by Pythagoras (Bauer-Petrovska 2012). Gold of pleasure (*Camelina sativa* (L.) Grantz.) belongs to the same family and is encountered in a rich concentration at Kastanas (Kroll 1983). In folk medicine, its oil has healing effects on ulcers and wounds and is still grown in some parts of Europe such as Slovenia (Rode 2002).

Many of the *Lallemantia* species are known for their medicinal properties. The leaves, seeds, and seed oil are used in traditional medicine in Iran (e.g., Samadi et al. 2007), Iraq, Afghanistan, Pakistan, and India (e.g., Abbas et al. 2012; Albayaty 2011; Khan et al. 2019). In a recent review of *Lallemantia iberica* Fisch. and Meye., the medicinal uses of the seed are mentioned for the treatment of colds, coughs, and abdominal pain while the mucilage contained in *L. iberica*'s seed is also mentioned for its uses in treating "nervous, hepatic and renal diseases and as general tonic" (Al-Snafi 2019:98). Recently, the possibility of introducing *Lallemantia* as a cultivated crop in Europe (again) was investigated, in light of the beneficial oil contained in the seeds (Strašil and Káš 2005).

In the remaining part of this section, terebinth and opium poppy are examined in more detail due to their special properties and potential contexts of consumption suggested by various lines of archaeological evidence.

Terebinth

Terebinth occurs in the archaeobotanical record of Greece, and it is reasonable to assume that its aromatic nuts and resin would have been noticed already in the early Holocene, when it is clearly present in pollen diagrams and the archaeobotanical record. Ancient recipes mention the use of terebinth, especially of resin, applied externally, together with other substances. It was used in the treatment of scorpion bites, skin ulcers, and pain, for example, in several preparations prescribed in the *Materia Medica* of Dioscourides (*De materia medica* 3.146.2.2). Terebinth resin was widely used in the past as incense for its aromatic and therapeutic powers, and the edible oil has also been used in traditional healing remedies (Gennadios 1997; Özcan 2004; Marcopoulos 1965; Simpson 2005). It, along with many other medicinal and/or aromatic plants, is still being used for the preparation of the Holy Myrrh of the Orthodox Church. Its preparation is laden with ritual (Menevisoglou 1972). Theophrastus, a Greek philosopher and botanist of the fourth/third century BC, praised it for its perfume and resin properties. These properties, found both in the nut oil and the resin of the plant's woody parts, continue to be recognized and used in modern times (Polunin and Huxley 1987; Huxley and Taylor 1989; Gennadios 1997; Özcan 2004; Marcopoulos 1965; Merousis 2016; Lardos et al. 2011). Both *Pistacia terebinthus* and *Pistacia atlantica* have similar properties and uses in the wider eastern Mediterranean (Nielsen 1986).

Terebinth is known among prehistoric archaeologists mainly for its resin, identified in the Uluburun shipwreck (e.g., Mills and White 1989; Hairfield and Hairfield 1990; see, for example, Peachey 1995 for an in-depth consideration of terebinth in the broad sense of *Pistacia* sp. in the Bronze Age and in antiquity). It seems that it was traded along the routes of the eastern Mediterranean at least during the Late Bronze Age (Nielsen 1986; Knapp 1991; Artzy 1994). It was probably imported at Amarna in Chanaanite amphorae and was burned in special bowls, locally produced, as incense (Stern et al. 2003). Terebinth was a favorite incense burned by ancient Greeks (Majno 1975); its nuts were also burned, as the smoke was believed by the Persians to have healing properties (Mohagheghzadeh et al. 2006). The healing properties of the smoke of terebinth resin are mentioned by Dioscurides (*De materia medica* 1.18.1.9). It may not be entirely fortuitous that terebinth is one of the resins identified by residue analysis in an incense burner at Dikili Tash (René Treuil, personal communication based on analyses by Nicolas Garnier).

D'Agata, in her review of incense and perfumes in the Aegean during the Late Bronze Age, considers that terebinth resin was likely a traded commodity in the wider context of aromatics being imported by the Mycenaean perfume industry, probably in Canaanite jars, reaching the harbors of Crete such as the town of Kommos (D'Agata 1997). Was the import of terebinth resin necessary given that terebinth trees were growing in prehistoric mainland Greece and the Aegean? Peachey (1995), on the basis of quantities of ki-ta-no mentioned in Linear B texts, estimated the area and the number or terebinth trees that would have been needed to have been available in Bronze Age Crete for the Mycenaean perfume industry. She points out that although terebinth was growing in the prehistoric landscape of Greece and small-scale resin extraction was possible, importing it would be necessary if production of perfumed oils and incense was large scale, as seems to have been the case in Minoan and Mycenaean times.

The association of terebinth with the perfume industry during the Late Bronze Age may have been related not only to its perfume and oil but also its resin and the medicinal properties it possessed. Perfumes and incense are important elements of healing practices in various cultural contexts, past and present (cf. Fappas 2008; Nielsen 1986; Menevisoglou 1972; Brun 2000). In terms of the archaeobotanical evidence, terebinth has been identified in a rich, more or less pure concentration at Neolithic Makriyalos and may have been associated with healing practices in light of its contextual associations (Valamoti 2013a).

Opium Poppy

Opium poppy is one of those special plants that could have had many uses due to the different products that can be extracted from it. Seed oil could have been used for cooking as well as for cosmetic purposes (Azcan et al. 2004; Özcan and Atalay 2006). Opium in many cases has been associated with medicinal and psychotropic uses due to the alkaloids contained in the sap of the unripe head. The alkaloids (e.g., morphine, codeine, papaverine) render the plant analgesic and hallucinogenic. It likely was used from very early on by the prehistoric inhabitants of Europe. Old and recent reviews of the archaeological evidence (Kritikos and Papadaki 1963; Merlin 2003) have highlighted the importance of this plant, while several overviews of the archaeobotanical evidence for opium poppy show its long use in Europe and the east Mediterranean (Salavert 2010; Salavert et al. 2020; Bakels 1996; Marinova and Valamoti 2014).

In Greece, opium poppy appears rather late compared with the Early Neolithic farming communities of central, northern, and western Europe known as Linearbandkeramik (Valamoti 2003). The spread of opium poppy among Neolithic communities of Europe is currently being investigated by directly dating a rich record of opium poppy finds from archaeological sites (Salavert et al. 2020). The recent evidence (Salavert et al. 2018, 2020) confirms earlier suggestions that opium poppy was first cultivated in the western Mediterranean, spreading to the Linearbandkeramik of western, northwestern, and central Europe during the Neolithic (e.g., Bakels 1996), then apparently spreading southward to the Aegean via at least one path connecting Macedonia with central Europe (Valamoti 2007b). The recent dating of prehistoric poppy seeds from twenty-two archaeological sites in central and western Mediterranean, northwestern temperate Europe, and the western Alps confirms that the crop spread northward from parts of the Mediterranean where it grew, gradually and at different paces during the course of the fifth millennium BC when its cultivation became widespread (Salavert et al. 2020). Such a gradual spread in the context of rituals and healing practices involving hallucinogenic plants makes sense as a new psychoactive plant (assuming that others were already in use, e.g., wild mushrooms) would have required time to become accepted and adopted (see discussion about wine consumption in prehistory in Chapter 6). Sherratt argued that, before north Europeans became familiar with alcoholic drinks such as wine and beer, they would have consumed opium in the west and cannabis in the east perhaps in special braziers where it would have been burned in ritual contexts as a means to enter other worlds (Sherratt 1991b, 1995).

Opium poppy is closely connected to Minoan worship as is eloquently shown by the presence of opium poppy heads emerging from the head of a figurine at Gazi in Crete (Figure 7.3) dated to the middle of the second millennium BC (Marinatos 1937; Kritikos and Papadaki 1963). In terms of its archaeobotanical distribution, opium poppy was cultivated during the Bronze Age in northern Greece, encountered primarily at Late Bronze Age sites. The relationship between opium poppy and religious worship is also evidenced since early historic times, one example being the ivory opium poppy heads found at the Heraion on Samos and dated to the eighth century BC (e.g., Kyrieleis 1983). The opium head is depicted in relation to the worship of Demeter, the goddess of cereals and crops in ancient Greece (for a recent discussion, see Voutiras 2016), while it is highly likely that the medicinal properties of opium poppy were widely known not only in ancient Greece but also in the wider area of Europe and the eastern Mediterranean in the past.

The widespread use of opium poppy in the eastern Mediterranean during the Bronze Age is inferred on the basis of artifactual and textual evidence, such as ancient Egyptian medical texts (Kritikos and Papadaki 1963; Nunn 2002; Rosso 2010), and it has been suggested that it was some sort of panacea, a cure-all, an equivalent to modern aspirin (Merrillees 1962).

Figure 7.3. Representation of a clay figurine found at Gazi, Crete, second half of the second millennium BC. (Drawing by Danai Chondrou. Courtesy of PlantCult.)

Ancient Greek texts clearly refer to opium under the name *opos*, meaning the sap of the poppy head, as can be seen in the works of Theophrastus (fourth century BC) and the Hippocratic corpus, in contrast to Bunimoviz and Lederman (2016). Theophrastus (*Hist Plant* 9.8.2.1 to *Hist Plant* 9.8.2.19) describes how the juice of opium poppy was extracted from the head of the plant, while in the Hippocratic corpus we find that opium juice was used to treat various health complaints (e.g., *De mulierum affectibus* 201).

Like terebinth, opium was probably traded in the east Mediterranean. It was most likely exported from Cyprus, placed in a particular type of ceramic vessel, leaving little doubt of its contents as its body closely resembles an

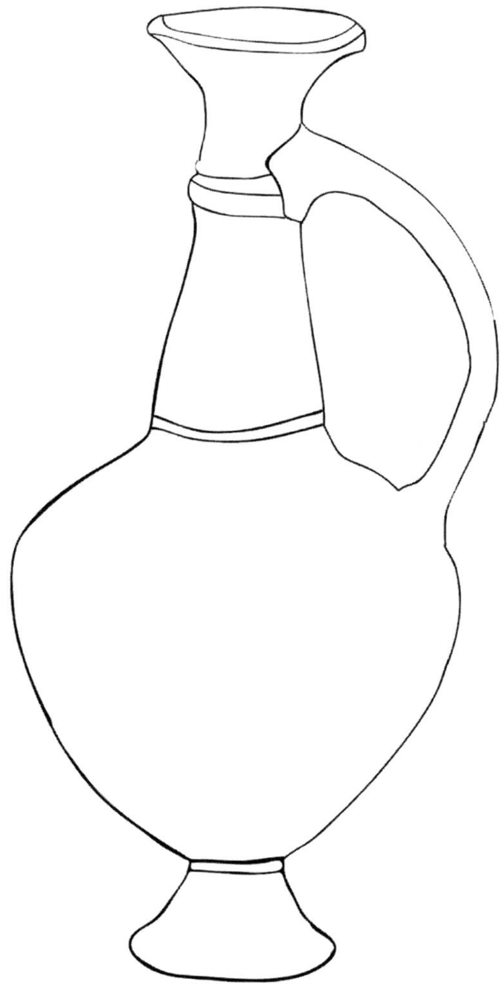

Figure 7.4. Bronze Age Cypriot ceramic vessel resembling a reversed opium poppy head. (Drawing courtesy of Panos Litsios.)

Figure 7.5. Opium poppy head, incised with sap oozing out. (Soultana Maria Valamoti)

inverted opium poppy head (Figure 7.4 and Figure 7.5), occasionally decorated by pale whitish lines, suggestive of the cuts performed on unripe opium poppy heads for the extraction of opium (Merrillees 1962). Moreover, analyses by infrared mass spectroscopy performed on the remains of such pots identified a component of the molecule of papaverine (Merrillees and Evans 1989). Further residue analyses performed on such vessels have questioned the widely held association between the vessel's shape and a content closely associated to this shape, that is, opium. Covachec and colleagues (2015) and Bunimovic and Lederman (2016) detected opium in only one of a series of such vessels that were analyzed, thus concluding that these vessels contained most likely oils rather than opium and suggesting that opium may not have been used at all during the Bronze Age and in antiquity (Bunimovitz and Lederman 2016). Yet, there is little doubt that given the good knowledge humans seem to traditionally possess of the properties of the plants surrounding them, the opium head sap and its properties would have been noticed. Although opium has not been detected in all analyzed but in some vessels (Chovanec et al. 2015 and more recently Smith et al. 2018), it is very likely that the Cypriot juglets reminiscent of the opium head contained probably both oils and opium, perhaps mixed together or leaving their trace after a sequential use in the same vessel. The connection between the container and a content that had opium as one of its ingredients remains strong, despite the failure to detect it in some of the residue analyses: the vessel shape is very eloquent in almost all of these cases (see also Sherratt 1995). Whether all of them contained opium or whether they always contained opium throughout their biographies is another matter, and it is always possible that a vessel initially prepared to hold a specific substance (opium) was later recycled to contain other substances, oil, for example. Perhaps the popular shape exclusively used for perfumes in ancient Greece, the *aryballos*, reflects such an association, its shape being very reminiscent of the opium poppy head; residue analyses have indeed identified opium in such perfumed oil containers (Jannot 2009). If plant oils and opium were mixed in the same vessel, this would turn out to be a very interesting association, potentially linking a special mixture of plant oils and opium to healing practices.

Contexts of Healing with Plants in Prehistoric Greece

Nowadays in the Western world, treatment of illness is achieved by virtue of a powerful substance contained in a pill or capsule, yet, among traditional societies, the plants themselves, used in healing, are often treated as being actively involved participants in the healing process (cf. Insoll 2011b; Insoll et al. 2012). For this reason, their harvest is regulated by strict rules such as the time of the day or the phases of the moon as was the case in ancient Greece with famous witches like Medea (Scarborough 1991). Although plants hold a key role in traditional healing practices, they are not the sole actors in rituals of healing. Depending on the cultural setting and/or the healing purpose, the ritual involved may simply require the intake of an herbal infusion or more complex practices and healing rituals. In some of these rituals, as can be seen in the ethnographic record, contact with the ancestors, the dead, spirits, and gods is required (Townsend 1997:431; Winkelman 1997; see Collard 2011 and Karatasaki 2017 for reviews of the literature), while objects such as bracelets and figurines may be required in healing contexts as well (Insoll et al. 2012; Kankpeyeng et al. 2011; Insoll 2011b).

Plants with healing properties cannot be considered medicines in the modern sense, isolating them from the whole healing process. Healers, depending on the specific cultural context, know the properties of plants and possess the knowledge and abilility to perform the rituals required to contact the world of spirits and remove illness from a patient (for example, see Heinrich 1994 on the Mixe of Oaxaca, Mexico). Healers or shamans interact with the spiritual world and have the knowledge and power to establish equilibrium between the world of the living and that of spirits/gods/ancestors (e.g., Saijirahu 2008; Kale 1995; Insoll 2011b; Insoll et al. 2012), and psychoactive plants and altered states of consciousness might also be used for this purpose (Eliade 1964; Sherratt 1991b, 1995; Rudgley 1993; Pearson 2002; Hayden 2003; Merlin 2003; Collard 2011). Moreover, plants have been used symbolically as offerings in an attempt to ease and please this invisible world, which was probably very present in prehistoric times as is in many contexts in the modern world. This process of protection is evoked when placing a *chimaki* (an amulet made of bamboo stems tied together) outside a door in Kyoto, Japan, or hanging a corn dolly for good luck outside a shop today (Figure 7.6). When food, mainly special cereal food preparations, is offered to the dead (e.g., Mesnil and Popova 2002) (see recipe section for recipe for *kolyva*), either when the body is buried or in commemoration of the departed soul, plants and plant foods mediate between the world of the living and the dead, unifying them and at the same time delineating their boundaries. The living prepare food for the dead and, through its consumption the dead souls are also fed (see Danforth 1982). In some parts of China a connection between souls and rice seeds exists on a symbolic level of regeneration (Leslie and Young 1992). It is as if the plant foods, by being symbolically fed to spirits, secure the continuous participation of ancestors in the world of the living.

Healing in prehistoric Greece may have been linked to the intake of medicinal plants, including some with special, mind-altering properties. The images that we have constructed of prehistoric communities of Greece are rather tame, avoiding entering the realm of senses or breaking the rules or exhibiting behaviors that might perhaps shock modern concepts of normality. The ethnographic record leaves no doubt that conditions diverging from daily normality are evoked on special occasions and may be rites of passage, special celebrations, or rituals (e.g., Staal 2001). A rich ethnographic literature shows how specific plants and/or plant-derived substances are used

Figure 7.6. Corn dolly, believed to bring good luck, in the tavern o Mageiras in the village of Agios Prodromos, Chalkidiki, northern Greece. (Soultana Maria Valamoti)

in rituals involving special states of consciousness. Examples include ayahuasca, soma, ephedra, and various mushrooms such as *Amanites muscaria* (see, for example, Michaelides 2014; Merlin 2018; Fitzpatrick and Merlin 2018 for a recent discussion on the use of such plants in recent and past times). A classic example from the Greek world is women followers of Dionysos, called maenads, who enter into an ecstatic and manic state (Figure 7.7) (cf. Bierl 2012; Goulimaris et al. 2005; Ustinova 2017). In archaeobotany, grape pressings and the identification of wine production reveal the special, intoxicating substance consumed in certain contexts.

On the basis of the above discussion prehistoric healing can be perceived as a practice that would have involved the use of plants in addition to a series of rituals establishing contact with other worlds, spirits, gods, and ancestors. Keeping well is not only about restoring good health; it is also about securing overall balance and protection from evil spirits, angry ancestors, and gods. A connection between medicine, ancestors, and food, grains in particular, can be seen in different cultures in the world (see Aijmer 1968

Figure 7.7. Maenads worshiping an idol of Dionysos. After the painting on a red-figure Attic stamnos attributed to the Dinos Painter (450–400 BC), Museo Archeologico Nazionale di Napoli 81674. (Drawing by Danai Chondrou. Courtesy of PlantCult.)

as an example from China). Senses are culturally prescribed as well as the ways plants for healing may be classified as such (e.g., Johns and Keen 1985; Shepard 2004; Pieroni and Torry 2007), thus the archaeobotanical record from Greece offers a basis for speculation in relation to similar uses in prehistoric times. It is impossible to track down the steps followed by a prehistoric healer who might have intervened to cure an illness or a wound in prehistoric Greece, yet we know that such practices were performed.

Inferring healing practices from the archaeobotanical record is a real challenge, as in most cases the remains of such activities can be hardly distinguished from the contents of a rubbish pit. Paleopathology offers an insight into wounds and diseases of which prehistoric people suffered, such as anemia, osteoarthritis, stress due to malnutrition and/or disease (see Papathanasiou 2011), and it is possible that some of the "special" plants encountered in the archaeobotanical record could have been used in healing contexts to alleviate pain, protect from infections, and ease recovery. An early indication for a wound that healed comes from a Neolithic skull from Alepotrypa Cave in Greece (Papathanassiou 2001) and could indirectly point to the use of some medicinal preparation used in this context (Figure 7.8). Robert

Arnott has reviewed the evidence for prehistoric medical practice, especially for Bronze Age contexts of southern Greece, that were at times successful, at times fatal (Arnott 2014). There is evidence for tripanism, that is, a kind of surgery performed to the skull (Crubézy et al. 2001), from Final Neolithic and Middle Bronze Age Crete (Mountrakis et al. 2011) as well as for wounds that healed (e.g., Chlouveraki et al. 2008; Papathanasiou 2001; Papagrigorakis et al. 2014). A careful and very skillful surgery, with very clean cuts, performed by means of tripanism has been detected on the skull of a male of 20 to 30 years of age buried in grave C of the grave circle B in Mycenae. The surgery was unsuccessful, as indicated by the lack of bone growth (Angel 1973). Bronze Age wounds that healed are visible in a female skeleton buried in circle B of Mycenae: arm bone fractures had entirely healed indicating special treatment (Angel 1973; Arnott 2014).

Artifactual evidence may imply contexts of healing. Pots, for example, might have been used in healing practices as suggested by Vitelli (1995) for the Early Neolithic of the Peloponnese. In the Bronze Age, special plant substances were detected in pots from Minoan and Mycenaean contexts on the basis of residue analyses, suggesting the combination of oils and iris

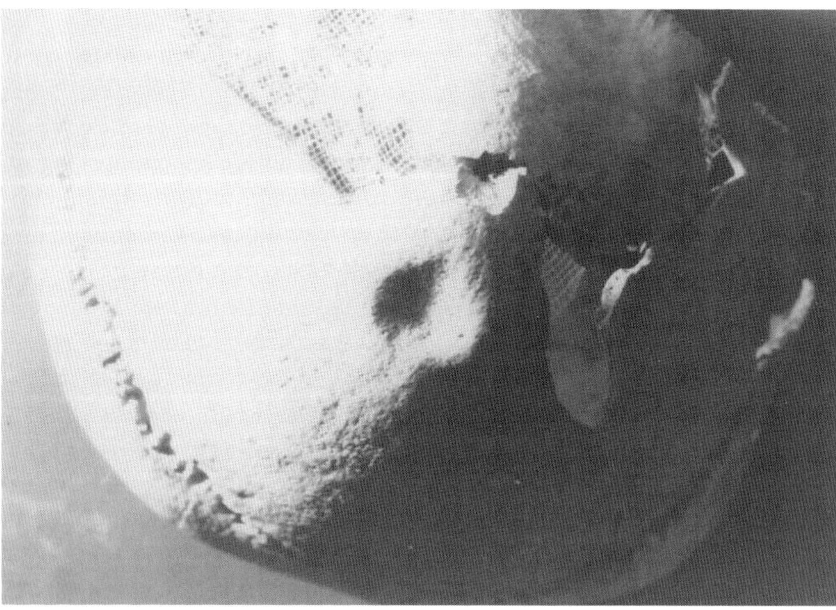

Figure 7.8. Wound on a neolithic skull from Alepotrypa cave, Peloponnese. (From Papathanasiou 2012, Figure 14.6. Photograph courtesy of Anastasia Papathanasiou.)

at Chamalevri on the island of Crete and rue mixed with wine at Mycenae (Andreadaki-Vlazaki 1999; Tzedakis and Martlew 1999:164). Such analyses, however, need to be treated with caution given the limited chances of preservation of such substances in the archaeological record (Evershed 2008). Vessels involved in healing practices could have contained ointments and aromatic oils (see Andreou et al. 2013) (see Figures 4.10 to 4.12). Bronze Age textual evidence on the healing use of oil and ointments is absent from prehistoric Greek sites, yet they are to be found in the Middle East (Fappas 2008, 2011). Vessels thought to contain oil/unguents are usually small in size with a narrow neck when containing oil, such as the stirrup jars of the Late Bronze Age. Ointments might have been contained in the kantharoi and amphorisks (small clay vessels) analyzed from Toumba Thessalonikis (Andreou et al. 2013). Amphorisks were also found at several other northern Greek sites, especially in funerary contexts.

Archaeobotanical indications for healing practices in Greek prehistory are indirect in most cases. Especially interesting in this respect may be a pit from the Late Neolithic site of Makriyalos. This pit contained an interesting combination of plants, most with medicinal properties that were not encountered in other pits of the site and thus special. In addition to cereal chaff and fig seeds, the usual component of the site's pits, it also contained hundreds of linseed as well as the only lentil concentration found at the site (Valamoti 2004). Lentil, although not widely known now as a medicinal plant, is mentioned as an ingredient of many medicinal recipes in the Hippocratic *Corpus* (see Valamoti et al. 2022a). The pit stands out also for its terebinth finds, present in a more or less pure concentration totaling 71 nuts. In addition, it contained a range of plants with known medicinal properties: *Hypericum* sp., *Anagallis arvensis, Sinapis/Brassica, Urtica* sp., *Trigonella* sp., *Medicago* sp., Solanaceae, Compositae. Of the plants found in the pit as seed, all belong to species or genera that comprise species widely used in ancient and traditional folk remedies (Gennadios 1997; Bown 1995; Brockhampton Reference 1996). Terebinth, as already described, was widely used in medicinal recipes in the past but also in the present for its aromatic resin, nuts, and oil. *Anagallis arvesis* is well known in folk medicine as an antipyretic, diuretic, and antidepressant. *Hypericum perforatum* is a widely known medicinal plant, used externally on wounds and as an infusion against depression and other related problems. *Hypericum,* although not identified to species level due to the large number of *Hypericum* species growing in Greece, could correspond to *Hypericum perforatum,* St. John's wort (see Figure I.8), yet another genus that points toward medicinal uses.

Flax's medicinal properties can be found in ancient sources as well as modern herbal remedy sources. Flax seeds are mentioned in various medicinal recipes, in the work of Dioscorides, a Greek physician, botanist, and pharmacologist who lived in the first century AD. They are mixed with other ingredients for the preparation of poultices. Flax seed, occasionally roasted and mixed with honey, vinegar, roasted figs, or with several other ingredients, was used to treat various problems, externally as poultices against pain, scorpion bites, vaginal infections, or internally to treat stomach and bowel problems (see Valamoti 2013b). Sometimes linseed is roasted and/or ground before being used. Publications of traditional plant-based medicinal remedies from Greece (Pyrros 1838; Gennadios 1997) and other parts of Europe (Bown 1995) and the Near East (Kislev et al. 2011) show the use of linseed in many preparations for the treatment of respiratory and digestive problems.

Although linseed and lentils found in the Makriyalos pit point to the direction of food, and a skeptic might interpret this particular assemblage mixture as corresponding to charred crops with their weeds, the whole assemblage is exceptional for the nearly 1,300 samples examined (Valamoti 2004). The large variety of wild species and genera, with most corresponding to plants with medicinal properties, signifies an exceptional context. The pit might have received the remains of healing practices involving the use of plants with special properties. The burning of seeds for medicinal purposes, recorded in the ethnobotanical record (e.g., Pennacchio et al. 2010; Mohagheghzadeh et al. 2006), would be a likely source of some of the charred plant remains from this Makriyalos pit. Avicenna, for example, mentions the use of smoke from the burning of terebinth nuts to treat sore eyes (Pennacchio et al. 2010).

Perhaps a Neolithic healer operated at the site. A use in the context of healing may be corroborated by other finds from the same pit, which is also special as it contained a large number of *Spondylus gaederopus* artifacts, especially remarkable for the highest occurrence of *Spondylus* bracelets for the whole site (Pappa 2008:283, 338; Pappa and Veropoulidou 2011). *Spondylus gaederopus* artifacts, according to their spatial distribution and contextual associations at Makriyalos, seem to be closely related to burial practices (Figure 7.9 and Figure 7.10) (Pappa and Veropoulidou 2011). A context involving healing rituals whereby medicinal plants, *Spondylus* shell objects, and ancestors came into play seems a possible interpretation.

Rituals involving ancestors and possibly linked to healing, not necessarily always involving plants, may be indicated by the contents of Neolithic pits such as those from the Late Neolithic Toumba Kremastis Koiladas. These

Figure 7.9. Burial of child with *Spondylus gaederopus* bracelet offering, Makriyalos, Late Neolithic (From Pappa and Veropoulidou 2011, Figure 3. Photograph Makriyalos archive. Courtesy of Maria Pappa and Rena Veropoulidou.)

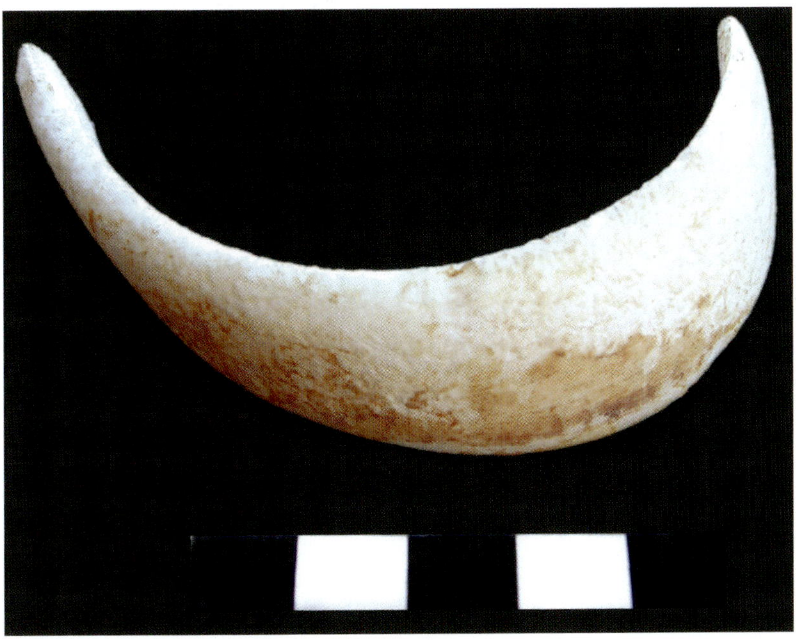

Figure 7.10. Bracelet from a child burial (see Figure 7.9) at Late Neolithic Makriyalos. (From Pappa and Veropoulidou 2011, Figure 6.g. Photograph by Nikos Valasiadis. Courtesy of Maria Pappa and Rena Veropoulidou.)

pits are among the most impressive in this respect, containing skeletal remains of humans and animals as well as a range of artifacts such as complete pots, house models, and a flute made from a human femur (Figure 7.11 and Figure 7.12) (Chondrogianni-Metoki 2009) as well as a broken large vessel full of charred grass pea seeds (Valamoti et al. 2011:Figure 2a; this book, figure 3.5). Similar connections may be reflected by the numerous *Spondylus* shell ornaments buried in specific pits at Makriyalos I and II, including those that contained the interesting combination of medicinal plants. Pit clusters encountered at several Neolithic sites such as Toumba Kremastis Koiladas, Apsalos, Makriyalos, and Koroneia in northern Greece might correspond to the remains of ritual practices placed inside the earth and sealed, marking in this way the completion of rituals and sacred contexts possibly related to the ancestors and healing (see Sherratt's interpretation of Neolithic "ritual" pits containing the remains of sacrificial meals, Sherratt 1991b:50). Feasting reported for the Kepele ceremony in Papua New Guinea included the symbolic feeding and appeasing of ancestors whereby human bones, plant foods, firewood, and pig fat were ceremonially cremated (Wiessner 2001:132–133).

Figure 7.11. Late Neolithic pits, Toumba Kremastis Koiladas. (Photograph courtesy of Areti Chondrogianni.)

Figure 7.12. Miniature clay vessels and human bone found in one of the Late Neolithic pits at Toumba Kremastis Koiladas. (Photograph courtesy of Areti Chondrogianni.)

Compared with the slim and ambiguous evidence for healing from Neolithic Greece, the Bronze Age appears to offer a wealth of indications for the use of plants that would have had primarily special uses. Plants with particular properties seem to proliferate in the Bronze Age, some corresponding to the indigenous flora, others being introduced from distant places such as opium poppy and *Lallemantia* (see Chapter 4 and Appendix A:Table 4.2). Others, like terebinth, although locally available, were probably insufficient for the needs of the elites of the palaces, leading therefore to its importation from other regions in the east Mediterranean.

The wide range of plants with medicinal and psychoactive properties that appear during the Bronze Age might point to more elaborate, ritualized healing processes, perhaps related to social changes and emerging hierarchies that might have brought about changes to the role and status of healers and healing practices. By the Late Bronze Age in the south, the royal healer had a special status and land ownership; it seems, however, that not everybody had access to special medical treatment, judging from the osteoarchaeological indications for injuries that were untreated (Arnott 2014). A tablet from Pylos (PY Eq 146) mentions the term *ijate,* which has been interpreted as "healer" (from ἰητήρ, ἰατρός) (Ventris and Chadwick 1973:547). This person seems to be one of the palace officials, yet ranked on the same level as the palatial potter (*ke-ra-me-wo*) (Arnott 2014:43–46). Perhaps a healer's kit is what corresponds to the grave goods, dated to Late Helladic II, which included bronze tools such as tweezers, knives, a large, indentated pair of scissors, and two grinding tools, perhaps for the preparation of medicines (Protonotariou-Deilaki 1973:92). Using the smoke of aromatic plants and resins might have been a way of healing, and such plants as terebinth and pine, detected in the archaeobotanical record, could have been burned for their aromatic smoke in healing contexts. Special vessels have been associated with the burning of aromatic substances, or even, in the context of northern Europe, with the burning of plant-derived psychotropic substances

Figure 7.13. Incense burner from Palaikastro. (After Georgiou 1979:Catalogue XIII, Heraklion 4536, Plate 62; Drawing by Danai Chondrou. Courtesy of PlantCult.)

Figure 7.14. Young priestess from the Western house. (Photograph courtesy of the Excavations at Akrotiri, Thera.)

like opium and *Cannabis* (Sherratt 1991b). Similar artifacts have been found in Neolithic and Bronze Age contexts in Greece. Several Cretan Bronze Age examples have been interpreted as incense burners, and they show a complex structure consisting of a bowl-shaped vessel with a perforated lid resting on top of the vessel (Figure 7.13). The burning of special substances in particular vessels (*pyraunoi*) to produce smoke may be depicted in a Bronze Age fresco excavated at Akrotiri on Santorini. A female figure is depicted holding a *pyraunos*, a brazier used here as a censer, to which she appears to be placing something, perhaps incense (Figure 7.14).

Snapshots of healing practices are also provided by micrographic representations of human body parts that have been found in peak sanctuaries, some exhibiting deformations, probably corresponding to that part of the body with a specific health problem for the cure of which divine help was probably sought (Morris and Peatfield 2014). These finds, occasionally perforated, closely resemble later practices of antiquity that are still encountered in modern Christian temples, with variants including body part offerings in the form of breads as those offered to St. Antonios in the Sphakia region on the island of Crete (Psilakis and Psilakis 2001). These bread-like offerings emphasize the strong connection between healing, food, and rituals (Psilakis and Psilakis 2001:423–429).

The preceding exploration of archaeobotanical remains from Greece, mainly prehistoric, reveals that many of the major plant ingredients used in later *Materia Medica* (pharmacology, medicines) were already available since prehistoric times, increasingly so during the Bronze Age. Healing in prehistoric Greece involved a range of "ingredients," rituals, and special contexts for which very little is known and mainly imaginative inferences can be made. Healing with plants would have involved a wide range of knowledge and skills as well as abilities of the people who had the role of healers. We may hypothesize the existence of the elusive healers of the Neolithic and the Bronze Age in Greece, those who knew the properties of plants and used them in special contexts. They would have harvested medicinal plants from the wild, their fields, or special gardens. Some were community members who, over time and in certain contexts, became official palace healers.

8

The Transformation of Plants into Dishes

Cooking for Daily and Special Occasions

Culinary Encounters between Nature and Culture

Cooking converts raw ingredients to cooked food, thus transforming nature (raw ingredients) into culture (cooked meal), as very accurately researched originally by Lévi-Strauss (1965) in a series of works including the *Le triangle culinaire* (see also Goody 1982:17–29; Hastorf 2017:42–47 for reviews of this discussion in the anthropological literature). There is little doubt that modern humans over the course of their presence on earth developed a wide range of methods for removing toxins, fermenting and storing plant foods, and converting ingredients into dishes, which were the outcome of a long interaction with their environment. Food preparation dates back to these Hominins who first mastered fire and gradually developed food preparation technologies. The rich cultural culinary heritage experienced today has its roots in the deep past of humanity. It has been shaped on the basis of available food ingredients and technologies of transformation into meals but also according to cultural classifications of what is "good to eat" (*bonnes à manger*) and "good to think" (*bonnes à penser*) at the same time, to repeat the classic association made by Lévi-Strauss between animal species as food and animals as totems (Lévi-Strauss 1962:128).

The debate regarding the role of plant foods and plant food processing in the evolution of the various species of the genus *Homo* is lively (e.g., Ungar and Sponheimer 2011; Revedin et al. 2010; Revedin et al. 2015; Wollstonecroft 2011; Pryor et al. 2013; Butterworth et al. 2016; Ragir 2000). Plant food processing appears to be crucial for human survival and the expansion of *Homo sapiens* in diverse environments (e.g., Jones 2007, 2009; Wollstonecroft 2011; Hardy and Kubiak-Martens 2016b; Arranz-Otaegui et al. 2016;

Power et al. 2015; Cristiani et al. 2016). The cultivation of wild plant foods leading to their domestication most likely relied on an early acquired knowledge of plant food processing (see Wollstonecroft 2011) and an excellent knowledge of plants and their properties (Abbo et al. 2011). Stone tools and "hidden plant foods" (see Cristiani et al. 2016), that is, starches and phytoliths trapped on grinding stone surfaces dating to the Paleolithic, are eloquent indicators of plant food processing through crushing or grinding (approximately 25,000 years ago) (Aranguren et al. 2007; Piperno et al. 2004; Nadel et al. 2012). Besides grinding, fermentation and cooking of plant ingredients would have further improved their digestibility and in some cases would have removed toxins, thus widening the spectrum of edible ingredients; in turn, food processing and cooking would have facilitated nutrient uptake by the human digestive system, thus contributing to an increase in brain size and the potential to sustain larger populations with all the socioeconomic consequences of such changes (e.g., Stahl 1989; Perlès 1977, 1999; Wrangham 2009; Groopman et al. 2015). This of course does not necessarily imply that the choice of specific processing procedures was always optimal or nutritionally beneficial or perceived as such. Cooking as a process, involving ingredients, equipment, gestures, and the senses, corresponds to that critical point where culture encounters nature through the "mediation of fire," the two containing each other, through the transformation of raw ingredients to cooked food and the creation of cosmologies (Lévi-Strauss 2001:185; Bourdieu 1977:143–158; see also Hastorf 2017:91–107, 117–128 for a recent discussion and overview of the relevant literature).

Several factors could have come into play leading to the first attempts to try various "recipes" in the course of human history, resulting in a wealth of diverse cuisines across the world. Initially, fermenting ground cereals into an alcoholic drink or baking fermented cereal dough might have been the outcome of an accident or curiosity: "Dough" in a more or less liquid form, left for several days in a warm, humid place, its smell changing to something that would have been perceived as pleasant or interesting or bearable, baked so as not to waste the food, ending in something palatable and acceptable for its organoleptic properties and "behavior" in the digestive track and therefore approved for consumption. If indeed the gatherer-hunter ancestors of early farmers baked ground wild cereal grains mixed with water (see Arranz Otaengui et al. 2018) or brewed them in some alcoholic beverage (e.g., Hayden et al. 2013; Liu et al. 2018), this would have been perhaps the initial way in which the first baking and brewing experiments of humankind were ever carried out (cf. Braidwood 1953; Katz and Voigt 1986).

It is far from certain whether the making of leavened bread would have required the availability of a hexaploid wheat with its special gluten characteristics ready to respond to an encounter with certain yeasts and/or sourdough, as well as the availability of such yeasts that would have led to the formation of sourdough (see also Hansson 1994). For these two (hexaploid wheat and yeasts) to work together into a risen loaf, specific conditions would have also been necessary, namely, a humid and warm environment (Katz and Voigt 1986). The where, why, and when this might have happened are yet to be determined, and it is very likely that it happened in different parts of the world where hexaploid wheats or wheat and barley were grown. Hexaploid wheats, however, are not a prerequisite for bread to have been discovered. In the wide sense of the term, mixtures of cereal meal and water can be transformed into a wide range of cereal foods such as cookies, porridges, and breads of different textures and elasticity, depending on the type of meal and liquid used as well as the presence or absence of yeasts in the mixture.

Grinding Plant Food Ingredients

The archaeobotanical evidence from Greece, explored in earlier chapters, shows that plant food ingredients were ground in various forms for the preparation of a wide range of foodstuffs: split pulses, coarsely and finely ground cereals, lumps of cereal fragments and meal have been found from contexts spanning the Neolithic and the Bronze Age. Unlike the Near East where numerous grinding stones, pestles, and mortars have been unearthed from settlements of the late Pleistocene gatherer-hunters who experimented with the processing and cultivation of wild plants (see, e.g., Wright 1991; Dubreuil and Nadel 2015), such equipment is nearly absent from the Greek Mesolithic (see Bekiaris et al. 2020, for a recent overview of the evidence). From the onset of the Neolithic, however, all excavated settlements of Greece demonstrate a range of stone tools such as mortars, hand stones, and grinding slabs that could be used for processing various substances including cereals or/and other plant foods (e.g., Perlès 2001; Demoule and Perlès 1993; see Ninou et al. 2022 for some interesting patterns in the Early Neolithic). Neolithic pestles and mortars could have been used for pounding spikelets for de-husking the glume wheats or for de-branning cereal grain and breaking clean grain into coarse fragments. Fresh or dried fruit, herbs, and medicinal and aromatic plants could also be crushed with pounding equipment as well as thoroughly mixed for the preparation of dried food, for example, that could be stored for a long time and consumed during journeys or in winter. Such stone tools, however, are less common in the Neolithic of Greece (see Bekiaris et al. 2020). Wooden ones, however, perishable under prevailing archaeological preservation conditions in Greece, could have been employed. The number of pounding tools increases in the Bronze Age, in the south of Greece, and this could be linked to olive oil production and the perfume industries of the Minoan and Mycenaean palatial centers.

Quern stones would have operated in a different way and might have been used as alternatives or complementary to pounding tools (see Bekiaris et al. 2020 for a review of the evidence for Neolithic and Bronze Age Greece). It has been argued that during the Early Neolithic, small-sized querns were not primarily used for grinding cereals into fine meal. Stroulia (2010), in agreement with Runnels's earlier views (Runnels 1981:250–251), suggested that, during the Neolithic, small querns, when used for food preparation, were most likely used for the grinding of grain into coarse groats rather than fine meal or, that grain was not ground but consumed in other ways. This view has recently been questioned as the function of these tools would have depended on a number of cultural and economic parameters (Valamoti et al. 2013; Stroulia 2018; see also Bekiaris et al. 2020).

Ethnographic studies have shown that grinding is among the most tedious and time-consuming tasks in plant food preparation, usually but not exclusively performed by women (Alonso 2019), sometimes visible in female human skeletal remains as is the case at Çatalhöyük (Sadvari et al. 2015). Pathologies related to grinding and gender have not been reported from prehistoric Greece (Anastasia Papathanasiou, personal communication), yet, long hours of grinding cereal grain into flour on small grinding stones would have been required (Valamoti et al. 2013; Chondrou et al. 2018). Experiments conducted on a larger scale and a more systematic way (Bofill et al. 2020) have shown that small-sized grinding stones (smaller than 30 cm) can yield a kilo of flour in approximately 3 1/2 hours. One very important observation of the recent experimental grinding sequences (Figure 8.1) was that the grinders became more skilled after the first couple of days, drastically reducing the grinding time required to turn grain into fine flour (Bofill et al. 2020). Thus, skill and experience would have certainly played their part in grinding productivity while the availability of small-sized grinding stones might have been related to the involvement of young members of the community in grinding activities.

One needs to ask how often such grinding of grain into fine flour or coarse fragments might have been required for different recipes and different

Figure 8.1. Experimental grinding with prehistoric replica grinding stones, PlantCult project. (Photograph by Vasiliki Kapetanaki. Courtesy of PlantCult.)

occasions. Grinding to generate these products would have varied according to the people involved and the context of consumption. Grinding could have been performed on the level of breaking the grain only coarsely into groats (Stroulia 2010), yet this does not preclude the use of the same grinding stones for producing fine meal, perhaps for special food preparations that required finely ground grains (Valamoti et al. 2013). The processing methods employed, aimed at different end products, could be further explored through the study of usewear patterns on the grinding surface (see, for example, Bofill et al. 2020). The combined experimental and archaeological investigation of usewear on grinding stone surfaces has indicated that some grinding slabs were used for processing cereals in their husk (Chondrou et al. 2021). As such, processing is not ideal for de-husking glume wheats (Meurers-Balke and Lüning 1992), it is more likely that grinding in the hull concerned primarily hulled barley. It is expected that future integrated approaches with the combination of usewear studies and plant microremains, experimental and archaeological, currently under study in the context of project PlantCult (Valamoti et al. 2020) will ultimately provide valuable insights on the issue of plant food processing with stone tools (for a highly informative north European recent example, see Hamon et al. 2021).

Figure 8.2. Antoni Palomo checking sandstone from a stream for the fabrication of experimental grinding stones for PlantCult project, October 2016. (Photograph by Vasiliki Kapetanaki. Courtesy of PlantCult.)

The preparation of the grinding kit would have required excellent knowledge of the territory as well as skills in determining the appropriate stone to be used. The whole procedure must have been a time-consuming process, not only for obtaining the raw material from riverbeds or other locations but also for working the surfaces that were to be used for grinding. Selection of the appropriate raw material for grinding stones was probably dependent on local availability but also on the mechanical properties of stone required for the purposes of grinding grains or other plant parts (Figure 8.2). The targeting for conglomerates at Late Neolithic Kleitos, for example, was probably a deliberate choice as it reduced the time and effort for regularly pecking the grinding stone: the stone would "peck itself" during grinding (Figure 8.3) (after Chondrou et al. 2018). This might have resulted in a higher presence of tiny particles of stone ending with the ground product, which in turn would have affected tooth surface, something observed in prehistoric Europe and elsewhere (Lev-Tov and Smith 2006; Smith 1984). Such particles are visible in remains of plant foods at some northern Greek sites (PlantCult material under study).

Figure 8.3. Conglomerates intentionally selected for making grinding stones at Late Neolithic Kleitos (From Chondrou et al. 2018, Figure 11. Photograph courtesy of Danai Chondrou.)

The placement of grinding stones and the areas of grinding would have probably formed areas of social interaction, especially if grinding was organized in communal spaces, shared by several households. Some houses seem to have contained a large number of grinding stones as is the case with the Early Neolithic house of Sossandra in northern Greece (Georgiadou 2015); others might have contained fewer or none at the time of destruction and/or abandonment, while in certain cases grinding stones were "buried" together with grain or other objects, complete or fragmented, perhaps as part of some ritual (Bekiaris 2020; Stroulia and Chondrou 2013; Bekiaris et al. 2020). Grinding stones can also appear to be organized in a space specifically designed for grinding, pressumably on a large scale, as is the case at Akrotiri on the island of Santorini (Moundrea-Agrafioti 2007:82–90). This permanent installation associated with milling reveals efforts to improve and organize more systematically these kinds of activities (Moundrea-Agrafioti 2007:82, 88) and seems to be a feature of the Bronze Age, mainly in the south of Greece (Bekiaris et al. 2020).

The grinding kit was not only a tool in the sense that a modern cook would use a food processor or a farmer may use a milling machine to produce flour. Archaeological and ethnographic evidence shows that grinding stones and the act of grinding have had a strong symbolic dimension, expressed in rituals for the living and the dead. Grinding stones symbolize the mother among the Mursi in Ethiopia (Robitaille 2016) and are used as elements of house foundation rituals, buried for protection, in Sihó, Yucatán (Fernández-Souza 2016:15–27). They play an active role in Kepele feasting among the Enga in Papua New Guinea, representing ancestors, the active stone corresponding to male, the passive to female. During the feast the ancestors (grinding tools) are fed (covered in fat) and buried, this guaranteeing prosperity of the living (Wiessner 2001:132, 141). Ritual grinding has been performed by women in different cultural contexts such as the Zuni of New Mexico in their *Ashiwanni* and *Shalako* ceremonies (Coxe Stevenson 1904:165, 229, cited in Mobley-Tanaka 1997) as well as ancient Athens, where noble virgins were selected to grind cereals for the preparation of ritual breads for the cereal goddess, Demeter (Aristophanes, *Lysistrata* 643). Ritual grinding for the preparation of special foods for ceremonies has been suggested for archaeological assemblages such as subterranean mealing rooms of the Anasazi (Mobley-Tanaka 1997).

A strong symbolic meaning of grinding stones might be implied by finds of grinding stones in burials or pits of Neolithic Greece (Figure 8.4) (Bekiaris 2020; Stroulia and Chondrou 2013; Chondrou et al. 2018). In such contexts, grinding stones are sometimes found intact with very little use prior to deposition and therefore still functional (e.g., Bekiaris 2020). Fragmented grinding stones were found in pits at Toumba Kremastis Koiladas in northern Greece. They were probably intentionally broken and buried, perhaps an act of "destroying the means of production," marking the end of their grinding

Figure 8.4. Grinding stone tool still functioning, buried in a pit at Late Neolithic Kleitos. (From Chondrou et al. 2018, Figure 12. Photograph Kleitos Excavations Archive. Courtesy of Danai Chondrou.)

cycle (Stroulia and Chondrou 2013). Similar practices have also been inferred on the basis of archaeological finds in Europe, the Near East, and the New World (Larsson 2003; Hansson 2011; Sadvari et al. 2015; Lidström-Holmberg 1998; Adams 2008; Joyce 2001). Metates/grinding stones were included in Maya burials (Scherer 2015), and Adams (2008) gives several such examples of object sacrifice as part of a funerary or commemorative mourning ritual. Ethnographic observations of practices related to the destruction either by breakage or burial of the grinding stone suggest that these objects, by being considered alive, ended their life with the death of their owner, their life being ritually terminated, or sacrificed through breakage, burial, and burning, or combinations of these acts (see, e.g., Adams 2008 and Joyce 2001). The association between death and broken grinding stones is reported for indigenous women of Mexico (Fray Bernardino de Sahagún, Fiorentine Codex, Burkhart 1997:42).

Tools for grinding and pounding thus emerge as an essential element of the culinary apparatus for prehistoric communities of Greece, the first step toward transforming plants into food. Tools were used in food preparation but also in social interaction and the construction of memory. They could have been used for the preparation of particular foods including alcoholic beverages (e.g., Hayden 2001:30; Brown 2001:378–379) consumed only during feasting, and specifically for the occasion (see Brown 2001:385), thus demonstrating little usewear on their surfaces. These tools were also loaded with a strong symbolic element, little explored in the literature, a role that can be detected in archaeological indications for the termination of the tools' life cycle, perhaps in association with their owners' death. Grinding and pounding as well as the tools associated with the process were not just activities for preparing daily meals and dishes for rituals and feasts in prehistoric Greece; they connected people to their landscape and each other, to the cycle of life and death.

Cooking Facilities: From Hot Embers to Complex Food Preparation

The cooking of food using fire has been considered an essential step in human evolution. Cooking with fire would have unleashed nutrients that allowed the development of larger brains in the genus *Homo* (e.g., Wrangham 2009), although the processes involved are still poorly understood (Wrangham 2017). There is no consensus as to whether the use of fire for cooking plants was a generalized practice among early humans (Brain and Sillent

1988; Goudsblom 1986; Henry 2017; Wrangham 2017; Perlès 1999). Cooking has been linked to identity formation (Perlès 1999; Hastorf 2017) and by consequence gender issues and power relations between the two sexes, on the assumption that the people in charge of cooking food hold the key to the well-being of those consuming it, the ultimate abuse of this power being poisoning. This "key" has been traditionally in the hands of women who are involved in plant food processing and cooking of daily food (see, for example, Stanley 1981).

Fire and containers for cooking constitute main transformation technologies in the preparation of food. Cooking can be performed without sophisticated apparatus; thus, neither pots nor specially prepared hearths are necessary to cook or roast. Cooking in pits (Wandsnider 1997) or with hot stones (Connors 1994; Balls 1962) or roasting and baking directly on hot embers (Figure 8.5) would have resulted in nothing more than an ashy layer, or a pit containing ashes and a concentration of stones, thus leaving hardly any diagnostic traces in the archaeological record. Ember baked bread traditions are known from different parts of the Old World (Figure 8.6) such as among the Bedouins who bake *abud* directly on embers (Iddison 1996), a

Figure 8.5. Experimental "roasting" of acorns by placing on embers, PlantCult project. (Photograph by Georgios-Vily Kapetanakis. Courtesy of PlantCult.)

Figure 8.6. Bedouin bread baked directly on ashes. (Photograph courtesy of Daniel Noll, UncorneredMarket.com.)

practice also encountered, for example, in Israel (Avitsur 1977), Afghanistan (Cousin and Monzon 1992), and Greece, modern and ancient (Psilakis and Psilakis 2001).

Of course, more sophisticated clay cooking facilities are known ethnographically and archaeologically, providing alternative ways to prepare a variety of recipes as are the *tanur*, *tabun*, and *saj* encountered in Anatolia, the Near East, and elsewhere (see, for example, Mulder-Heymans 2002; Lyons and D'Andrea 2003). Prehistoric farmers of Greece in addition to the ashes of open fires made use of a varied culinary kit consisting of cooking pots and thermal facilities that could have been combined in different ways for different occasions and numbers of people.

Hearths and Ovens

Hearths are present at Paleolithic and Mesolithic sites of Greece (see, for example, Galanidou 1997; Koumouzelis et al. 2001; Karkanas et al. 2004; Darlas and Lumley 1999; Kyparissi-Apostolika 1999), and one can assume that, besides other purposes, they would have served as a means to cook food, including plant foods. The first farmers that built the early Neolithic villages of the region were mastering cooking techniques as is evident from a wide variety of cooking installations retrieved archaeologically. This variety has been related to variability in cooking practices (Perlès 2001:197), suggestive of different groups and culinary traditions or different ways to cook in different seasons of the year or on different occasions.

Cooking installations could have been fixed or portable, inside or outside houses, in clusters or isolated, depending on the site, period, and cultural context (Papadopoulou 2010; Prévost-Dermarkar 2002; Papadopoulou and Prévost-Dermarkar 2007). When found in external spaces, it is reasonable to assume that they were under some kind of roofed space as they would have been weathered down if left exposed to weather conditions such as snow, wind, and rain. Thermal structures have been found directly on the ground surface, below or above ground level, inside or outside buildings, open, surrounded or not by a rim or covered by a dome (Prévost-Dermarkar 2002). Kalogiropoulou (2014), in her discussion of Neolithic thermal structures from northern Greece, observes four types of built cooking installations (two types of ovens, two types of hearths) and a very simple type consisting of a layer of ashes and charcoal remains (Kalogiropoulou 2014:76–79). One criterion for classifying thermal structures is the firing environment: open in hearths and fully or partly closed in ovens (Papadopoulou 2010:92). In this respect the "oven" with a round opening on top (e.g., Stratouli et al. 2015:Figure 12) would have functioned as a hearth with raised walls or an oven if all openings were closed during cooking. The basic categories for prehistoric Greece remain the oven, the hearth, and the open fire, and all three may be encountered at a single site (e.g., Perlès 2001; Kalogiropoulou 2014).

A complex cooking structure identified at Archondiko in northern Greece (see Figure 6.8) appears to have been more widely used in Greece during the third/second millennia BC as noted by Papadopoulou (2010:429–432). The low temperatures experimentally observed in its back chamber (Papadopoulou 2010) raise the possibility that it might have been used, among other things, for preparing the wort prior to fermentation (see Valamoti 2018).

Overall, ovens are distinguished from hearths on the basis of how cooking is being performed. In hearths, cooking is done with the heat underneath the food and a medium, such as a pot, a pan or a griddle, often needs to mediate for the heat to reach food, although for some forms of cooking this can be done directly on embers. In ovens, the heat circulates above the food and on the floor without the necessary need for a container, unless food is in liquid form (Papadopoulou and Prévost-Dermarkar 2007). The two, hearth and oven, can also work together: hot coals from a constantly burning hearth could be used to feed an oven with fuel or vice versa (cf., e.g., "stove" and "griddle" in Ethiopia) (Lyons and D'Andrea 2003).

Hearth and oven floors were made of clay and pebbles or fragmented sherds that acted as heat insulators for household hearths of the Neolithic (Germain-Vallée et al. 2011:41–63; Papadopoulou and Prévost-Dermarkar 2007:123–135; Prévost-Dermarkar 2003:215–223) and the Bronze Age as is the case with Agios Athanasios (e.g., Mavroidi 2012). This applied both to small, household installations and to large ones intended to serve a feast or large-scale cooking as was the case with a large open-air hearth from Late Neolithic Kleitos (Figure 8.7) (Ziota 2014) and late Bronze Age Tiryns (Figure 8.8) (Maran 2012). Although on mainland Greece the lining of pebbles or sherds as insulation beneath the hearth floor is widely used since the Neolithic, this does not seem to be the case on the island of Crete (Borgna and Levi 2017). This lining for the south of Greece is considered a later feature, associated with big regional centers like Mycenae and Tiryns (Tournavitou 1999; Borgna and Levi 2017). Such hearths are considered a Mycenaean influence on the island of Crete and the east Mediterranean in general (Borgna and Levi 2017). Besides clearly identifiable hearths, at some sites on Crete, hearths are indicated only by areas with burning debris (Borgna and Levi 2017). Domes were probably made with clay and straw or chaff and were lighter than the base of the oven, as is indicated by thin sections performed on hearths from Dikili Tash (Germain-Vallée et al. 2011). The construction of ovens in particular required good skills and a mastering of the clay. Sometimes hearths and/or ovens were decorated both in the Neolithic (Figure 8.9) and the Bronze Age (Figure 8.10) (e.g., Prevost-Dermarkar 2002; Papadopoulou 2010).

In the south of Greece during the Bronze Age, both fixed and portable hearths are known. In mainland Greece, most hearths and ovens were permanent installations, whereas portable ones, in the form of braziers, dominate Crete in the Bronze Age (Muhly-Metaxa 1984; Borgna and Levi 2017). On Crete outdoor kitchens or kitchen complexes are found in the periphery of large habitation spaces (Borgna and Levi 2017:121; Muhly-Metaxa 1984:119). More robust hearths on Crete, with insulation on the floor, when found in households, have been linked to ritual practices; this, however, need not have precluded a culinary use (Borgna and Levi 2017 for a recent review of the evidence). Hearths with a central depression are encountered on Crete, either portable or fixed in Minoan times (Muhly-Metaxa 1984). This depression is also present in some hearths on mainland Greece dated to the Late Bronze Age (Maran 2012) (see Figure 8.8).

Figure 8.7. (*above*) Open-air large hearth from Kleitos, Late Neolithic, Ephorate of Antiquities, Kozani. (From Ziota 2014, Figure 4. Photograph courtesy of Christina Ziota.)

Figure 8.8. (*left*) Sherd pavement of Late Bronze Age (LH IIIC) clay hearth in the northeastern Lower Town of Tiryns with round opening in the middle. (From Maran 2012:128, Figure 11.8. Tiryns Archive. Photograph courtesy of Joseph Maran.)

Figure 8.9. Decorated hearth from Kleitos, Late Neolithic, Ephorate of Antiquities, Kozani. (Photograph courtesy of Christina Ziota.)

Figure 8.10. Decoration on a fragment of a rectangular fixed hearth with a cavity from Phaestos. (From Muhly-Metaxa 1984:110, Plate 26, Figure 10. © Archivi SAIA, B/16059.)

At Late Neolithic Dikili Tash, where the excavated space corresponds to houses, hearths and ovens are primarily located inside houses (Figure 8.11) with each internal partitioning possessing its own thermal structure (Tsirtsoni et al. 2018:Figure 18.7). At Makriyalos, an area with several hearths/ovens clustered together existed during phase II without indications of built space containing them (Pappa 2008). Clusters of hearths or significantly large hearths could point to large-scale feasting occasions, especially so in the case of outdoor ones.

A study of Early Bronze Age clay structures from the site of Archondiko reveals great variability in cooking installations and the presence of complex morphological types within the same settlement (Papadopoulou 2010:165–180). The portable *pyraunos,* whereby a hearth and a pot

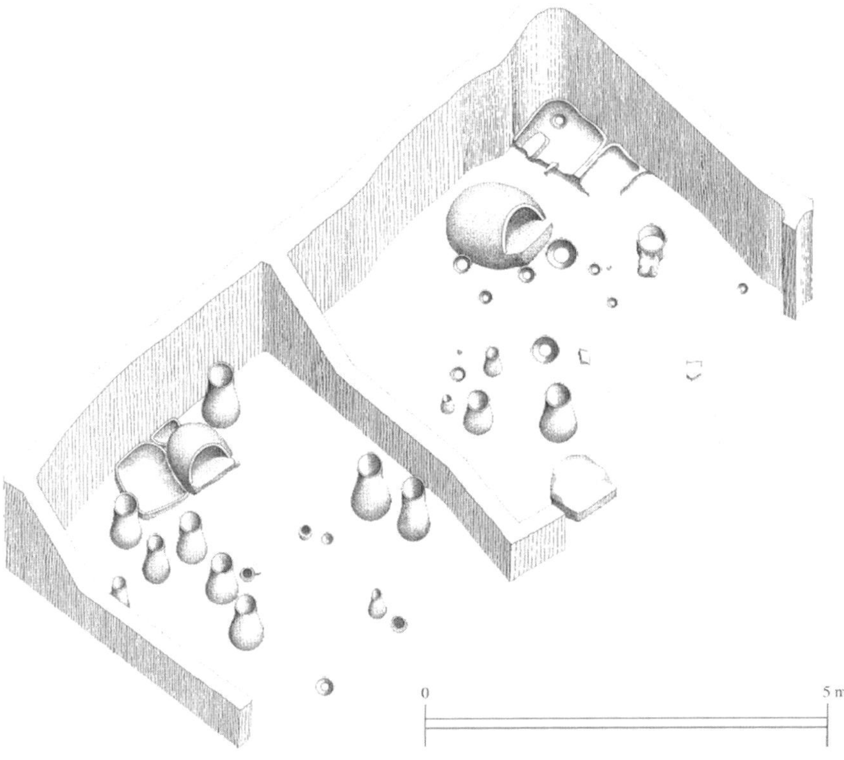

Figure 8.11. Dikili Tash, House 4, reconstruction showing two rooms, both with thermal structures in them. (After Tsirtsoni et al. 2018:Figure 18.7b. Drawing by Charalambos Romanidis. © Dikili Tash Research Project.)

are fixed in a single cooking apparatus, is discussed in the next section. As will be seen, it corresponds to an interesting Bronze Age culinary innovation with a specific geographic distribution that may be related to the emergence of new culinary traditions in the region. Papadopoulou in her thorough examination of fire structures and their temperatures concludes that a wide range of cooking techniques took place at Archondiko: roasting, baking, boiling, and smoking (Papadopoulou 2010:339–343, 455–457; Papadopoulou and Maniatis 2013). Different types of thermal installations coincide in the same spatial unit, suggesting that cooking of different types of food was taking place at the same time, or/and that perhaps large quantities of food or drink were required and thus prepared at a given moment (Figure 8.12).

The cooking spaces at Archondiko might have been only partly walled, with one side open for good aeration and access (see modern examples from Afghanistan further on). It is difficult, however, to infer the scale of food preparation by the arrangement and variety of hearths alone. Two similar or different facilities may be placed together to serve the needs of a family, allowing for the preparation of different dishes at the same time. Such cooking

Figure 8.12. Cooking installations from phase IV at Archondiko Giannitson, 2100–1900 BC. (Photograph courtesy of Aikaterini Papanthimou.)

Figure 8.13. Summer kitchen spaces under the same roof, village of Shamarq, Baghlan Province, Afghanistan, summer 2013. (Photographs courtesy of Shafiq Ahmad Sedeqy.)

facilities organized in clusters are still being used in different parts of the world. In the village of Shamarq in the region of Baghlan, Afganistan, village kitchens include a summer and a winter kitchen. In the former, there is a variety of cooking installations with different sizes/types of hearth and oven and one *tabun*, similar to those excavated at Archondiko, arranged in clusters (Figure 8.13). In the latter, only a tabun is placed, in the center or in one side of the room, and it is used for all types of cooking (bread, local dishes, and tea) (Shafiq Ahmad Sedeqy, personal communication). There the fuel used is mainly dung and wood. Different types of cooking facilities could be used simultaneously for the preparation of different dishes, for example, frying and baking as can be done on a baking dish or boiling and baking.

The available literature on cooking installations shows that some features remain constant from the Neolithic through to the Late Bronze Age, whereas others are regionally and/or temporally variable, for example, fixed or portable hearths. Some of these features seem to generate a "community" of shared culinary equipment as is the case of Crete where different cooking traditions from those of the mainland seem to prevail. Crete seems to share more common features with parts of the east Mediterranean (Borgna and Levi 2017).

Pots for Cooking

Pots, a stereotypical element of the Neolithic way of life, despite their frequent association with domesticated food ingredients, may not initially have been made for cooking in prehistoric Greece: their small size, low number, and lack of indications for contact with fire in Early Neolithic Peloponnese led Vitelli to suggest that they were not used for cooking among early farmers; instead, she saw a strong symbolic element in the process of pottery making itself, that of transforming clay through fire, a more or less magic process. According to her, the knowledge of ceramic making was in the hands of those who knew the properties of plant ingredients, including medicines, and were responsible for cooking and preparing healing remedies (Vitelli 1989, 1993, 1995).

As high temperatures will tend to eliminate sooting clouds produced during cooking, the issue of recognizing traces of contact with fire remains largely unresolved. Pots need to possess certain properties to be used for cooking. The main problem that had to be solved in prehistoric times would have been the fabrication of pots suited for easy and even conduction of heat in a way that would lead to the cooking of food (Skibo 2013). For Greece, there is no consensus about whether Early Neolithic ceramic vessels were indeed used for cooking, with some arguing that cooking was unlikely (Nea Nikomedeia, Yiouni 2004) and others recognizing traces of cooking on pot surfaces (Early Neolithic Knossos on Crete, Tomkins 2007). For western Macedonia during the Early Neolithic and a large part of the Middle Neolithic, "it appears that pottery, for a long time after its appearance, was still not regularly used to fulfill everyday needs such as cooking" with one exception, however. At Ritini "at least some pots were used over fire" (Urem-Kotsou et al. 2014:510; Urem-Kotsou et al. 2018:329), suggesting that the use of pots for cooking over fire should not be completely ruled out, perhaps practiced among certain early farming communities or in very specific contexts of pottery use. Yet, a systematic investigation of this open issue, combining macroscopic and petrographic features related to cooking on a large dataset from several sites, is still lacking for Early Neolithic Greece.

Coarse cooking ware increases with time from the Final Neolithic onward and even more so during the course of the Bronze Age (see, e.g., Perlès and Vitelli 1999; Halstead 1999). Unlike the Neolithic, for which little comparative work has been done regarding cooking ware across regions, the

Bronze Age and the second millennium BC in particular seem better investigated. Bronze Age cooking ware from Akrotiri on the island of Santorini was investigated by exploring the materials of different categories of cooking pots, showing that different firing techniques had been employed for tripodic vessels and cooking jars, the former being more resilient to higher temperatures and therefore suited for boiling (Müller et al. 2013). Differences in cooking styles (boiling vs. simmering) have been inferred on the basis of different materials (Müller et al. 2013) and different shapes (Borgna 2012) in the prehistoric Aegean, reflecting perhaps different culinary practices and identities.

Borgna (1997), in her study of Late Minoan cooking pots from Phaistos in Crete, observes great variety both in the forms of cooking pots and their capacity, the dominant shape being that of the "tripod globular jar with restricted mouth" (Borgna 1997:191), the shape of the tripodic jar having a long tradition on the island, spanning the Bronze Age (Borgna and Levi 2017:119; see also Alberti 2016) (Figure 8.14). The large-mouthed cylindrical jar has been linked to heating and rapid boiling of foodstuffs, the round base securing even distribution of heat when placed over the embers of an open fire and the wide mouth preventing boiling over (Borgna 1997:204). This could correspond to open-air cooking, especially but not exclusively in ritual, communal food consumption events (Borgna 1997:202). The closed-mouth jar with a flat base (Figure 8.15) is seen as more appropriate for slow boiling (Borgna 1997:204). Open-mouthed vessels are traditionally used for parboiling bulgur (Valamoti 2011b), while experimental cooking with Neolithic replica pots showed that closed-mouth jars result in the faster cooking of grass pea fava compared with tripodic, open-mouthed pots (Dimoula et al. 2020). This highlights the need for systematic cooking experiments that could help researchers better understand the variety observed in cooking pots in the archaeological record of prehistoric Greece.

The open cylindrical tripod jar with a flat base appears to be the characteristic Minoan cooking pot, the type of cooking vessel known from Crete, where a "rich set of different vessels, including trays, dishes and different kinds of pots, such as low-walled jars or pans . . . point to the development of that elaborate cuisine in which the diversification of cooking procedures, involving roasting, boiling, simmering, frying . . . , directly points to the complexity of social relations" (Borgna 1997:205). Based on the study of the Petras cooking ware, it has been suggested that a Minoan set of cooking vessels might have existed, consisting of cooking pots (a large-mouthed and a narrow-mouthed), a pan as well as a few plates, trays, cooking dishes, and a fire stand (Alberti 2016).

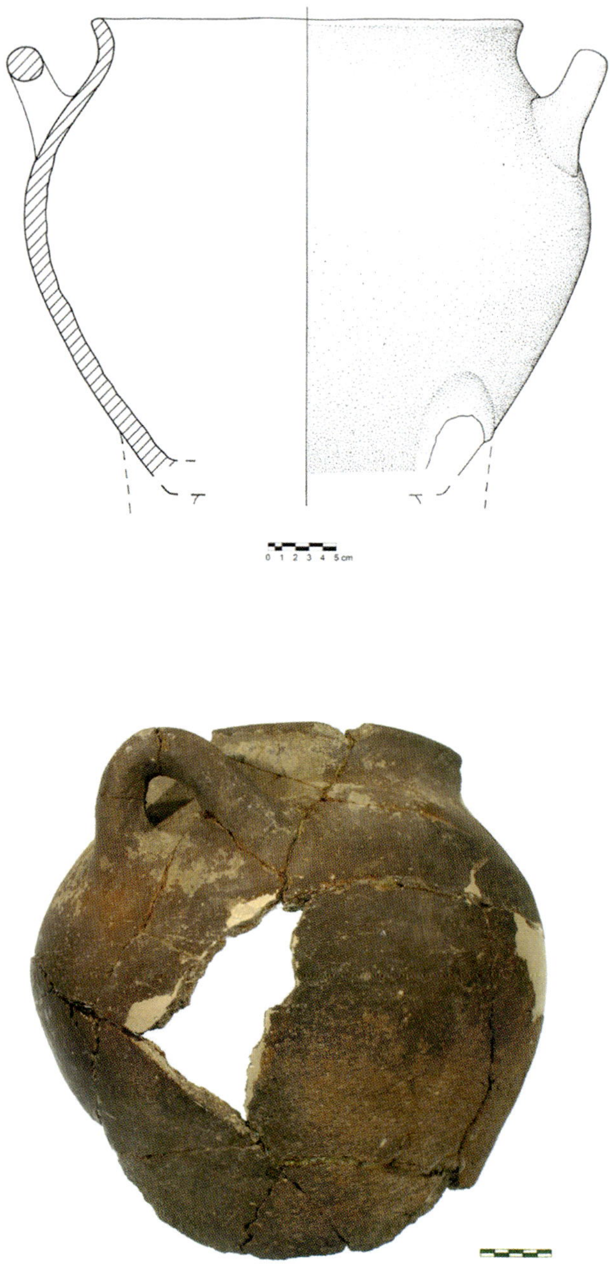

Figure 8.14. Tripod globular jar with restricted mouth is a typical cooking vessel for Minoan Crete. (Drawing by G. Merlatti. Photograph by Elisabeta Borgna. © Archivi SAIA, N.I.G. 6010 [*top*] and U/1179 [*bottom*].)

Figure 8.15. Cooking jar with a flat base, a typical cooking vessel from Bronze Age southern Greece. (From Borgna and Levi 2017, Figure 10.4. Photograph by Elisabeta Borgna. © Archivi SAIA, B/11.819.)

Culinary diversity in terms of available equipment and potential ways of transforming ingredients into dishes is manifested for the entire prehistoric Greece through the different types of cooking pots and other types of cooking vessels during the Neolithic and the Bronze Age as well as later periods (Gimatzidis 2017; see Dimoula et al. 2022a for a recent synthesis). Baking dishes and trays would have been ideal for multipurpose cooking, for example, baking pancakes or griddle breads, cooking meat and vegetables, or keeping a pot of water warm. Deep impressions on the surface of certain cooking vessels, including baking trays (Figure 8.16) from the Bronze Age onward, have been interpreted as a way to achieve a better conductivity of heat (Borgna and Levi 2017). This type of griddle with depressions could have been used as baking pans for thick doughs (Hruby 2017). This feature is observed on sherds interpreted as corresponding to pots, baking dishes, and portable hearths found, for example, in Thessaly (Iolkos, Adrymi 2019), Crete (Borgna and Levi 2017), and northern Greece (Mavroidi 2012; Gimatzidis et al. 2014). The *pyraunos*, a fusion between a portable hearth and a pot (Figure 8.17), was used in the Bronze Age and underscores the great variability and inventiveness in cooking facilities available during different periods and

Figure 8.16. Shallow depressions on the base of a baking dish. (From Gimatzidis et al. 2014:Figure 3. Photograph courtesy of Stefanos Gimatzidis, Yannis Karliampas, and Stavros Kotsos.)

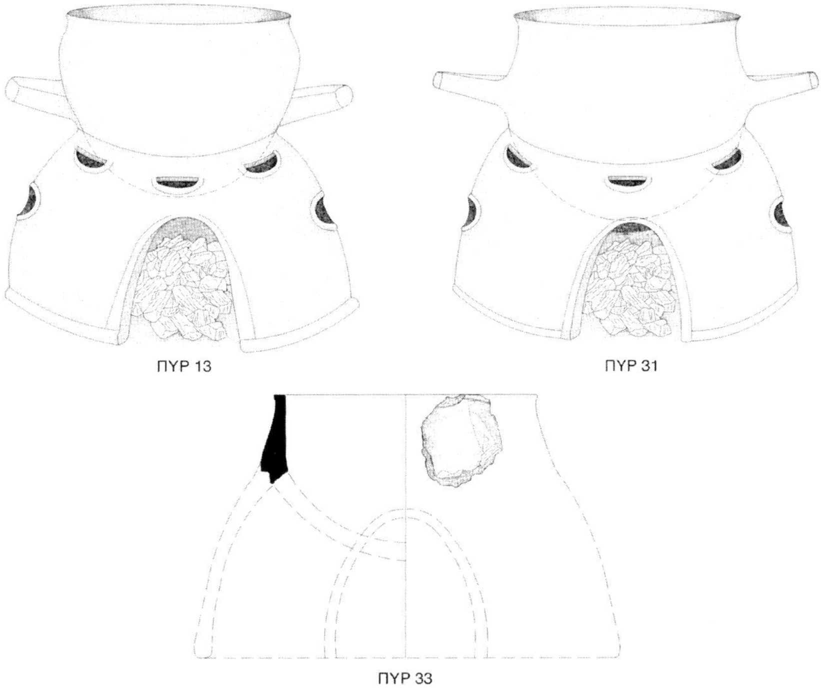

Figure 8.17. *Pyraunos* from Apsalos, Late Bronze Age. (Reproduced from Anastasia Georgiadou, *Apsalos Almopias: The Late Bronze Age Settlement*. University Studio Press, Thessaloniki, 2021, Figure 1. Courtesy of Anastasia Georgiadou and University Studio Press.)

in different regions of prehistoric Greece. Such vessels were widely used in the north of Greece and parts of the Balkans throughout the Bronze Age and during the Iron Age and antiquity in northern Greece (Wardle 1993; Gimatzidis 2017; Savvopoulou 2012; Horejs 2007).

Borgna sees in the variety of shapes in Cretan ceramic cooking pots a more elaborate, complex cuisine, requiring different forms of cooking, including baking, roasting, simmering, boiling, while for the Mycenaean period she interprets the shapes as indicating simpler, more "practical, utilitarian concerns" (Borgna 1997:205). In the Mycenaean world, metal cooking vessels are considered as the main vessel for preparing the elaborate meals of the elites of the mainland, while the flat-bottomed ceramic vessels were used for private, everyday cooking (Borgna 1997). There is no consensus on this, however, as Lis (2006) and Hruby (2006, 2017) argue that Mycenaeans developed an haute cuisine, too, on the basis of a wide range of ceramic cooking vessels suggestive of a large variety of dishes prepared. Bronze Age cooking pots from southern Greece, Crete, and the eastern Mediterranean are suggestive of cultural affinities between specific cooking pots, cooking installations, and regional culinary traditions, and by the Late Bronze Age, cooking vessels and cooking methods seem to acquire a characteristic typology with a widespread geographical distribution (e.g., Borgna and Levi 2017). Mycenaean cooking jugs and amphorae, together with hearth platforms, penetrate local culinary traditions in Cyprus (Jung 2012) and Palestine, partly or exclusively (Jung 2012; Yasur-Landau 2012).

The wide range of cooking vessels produced during the Final Neolithic and especially the Bronze Age shows a variety in cuisine, possibly more elaborate compared with that of the Early–Late Neolithic. This observation might in fact undermine the widely accepted suggestion put forward by Halstead for a change from the Early Neolithic to the Early Bronze Age toward a more introvert cuisine, confined within the household, gradually distanced from food sharing and hospitality over time (Halstead 1999:80–81). The variety inferred both from the available ingredients and cooking pots might actually be pointing to the opposite, that is, a more sophisticated cuisine with elaborate meals prepared for communal display in the context of feasts and food sharing, the emphasis now being perhaps placed on the quantity and/or the variety of cooked food rather than the vessels used for display as appears to have been the case in the Neolithic. The higher the elaboration the greater the opportunity for status display, and as Hruby (2017) observes, the appearance of new types of cooking vessels during the end of the Late Bronze Age suggests that status display was achieved through the ability to offer an elaborate cuisine in feasting contexts. In any case, a detailed spatiotemporal investigation of cooking technologies, cooking capacities, and the organization of cooking spaces in prehistoric Greece might reveal more nuanced commensal practices and gastropolitics (see Hastorf 2017 for a thorough discussion of the term) reflected in the archaeological record.

Cooking Places

A rich anthropological literature highlights the strong cultural significance of cooking and the symbolism associating cooking (dishes, seasonality, facilities, agencies) with the cosmos of each society. Cooking in a way forms the arena where social relations are reproduced and potentially negotiated through the daily repetition of tasks, words, songs, positions, and silences. Cooking indoors or outdoors, the form of cooking, and who is performing, each has its own cultural, symbolic, and utilitarian aspects, all closely interwoven. Bourdieu (1977) provides an illuminating example of this based on his work in Kabylia, Algiers: dry foods and roasting are associated with male, outdoor cooking and summer, whereas fermenting, boiling, and liquid foods are associated with the female. In this general framework a cereal food in the form of gruel has a female nature while dry wheat cake is considered as the male food par excellence (Bourdieu 1977). For different cultures across the world, a firing pit, the oven, or the pot symbolizes the womb or parts of the human body while the food prepared in it nurtures life (see, e.g., Jacobson-Widding 1992; Feldman-Savelsberg 1994; Boeyens et al. 2009; Gosselain 1999; Miller 2000; Johns 1998). Similar associations have been put forward in the interpretation of archaeological contexts (see, e.g., Bacvarov 2007 for examples from southeastern Europe and western Asia).

For prehistoric Greece, archaeological discourse on the social dimensions of cooking spaces has focused on hospitality and reciprocity among households and settlements. The presence or absence of decoration of ceramic vessels as well as the spatial arrangement of hearths have been the subject of discussion, especially since the suggestion by Paul Halstead (1999) that during the Neolithic, hearths are located outside houses, whereas toward the end of the Neolithic and during the course of the Bronze Age, they are located inside (Halstead 1999, 2012). Halstead argues that this is an indication for changes in hospitality and reciprocity levels between households and communities, with the household gaining a more prominent role rather than the whole community, in the course of time. This suggestion has received wide acceptance by prehistoric archaeologists working in Greece, even in cases where the evidence points to the contrary, as is the case in a recent examination of Neolithic cooking installations of northern Greece (Kalogiropoulou 2014; for a critique see Valamoti 2018). Despite this widely held view, the available data do not seem to support such an overarching pattern, whereby there is a linear evolution from the "communal" and reciprocity on one hand to the individual household and hoarding on the other, reflected in the archaeological record by a shift in the positioning of cooking installations over time, from open to roofed and walled spaces.

Cooking installations in fact have been found both inside houses and outside at the same settlement and/or the same habitation period. Open-air hearths have been found throughout the Neolithic and the Bronze Age: at the Late Neolithic sites of Makriyalos (Pappa 2008) and of Kleitos in western Macedonia (Ziota 2014) and at Early Bronze Age Agios Athanasios in the north (Mavroidi 2012, 2014) and Late Bronze Age Tiryns in the south (Maran 2012). At Kleitos, hearths are present both in house interiors and outside in what may have been clusters of hearths to be used on special occasions (Ziota 2014). Size-wise, those located outside are decorated and large, underscoring their special and communal character. Ethnographic studies on feasting in different parts of the world associate clusters of installations such as hearths with the preparation of large quantities of food or drink required for the feast (e.g., Hayden 2001:50, 55, Figure 2.3; Kirch 2001:175–176; Brown 2001:381–383, 385). At Makriyalos I, the available evidence for large-scale feasting during the early phase of the site (phase I, Pappa et al. 2004) raises the possibility that the cluster of hearths of phase II, despite their interpretation as installations for the preparation of daily meals (Urem-Kotsou and Kotsakis 2007), might have in fact served similar, large-scale feasting events continuing during phase II of the site.

Thus, the coexistence of hearths outside and inside houses may reflect the complex interplay of social and economic organization as well as the seasonality of cooking in the preparation of daily meals, rather than a straighforward, linear evolution from the communal in the Neolithic to the rise of individual household and elite power in the Bronze Age, which is a prevailing narrative for Greek prehistory (Halstead 1999; Urem-Kotsou and Kotsakis 2007).

The spatial arrangement at Archondiko has been interpreted as indicative of houses running parallel to each other (Papaefthymiou-Papanthimou et al. 2013; Papaefthymiou-Papanthimou and Papadopoulou 2014), each containing a set of cooking installations. The thermal structures clearly demarcate spaces of specific household activity related to food preparation. Likewise, kitchens in a row, perhaps related to brewing, were found at Early Bronze Age Argissa. Both examples could signify specialized spaces within a settlement allocated to specific activities during parts of the year, such as brewing beer. The spatial arrangement of the complex cooking structures found at Archondiko could indicate the preparation of elaborate dishes and/or drinks (e.g., beer) or ingredients (e.g., oil), and the execution of complex recipes, requiring perhaps more sophistication than the recipes prepared during the Neolithic.

Muhly-Metaxa (1984), in her detailed overview of hearths in Minoan Crete, points out that in palatial complexes, where a large number of people lived and worked, more specialized space would have been required for cooking. Circular or rectangular hearths in the form of fixtures of clay with a circular depression in their center constitute the basic type of Minoan hearths with parallels both in the eastern Mediterranean and the Aegean islands and mainland Greece (Muhly-Metaxa 1984). Cooking spaces in Bronze Age Crete were often in the form of sheds, open on some sides but roofed, with fire being placed directly on their floors, with the mainland tradition of fixed hearths with pebble or potsherd floors being later introductions to the island (Borgna and Levi 2017).

Ethnographic observations suggest that cooking on a household level usually takes place inside during the cold months of the year and outside in the summer. A house can have the summer and winter cooking spot (see Figure 8.13). The Late Neolithic site of Kleitos is very interesting in this discussion of placement of cooking installations. The cooking installations are located both inside rectangular houses and in open spaces with some of them being impressively large, probably used to feed large numbers of people in contexts of feasting (Ziota 2014). Their much larger size, compared with those thermal structures found in house interiors, suggests contexts of consumption involving large quantities of food and large numbers of participants in a meal cooked in the open air.

The placement of cooking installations and the act of cooking inside the house or in open-air locations could reflect seasonality in cooking and/or cooking in external, open air spaces for special occasions. Ethnographic evidence also highlights the complex factors determining cooking spaces such as the marital status of a woman or perceptions about exterior and interior spaces and the location of cooking facilities. Among the Dowayo potters of Cameroon, for example, cooking is always performed in interior spaces even in the summer, except for special occasions that are associated with the dead and rites of passage (Barley 1984). Among the Bangangté of Cameroon unmarried women cook outside in the open. Married women, however, cook inside, in the hearth of their husband's hut (Feldman-Savelsberg 1994). Thus, cooking inside or outside is subject to the different ways of structuring the world among different communities, alternating on the basis of seasonality, cooking cycles, the contexts in which food is consumed as well as gender and marital status.

From Daily Meals to Cooking for Special Occasions

In Neolithic and Bronze Age Greece the basic plant ingredients would have been transformed into a variety of dishes for daily meals but also for special occasions marking life cycles. The special occasions were closely linked with specific dishes and food consumption rituals with ancestors being also part of the process, at least in certain contexts. Plant food ingredients chosen for daily and special dishes may have varied. Yet, it is reasonable to assume that some basic staple food ingredients could have also been essential for ritual food preparations. There is abundant ethnographic evidence to suggest the ritual use of cereal and pulse seeds in modern Greece, for example. What might have differed for special occasions could have been the quantity, the method of preparation, and added ingredients and decoration. Most significantly, an occasion would have been made special, beyond the food itself, through rituals (see various examples in Dietler and Hayden 2001a).

Throughout prehistory, Greek farmers could grind their plant ingredients finely or coarsely, depending on the type of ingredient they needed. Grinding would have generated the raw ingredients for "breads," porridges, mashes, and pies and for fermenting plant-based liquids into alcohol. Medicinal plants or oil-rich seeds could be crushed for special occasions. Cereals were ground in two ways: a meal consisting of fine and coarse particles as well as coarsely ground fragments. Meal is indicated by lumps of "food" that have been analyzed, showing a homogeneous mass with some features of cereal components such as aleurone layers and occasionally larger identifiable cereal fragments (see, for example, Valamoti et al. 2019). Grinding into coarse fragments is clearly indicated by Neolithic ground barley fragments from Vasilika Kyparissi (Lathiras 2020) and Bronze Age cereal fragments from Archondiko and Messimeriani Toumba (Valamoti 2002a, 2011b).

Plant ingredients transformed through milling and crushing underwent the process of cooking using heat with the aid of a wide range of pots and cooking installations. Coarsely ground grain, possibly parboiled in water or milk and dried for later consumption, encountered in the form of cereal fragments, could be pointing to changes in the range of ingredients and in more sophisticated ways of transforming cereals into some form of precooked cereal. The wheat fragments from Mesimeriani Toumba, ground prior to charring and either boiled in water or milk, correspond to a complex recipe. The fine grade of fragments of similar size indicates good mastering of the grinding and sieving processes with a standardized mesh, which was designed to obtain this specific preparation, likely some form of bulgur or *trachanas* (Valamoti et al. 2021). At Archondiko, grinding might well have involved grinding of malt, probably for fermentation.

Pulses could have been consumed whole or split as the finds of bitter vetch from Agios Athanasios suggest. Acorns might have been ground, boiled, and roasted. Fruit would have probably been dried for storage and/or crushed to extract juice. Boiling down fruit juice to produce a syrup would have led to yet another produce with long storage life, probably a rare delicacy given the relatively limited scale of viticulture and the lack of sweet foods in prehistoric times. Simmering dried fruit would have thus added variety and taste in the cereal and pulse dishes.

Foodstuffs such as fermented drinks and breads would have required more sophisticated knowledge and mastering of vessels, temperatures, and processes. There are various cultural proscriptions against fermented foods that can either be totally excluded from diets, used, or avoided in specific contexts. It has been suggested that the emergence of fermented cuisines was the inevitable outcome of a continuous activity of microbial fermenters closely linked to the indigenous microbiota of humans: the bacteria and fungi present among different communities of people found their way into the raw ingredients available, be it milk, grain, or fruit, while people were prepared to eat these foods "modified" by microbial attack (Brumberg-Kraus and Dyer 2011). Bacteria and yeasts responsible for the wide range of fermented foods, consumed worldwide today, have their own fascinating history, and recent biomolecular analysis has revealed a previously unimagined wealth of variability even within the same broad food category (e.g., Sicard and Legras 2011; Lewicka et al. 1995; Galone et al. 2016; Romano et al. 2006; Guyot 2012; Papadimitriou et al. 2015).

Certain recipes may have been easy to reproduce on a daily basis throughout the year; others, however, may have been season specific, the complex outcome of ingredient availability, prevailing temperature, and weather conditions. Bulgur, *trachanas*, or fruit syrups or the drying of plant foods such as greens and fruit would have been such activities, requiring heat, dry conditions, and sunshine as well as the availability of raw ingredients, fuel, processing, and cooking equipment. De-husking of the glume wheats and hulled barley most likely took place piecemeal. Thus, grinding of cereals would have also been done on a similar basis (Valamoti 2004). Bulk processing may have been practiced in certain contexts for which the evidence is as yet lacking, such as, for example, when food for large numbers of consumers had to be accumulated and transformed into food for laborers or for a special dish for a feast. Seasonal preparations such as bulgur and/or *trachanas* would have been ideal ways to combine cereals and animal protein into a nutritious food/ingredient. Some seasonally prepared ingredients, such as *Grünkern* and *freekeh*, made of the grain obtained from parched, unripe ears of cereals, if at all produced in prehistoric Greece (no such evidence is available at present), might reflect a combination of taste and environmental parameters. Harvesting unripe grain and roasting/parching might have corresponded to a need to harvest unripe grain, when last year's stores were already depleted or when conditions risked the successful harvest of the grain (e.g., a wet summer with much rain); thus, practical reasons may have underlaid such a culinary choice (Berihuete-Azorín et al. 2020). Alternatively or, in addition, a desire for a particular taste, that of roasted, unripe grain, could have led to such preparations in prehistoric times.

Most of the recipes just described require great labor input and careful monitoring of the processes, in a rather short period of time. However, they could have provided food ingredients with a long shelf life that could be consumed piecemeal throughout the year, which is very important in a world functioning with no electricity for cooking and freezing and no canning technologies. Prehistoric societies largely depended on generating foodstuffs that would last long, preserve their taste and nutrients, as well as be easily transformed into a meal using dung and wood as fuel. Alcoholic beverages and preparations including hallucinogenic or medicinal substances would have corresponded more to culinary elements of special contexts of consumption (e.g., rites of passage and different types of feasts) rather than to the components of daily meals.

Differences in food textures could have been achieved through boiling, simmering, and parboiling, combining in different ways the wide range of cooking pots and thermal installations identified in the archaeological record. Thus, a thick batter of ground cereals, or pulses or even acorns, could be quickly baked on a baking dish/pan in the form of a crepe, flat bread, or griddle cake, being tastier with the addition of milk and the use of fat on the cooking surface, and dipped in honey or grape syrup for a special day. At the same time, breads could have been baked in an oven, and a soup containing cracked wheat or bulgur or bulgur simmered in water could generate a variety of dishes, all at the same time, especially in kitchens where a wide range of facilities coexisted, as is the case of Archondiko in northern Greece and Minoan and Mycenaean kitchens in the south. On a mundane, daily basis, larger families would have required the provision of more food; therefore, ingredients such as bulgur and/or *trachanas* would have enabled the fast conversion of ingredients into a warm, daily meal, rich in nutrients, suitable for all ages, including more vulnerable groups such as very young children and elderly people with difficulties in chewing.

The same ingredients could have been used for the preparation of meals for feasts, too. Although meat is the main dish during religious *panygyria* in modern Greece, a cereal component is also involved, boiled in the meat broth (Bozi 1997:28, 144). At Makriyalos, despite the impressive quantity of meat apparently consumed during one or several feasts, plant foods are considered as a significant component (Pappa et al. 2004:35). Such a

Figure 8.18. Foundation offering from Plateia Magoula Zarkou interpreted as reflecting members of a Neolithic household. (© Austrian Academy of Sciences, Austrian Science Fund Project P27159, Platia Magoula Zarkou in Thessaly. Photograph courtesy of Eva Alram Stern.)

combination seems to be confirmed by residue analyses conducted on some of the Makriyalos pots showing the cooking of both meat and plant food ingredients (Urem-Kotsou and Kotsakis 2007). In the Neolithic and Bronze Age contexts explored here, coarsely ground cereal grains could have been such an element, although no such evidence has been detected from the large pit containing the remains of the feast at Makriyalos.

The variable sizes of grinding stones, cooking pots, and cooking installations were probably closely linked to various needs for quantity and types of cooked food, depending on family size, number of people needing food provision, or requirements for a feast. Aspects of social organization such as family size (Figure 8.18) would have affected the quantities of food needed on a daily basis as well as the potential to produce certain ingredients in bulk, such as bulgur and *trachanas*, that require significanl labor input on the hottest days of the year. Of course, larger families would have required more food and at the same time would have been in position to generate it. Provisioning for food for the members of egalitarian communities or institutional elites with dependent laborers might have influenced the need for specific recipes that would be nutritious and easily converted into a dish. Variations in the quantities of food prepared, whether for daily dishes or for feasts, would be reflected in grinding tool and pot sizes.

Cooking pots of different shapes as well as baking dishes, which were common elements of the cooking equipment of prehistoric Greece, could have been placed on hearths and used for cooking either for daily meals or special occasions. In the latter case, many pots and dishes could have been pooled together by the whole community or those controlling access to the means necessary to prepare the feast. Sometimes, as is the case in Late Bronze Age Crete and the Peloponnese, it is possible to identify "kitchen spaces" associated with the preparation of food for feasts taking place nearby. This identification is based on the coexistence of different food preparation and consumption activities indicated by equipment and installations related to grinding, cooking, and baking (e.g., Muhly-Metaxa 1984; Borgna 2012). Preparing food for feasts is often closely linked to the ability to mobilize labor for the preparation of the food served in that context, with gender playing an important role, elusive though it may be for prehistoric communities. In the case of Luo feasts in East Africa, for example, there is a close relationship between the number of wives of a man and the size of the feast he can organize. The more wives one man has to prepare the food, the more food he can offer in a feast and thus the greater number of people who can be fed in this context and in so doing acknowledge the host's status (Dietler 2001; Dietler and Herbich 2001).

One can assume that large quantities of food would have been required in both Neolithic and Bronze Age feasting contexts. At Late Neolithic Dikili Tash, large cooking pots (Dimoula et al. 2020) would suggest either large families and/or vessels for preparing food for special occasions requiring the feeding of many guests. The available evidence from Neolithic Makriyalos suggests that food was probably cooked in ceramic vessels most likely in rather small batches for consumption by small numbers of people, though no indications for their capacity are provided (Urem-Kotsou and Kotsakis 2007). Although the Makriyalos cooking pots have been associated with daily meals for households or other small social groups (Urem-Kotsou and Kotsakis 2007), an alternative, albeit provocative, interpretation could be that these pots were used for cooking and serving the food for the feast(s) at the site, which may have primarily been a gathering place/space for performing specific rituals for feasts, including healing practices and ancestral worship. Cooking for the feast may also be indicated for the second phase of habitation at Makriyalos by the cluster of outdoor hearths.

Clusters of cooking facilities in the archaeological record have indeed been considered as indicative of feasting (Hayden 2001:40), marking the special context of food preparation and consumption and providing the larger quantities of food to feed the participants (e.g., Mills 2007; Hastorf and Johannessen 1993). Cooking facilities may have been prepared and/or constructed for exclusive use for feasting occasions, as encountered in recent and older feasting contexts in different parts of the world (e.g., Mills 2007 for Pueblo Indians of the United States). Examples of such clusters of cooking facilities are encountered at Greek *panygyria*, religious festivals taking place in different parts of the year (Figure 8.19). At Kleitos, the large size and evidence for repair of some of the hearths suggest reuse (Ziota 2014), which in turn could point to events repeated in the course of time. Similar rebuilding

Figure 8.19. Clusters of pits prepared for cooking the festive meal at the *panygiri* of Agios Prodromos, at the homonymous village of Chalkidiki. (From Katsagelos 1995, p. 43. Photograph courtesy of Georgios Katsagelos.)

is observed in a much later open-air hearth in the lower town at Tiryns, which is interpreted as evidence for intensive open-air cooking (Maran and Papadimitriou 2006). Wine for a large number of people (depending on individual servings) is evidenced from end of fifth millennium BC Dikili Tash, and if beer was indeed produced in the numerous open-mouthed jars found at Archondiko, their large number could suggest significant quantities produced for special occasions.

For Neolithic or Bronze Age communities, it seems that the location of cooking facilities, either open air or within houses, depended on the contexts related to their use, that is, the different ways in which spaces were involved in food provision for daily requirements and feasts. At Makriyalos the open-air cluster of hearths and at Kleitos the large open-air hearth could be pointing toward feasting under the sky in the Neolithic. Large open-air hearths have also been associated with feasting in a very different context from that of the Neolithic cases discussed: the large Mycenaean hearths found at palatial centers such as those excavated at Tiryns are closely linked to feasting activities, yet their placement shifts through time from interior spaces of the palace to open air locations (Maran 2012). The feasts held in each case might have been different in terms of participants, structure, and overall contexts of participation and goals (see Dietler and Hayden 2001a for an overview of different types of feasting occasions). The late Bronze Age examples of outdoor hearths from strongly hierarchical social contexts show that the location of hearths in interior or exterior spaces per se cannot be used as a basis to infer social organization and levels of reciprocity. Open-air hearths of the Neolithic, when operating, could still be used in contexts that would have underscored different status within a community or on a regional scale, as might have been the case at Late Neolithic Makriyalos.

In a palatial context from southern Greece, similar variability is observed in the placement of cooking facilities with a tendency to have hearths in open spaces and areas partly open at the margins of palace agglomerations, with portable hearths available as pot hearths at the same time (Muhly-Metaxa 1984). This is the case with the braziers of southern Greece but also the *pyraunoi* of northern Greece. The *pyraunos* is a vessel that is encountered in parts of southeastern Europe and the Danube throughout the Bronze Age, in some cases continuing into later periods (Gimatzidis 2017 for a recent review). With such portable cooking facilities available, cooking space becomes more flexible and can be placed inside or outside or in different rooms, involving few or many people, depending on occasion, season, and context of consumption.

The available evidence for cooking spaces reviewed here, as well as insights from ethnography, suggest that the prevailing narrative of a gradually less reciprocal society during the course of the Late Neolithic and the Bronze Age, reflected in an alleged temporal shift in the location of hearths from exterior to interior spaces, needs careful reconsideration. As discussed, a wide range of factors may influence the location of hearths in interior and exterior spaces, including hospitality and food sharing as well as seasonality in cooking and specific contexts of consumption for which cooking installations were intended. Rather than adopting generalizing, uniform, or evolutionary approaches to the placement of cooking installations, we need to carefully examine each site in a contextualized way as people in even neighboring sites might have had different ways of structuring their world, their bodies, and cooking spaces.

There is yet to be systematic study of cooking spaces for the whole of prehistoric Greece (work in progress, project PlantCult). The close relationship between cooking spaces and level of commensality between the household and the community and the dialectics between these two entities emerge as constant elements of both the Neolithic and the Bronze Age rather than as elements marking a temporal evolution to reduced reciprocity. Different contexts of cooking can be envisaged: (1) a daily one closely linked to fulfilling daily household needs, and (2) cooking for special occasions, in the context of feasting or special communal events, at times, at least, taking place outdoors.

The examination of the available evidence presented here suggests that the significance of open-air cooking can acquire different meanings and readings depending on the specific cultural context. In all cases it would have generated a powerful sense of community, regardless of whether this was structured on the basis of reciprocity or hierarchy and inequality.

This chapter provided a brief overview of various ways of cooking and consuming plant foods among farming communities of prehistoric Greece, as is evidenced through the processing equipment and cooking installations including hearths and pots. More often than not, the material remains of cooking offer a basis for speculation and hypotheses related to daily food preparation, collective food consumption, hospitality, food sharing, rituals, and commensal politics. This coalescing of different lines of evidence related to cooking practices is far from being complete, yet it has hopefully highlighted the potential of integrated considerations of prehistoric cooking technologies, bringing together ingredients, processing, and cooking equipment toward an understanding of past culinary practices and their changes over time. As evidence is gradually being pulled together on plant food ingredients, grinding and cooking equipment, insights about "regions" of culinary practices, and changes in time can gradually begin to emerge. Such integrative approaches hold the key toward enhancing our understanding of culinary identities and cultural change in prehistoric times.

9

Plant Foods and Identity in Prehistoric Greece

In the anthropological and archaeological literature, food has been closely associated with the formation and expression of cultural identity. Modern regional "microcuisines" are the outcome of processes operating over millennia, whereby recipes and ingredients have been adopted from the wild and/or introduced from distant lands, shaped over a succession of shifting cultural boundaries, changing economic systems, introductions, and/or modifications of ingredients and cooking techniques together with the movement of people and changing political and ethnic entities. At the same time, they may defend heterogeneity in the context of national states (e.g., Ayora-Díaz 2012; Appadurai 1981) and a modern globalized culinary environment. The cultural significance of taste and smell in everyday life, contributing to memory and "socially meaningful features of people's lives" (Walmsley 2005:44), has been highlighted by anthropologists, archaeologists, and psychologists (e.g., Lévi-Strauss 1964; Douglas 1982; Howes 1987; Seremetakis 1994; Walmsley 2005; Hamilakis 2014; Hastorf 2017). Taste and smell were in all likelihood significant elements of prehistoric meals, and this chapter uses the plant food ingredients and recipes from prehistoric Greece to approach culinary identities and social practices related to plant foods across space and through time in a region that was involved in complex contact networks with different cultural settings in the Balkans and central Europe as well as the east and west Mediterranean.

There is a large scope for generating culinary variability and expressing multiple, different identities through food preparation and consumption. This can be achieved through the ingredients and the recipes, the equipment used for its preparation, and the cooked meal. It can also be expressed through the etiquette followed during food consumption in regard to gender, timing, location, containers used, and the organizers and the participants to a feast (see, e.g., Hayden 2001; Dietler 2001; Hastorf 2017; see Zubaida 1994 for modern examples from the Middle East). Many dishes of modern Greece typically considered as ethnic cuisine transcend in fact the boundaries of the modern state. For example, consider the variations on phyllo pastry recipes known as *pita, buğatsa, banitsa,* or *börek,* encountered throughout the Balkans, the eastern Mediterranean, and parts of the Black Sea coastal areas (Kiziridou 2002; Krăsteva-Blagoeva 2008; Piskov 2010; Psilakis and Psilaki 2001). Thus, as is the case elsewhere (Zubaida and Tapper 1994), plants in modern Greece are cooked in a variety of ways, sharing culinary elements from different regions, with common traditions over a large area but at the same time with a wealth of regional variations expressing regional culinary identities (e.g., Voutsina 2008; Delatola-Foskolou 2006).

When exploring culinary identities in prehistoric Greece through archaeobotanical remains, there are too many unknowns and limited, ambiguous data, subject to alternative interpretations. Moreover, there is a danger to project modern trends such as "regional cuisines" (Ayora Diaz 2012) and culinary nationalisms (see Parkhurst-Ferguson 2010) to a place and time where the sense of community and boundaries was very different from the present situation. Attempting to decipher the prehistoric cuisine of a region runs the risk of generating an artificial stereotype of a past "foodscape" that never really existed in prehistory, in this process, projecting modern preconceptions about the culinary characteristics of the present to the prehistoric past. Stereotypes such as the concept of the "Mediterranean diet" (e.g., Nestle 1995; Manios et al. 2006; González-Turmo 2012) or "Mexican cuisine" (e.g., Pilcher 1996) have forged associations between ingredients and recipes and specific regions, generating modern culinary identities that may claim their justification to real or supposed ancient practices.

Relevant to our study area, the concept of the "Mediterranean diet" and its presumed *longue durée* from prehistory to the present, a concept famous for its alleged benefits to health all over the world, seems to correspond to a skewed reinterpretation of twentieth-century cuisine on the island of Crete (e.g., Matalas 2006a and b). Thus, the prominent position of olive oil as an element of Mediterranean cuisine in recent times has been projected to the prehistoric past of the region (e.g., Renfrew 1972; Zohary et al. 2012; Riley 2002; Tzedakis and Martlew 1999).

An arbitrary emphasis on cereals and wine, elements closely linked to the concept of the "Mediterranean diet," marginalized pulses for a long time (see Renfrew 1972; Sarpaki 1992a; Halstead 1981 for their importance in prehistoric diets of the region), and wild, gathered plants (cf. Forbes 1996; Valamoti 2015a) in archaeological discourse. Yet, both wild plants and pulses form important elements of countryside cuisines of modern Greece as well as elsewhere in the world, and it is reasonable to assume that this was also the case in prehistoric times.

Attempting to generate overarching patterns for prehistory that project modern perceptions results in a poor representation of a prehistoric culinary past that was most likely highly variable in choices of ingredients and culinary practices. Here I read through the lines of the available evidence and explore nuances that might imply variable culinary identities.

The available archaeobotanical record allows more conjecture than demonstration of the availability of ingredients on the basis of solid, statistically significant data. This chapter is an exercise in questioning the archaeobotanical data to extract fragments of information that might allow a rough picture of what a culinary landscape would have been like for Neolithic and Bronze Age Greece in different regions and contexts. Archaeobotany offers those analytical tools to explore the long temporal trajectory that plant food ingredients have followed from the first farmers to the emergence of the first cities in the region (see Valamoti et al. 2017). Aspects of a very slow process of "food globalization in prehistory" (see Jones et al. 2016; Liu and Jones 2014) are explored in the following pages, placing Greece and southeastern Europe in a wider nexus of interaction with wider regions.

The Construction of a "Neolithic Farmer" Identity

When the first farmers established their villages in the plains and low hills of mainland Greece and some of the islands, they must have encountered plants that they were already familiar with, plants that grew wild along the coasts of the eastern Aegean or farther inland, some with edible fruit like acorns, almonds, pistachios, and figs. Clearing woodland to open their fields would have familiarized them with the local flora. Although an encounter with local Mesolithic people is probable but not much evidenced in the archaeological record, farmers at the beginning of the Neolithic might have acquired knowledge on local plants, edible, medicinal and other special plants through their contacts with those "invisible gatherer-hunters" who were active in southeastern Europe during the first few millennia of the Holocene. The gatherer-hunters' knowledge of the terrain and its plant resources was probably the outcome of an accumulated knowledge acquired through millennia of inhabiting the area. Encounters between gatherer-hunters and farmers might have introduced different plant food ingredients and/or different ways of cooking/eating them in the culinary landscape of Early Neolithic Greece. A certain degree of familiarity with wild vegetation for farmers arriving from the east must be assumed on the basis of recent paleovegetation studies for the wider area of the Aegean that show that a significant number of shared plant species was growing on either side of the Aegean and people had access to them and most likely used them as food and fuel (Asouti et al. 2018).

The archaeobotanical record of the Mesolithic is quite scarce and shows the use of wild plants that managed to survive in archaeological deposits. Despite a hypothesized limited or nonexistent use of wild resources at the beginning of the Neolithic (Perlès 2001), these first farmers should not necessarily be perceived as "symbolically" excluding wild plants from their diets and/or settlements. This is probably a hypothetical scenario whereby farmer identity was constructed on symbols related to the domesticated part of nature, Hodder's *Domus-Agrios* opposition (Hodder 1990). This supposedly distinct "domesticated nature-farmer" concept, as opposed to that of an "unmodified nature and gatherer-hunters," may be more of a projection of what is now, millennia later, perceived as the outcome of the shift to food production, an artifact of archaeological approaches to the transition to food production and the Neolithic way of life rather than actual perceptions of the first farmers about plant food resources. Judging from what followed a couple of millennia after the first farmers appeared in Greece, it was agriculture that became established, marginalizing or integrating aspects of the gatherer-hunter ways of living or perhaps transforming them to encompass the new way of dealing with natural vegetation and wild plant food resources.

A recent reexamination of the archaeobotanical evidence from western Asia shows that there is no reduction in the use of wild plant food resources with the onset of farming in the area (Wallace et al. 2018). Farmers were gatherer-hunters before shifting to agriculture, and wild plants were part of their lives during preparing their fields or taking their animals for grazing. Wild plants in the form of seeds, fruits, nuts, roots, shoots, leaves, and flowers were absolutely necessary as food and fodder, fuel, timber, and medicines. Even in our modern globalized food environment, wild plants are still harvested and consumed (e.g., Pieroni 2021) alongside imported exotic foods. In a region such as Greece with 6,500 plant species, with *refugia* during the Pleistocene (Asouti et al. 2018), it is very likely that many plant food ingredients of the Mesolithic gatherer-hunters became part of the farmers' cuisines in the seventh millennium BC, despite the low presence of wild plant species in the archaeobotanical record. This limited presence may well be an artifact of limited excavations and sampling. One example is at Mavropigi-Fillotsairi, where the Cornelian cherry stones are quite regular in the otherwise poor archaeobotanical samples (Valamoti 2011a).

Differences in the selected ingredients, if they ever emerge on a sufficient level for generalizations, would have been influenced by what was locally available, depending on aspect, terrain, distance from the sea, altitude, and climate, with southern regions having Mediterranean vegetation and northern ones, with deciduous forests, under the influence of a continental climate. Yet, the anthropological literature teaches us that not all edible plant species are considered as food and thus consumed: culture shapes nature, naming and taming it through cooking and eating. This aspect of prehistoric cuisine remains elusive for the first centuries of the Neolithic (seventh to mid-sixth millennium BC).

It has been argued that the environmental gradient played a significant role in the spread of the domesticated crops of the Neolithic northward across the Balkans, influencing cereal crop choices (Ivanova et al. 2018). Yet, the cultivation of wheat species that are generally hardy during the Early Neolithic, and already adapted to the cold winters of continental northern Greece (e.g., Mavropigi-Fillotsairi), would not have been impeded from spreading north. Besides, these wheat species had already been under cultivation in the cold plateaus of central Anatolia from very early onward without any indications that climate acted as a filter against the different glume wheat species (e.g., Bogaard et al. 2017). Cultural choices related to ancestral crops and the origins of the first farmers settling new territories may have in fact been the decisive factors in this process (Valamoti 2004; 2017).

New genetic and dating evidence from the Aegean regarding the appearance of the first farmers in the region shows that, during the seventh millennium BC, different parts of Greece and western Turkey were being inhabited at a varying pace, with sites in the northeast being especially early in their establishment. In light of these observations, the archaeobotanical evidence from the Neolithic of Greece confirms the initial suggestion (Valamoti 2001a, 2004) that the preferences in wheat species were probably unrelated to environmental conditions but rather to origins of the first farmers. These farmers would have selected to cultivate the cereals that they were most familiar with, bringing grain from distant places where their ancestors first cultivated wheat. If emmer made it to the Neolithic United Kingdom (van der Veen 1992) and was a regular component of central and north European Neolithic communities such as the Linearbandkeramik settlements (e.g., Karg and Märkle 2002) and the Michelsberg culture (Kreuz et al. 2014), then there is no convincing argument for emmer being underrepresented at some of the northern Greek sites for environmental reasons. While einkorn, emmer, and *T. timopheevii* prevail in the north, in the south of Greece, the predominant wheat species, emmer and free-threshing wheat seem to resemble those of Italy and the Iberian Peninsula (Peña-Chocarro 2007; Rottoli and Pessina 2007). Einkorn, emmer, and free-threshing wheat, the first wheats of the Neolithic, are not similarly distributed along the paths through which agriculture appeared in Europe. In the western Mediterranean, the Iberian Peninsula, and the south of France and Italy, emmer and free-threshing wheat dominate the archaeobotanical assemblages of the Neolithic (Peña-Chocarro 2007; Rottoli and Pessina 2007; Antolín 2016).

This emerging pattern, if corroborated by further evidence in the future, might reflect the different itineraries that farmers or a farming way of living followed, as they spread westward and northward from the shores of western Asia. Different groups of settlers, from different parts of the broader region of the Fertile Crescent and Asia Minor, brought along with them their preferred cereals, establishing them in the new territory where they settled. Thus, different combinations of domesticated plants spread to the Mediterranean shores of Europe and the plains and river valleys of southeastern Europe. The new plant food ingredients as elements of novel culinary practices may have spread westward, together with people who moved, in a way similar to that proposed for the spread of spices and special plants for Latin America (Hastorf 1998): women moving away from their birthplace to establish families, bringing along seeds from their homelands to grow in their new garden plots, might be a plausible scenario, fitting for the spread of crops in a Neolithic European context. Isotopic studies on Neolithic skeletal remains indicate the movement of people in different parts of Neolithic Europe (Talheim and Vaihinigen in Germany; Bentley et al. 2003; Price et al. 2006) and this may well have been the case also for the Aegean on the basis of recent genetic evidence. Of course, it is unreasonable to think of the spread of agriculture as a unidirectional and static flow of ideas, people, and ingredients. The elusive gatherer-hunters of the Aegean knew well the land and its wild food-bearing niches, a knowledge that might have been negotiated with farmers, newcomers to whom the land was unexplored, or little-known territory. Recent research on the isotopic signatures of the diets of early farming communities of Greece shows no clear dividing line between the two ways of life in terms of food consumption (Sebald et al. 2022). This interaction between gatherer-hunters and early farmers remains in the realm of hypothesis for the Aegean and mainland Greece, and future research may hold interesting surprises regarding contacts between indigenous gatherer-hunters and farmers (see also Reingruber 2018).

Parts of the landscape such as upland zones with rich seasonal pasture, or areas close to lakes, might have been better known and controlled by gatherer-hunters. These communities, although still elusive, may have played a significant role in the emergence of Neolithic habitation around lake environments or in the development of pastoralism involving mobility with herding, away from settlements: they were the ones with good knowledge of the terrain, an advantage compared with the farming newcomers. In a context of cultural interactions between gatherer-hunters and early farming communities, expressing different identities through food could have been a powerful element for constructing social practices. Neolithic lakeshore communities recently unearthed in the region of western Macedonia (Chrysostomou et al. 2015), away from the coastal shores that had a long tradition of contacts across the Aegean Sea, might correspond to hybrid communities whereby elements of the two ways of living merged successfully.

From Plant Food Ingredients to Culinary Identities

Domesticated wheat was probably a special ingredient in the first centuries of farming in Europe, its grain traveling alongside paths that people followed involving a combination of maritime and overland routes. When farming communities appear in the Neolithic of Britain, dietary changes evidenced on human bones suggest variable diets and possible new food taboos, while cereals might have corresponded to special food for special occasions (Thomas 2003). The finds of wheat grain as burial offerings at Early Neolithic Mavropigi-Fillotsairi highlight their significance not only as staples closely linked to farming life but also as an important element of symbolic expression and ritual practice (Valamoti 2011a). Unlike past models explaining the emergence of agriculture as an inevitable choice imposed due to changes caused by environmental factors or population pressure, the process of transition is now better understood as a complex interplay of environmental and cultural factors with great variability in what was formerly described as the "Neolithic package." In this context, different cereal species might have played the role of identity signifiers, representing the cereals of ancestors or ancestral places from which people and/or crops set off to the lands of Europe.

In Greece, the archaeobotanical evidence available for the Neolithic, although not equally representative of the different parts of the country, shows variable crop choices possibly related to cultural factors. One way to explore culinary identities of the Neolithic of Greece might lie in the three glume wheats found at Neolithic sites: einkorn, emmer, and the new type glume wheat resembling *T. timopheevii*. For a long time these cereals had been lumped in one big basket of what has been described as the Neolithic "crop package" (cf. Colledge et al. 2004; Zohary 1996), only to gradually find out that it was several packages that may have been generated in the wider area of the Fertile Crescent (e.g., Asouti and Fuller 2012; Willcox 2013), which, in turn, would have had consequences for the range of crops that spread from the Near East westward to Europe (see, for example, variations in European Early Neolithic crops in the contributions in Colledge and Connoly 2007). Their geographic distribution in Greece as well as across the Mediterranean and southeastern Europe, along the paths through which a farming way of life gradually became established in Europe, may be revealing of regional preferences and cultural connections.

The richer and thus more reliable archaeobotanical record of the Late Neolithic (late sixth/fifth millennia BC) allows the observation of differences in the choice of wheats that were cultivated and presumably eaten. Emmer wheat and einkorn wheat are both present at nearly all settlements, yet, if one focuses on rich concentrations of grain or chaff, einkorn seems to predominate in the assemblages of several sites from the north of the country where retrieval has been systematic and therefore a reliable sample is available (Appendix A:Table 2.1). Einkorn, however, does not dominate in all northern Greek sites, and considerable variability is observed both within settlements and among different sites. At the end of fifth millennium BC Mandalo, emmer grain is dominant in grain-rich assemblages from house interiors. At a slightly earlier Makriyalos, the rare cereal grain concentrations identified in samples from Ditch Γ corresponds to a two-seeded wheat, emmer, or the new glume wheat type grain. This is in sharp contrast to the chaff-rich pits from the site, where einkorn predominates (Valamoti 2004:Figure 7.25). At Mavropigi-Fillotsairi, the cereal grains associated with an Early Neolithic burial correspond to einkorn and a two-seeded wheat, either emmer or *T. timopheevii* (Valamoti 2011a; Karamitrou et al. 2013). At Dikili Tash, both einkorn (Valamoti 2004) and emmer (Kokkidou 2018; Matterne 1993) dominate the assemblages.

In the south of Greece, the emerging picture is somehow different. Emmer and einkorn are present, but emmer appears to be more frequent than einkorn (Megaloudi 2006), while free-threshing wheat has been found as a rich concentration only in sites of the south (Megaloudi 2006; Valamoti

2009). Emmer and free-threshing wheat have been identified from the Late Neolithic site of Koufovouno in the Peloponnese (Vaiglova et al. 2014), while at Franchthi mainly emmer wheat is the most frequently encountered in the samples (Hansen 1991). At Early Neolithic Knossos, free-threshing wheat has been identified as a crop (Appendix A:Table 2.1). Until now, *T. timopheevii* has not been identified from sites in the south. These differences between the north and south in terms of prevailing wheat species are suggesting regional differences in plant food preferences most likely related to local identities and origins of the first farmers. Taking *T. timopheevii* into consideration in this discussion, the picture becomes more complex and nuanced than the north–south difference suggested here. This wheat appears in very low percentages at Makriyalos, in a few of the pits (Valamoti 2004). It is hardly present at sites such as Mandalo and Dikili Tash. Yet, recent evidence from the region of western Macedonia clearly shows that it is as important as einkorn in chaff-rich assemblages from the Middle Neolithic site of Apsalos (Valamoti 2006a) and the Late Neolithic site of Kleitos (Valamoti and Stylianakou 2015). *T. timopheevii* is also prominent at Vasilika Kyparissi in the Late Neolithic (Lathiras 2020). Regional variation in wheat species dominating Neolithic food ingredients of Greece could have acted as identity markers. Variation within the same community or gathering place for large-scale feasting regarding species' choices might alternatively have reflected status as well as complex networks of connecting people with other settlements. At Makriyalos, *T. timopheevii* appears only in a few pits that were rich in plant remains, as a refuse component. Only in one out of 31 samples does it hold a prominent position in the assemblage (Valamoti 2004:Figures 7.28, 7.29). Could this suggest differences in cereal species preferred between different families inhabiting the site or visiting to participate to a feast? Could it correspond to some gift brought along with a visitor, perhaps from the region of western Macedonia, where *T. timopheevii* prevailed, a contribution to a feast or a ritual the participants of which were pooled from a wider region? Alternatively, it could represent differences in preferred crops among the inhabitants of the same community.

In the wider area of western Asia and the Balkans, einkorn and emmer are the main wheats in Serbia (Filipović et al. 2018) and Bulgaria (Kreuz and Marinova 2017). *Triticum timopheevii* is considered a rare find in Bulgaria (Kreuz and Marinova 2017) and therefore not a crop in its own right, while recent evidence from Serbia (Filipović and Obradović 2019), Turkish Thrace, and western Asia (Ulaş and Fiorentino 2020) shows that in some locations this wheat was a major cereal cultivated by Neolithic farmers of western Asia and southeastern Europe. In light of the preceding discussions on first farmers, the reasons underlying preferences in different cereal species by these Neolithic farmers might be more telling about farming origins than thought. The connection between plants and ancestors is well documented in the ethnographic record, an interesting example offered by Sherratt (1995) in relation to the cultivation and use of coca varieties in Amazonia. Likewise, gourds and different varieties of chili pepper are associated with offspring and different families and lineages within communities of parts of Latin America (Hugh-Jones 1995 cited in Hastorf 1998). Similar perceptions can also be gleaned from ancient Greek texts in relation to the grapevine varieties (Valamoti et al. 2022a:146).

The choice of wheat species by Neolithic farmers of Greece was probably related to cultural preferences for specific cereals that generated the regional variability observed in the archaeobotanical record of the Late Neolithic (Valamoti 2001a; Valamoti 2003). Caution, however, is required when generalizing on the basis of a house or a settlement, and future investigations may reveal an even more complex picture. A first attempt to systematically compare choices in cereals and pulses from Neolithic northern Greece reveals the complexities involved in the ability to detect patterns related to crop choices rather than to different types of contexts and the predepositional and depositional processes that generated the archaeobotanical record depending on how the data is used and compared among different sites (Valamoti et al. 2022b).

In contrast, barley is present everywhere in the region during the Neolithic and in the Bronze Age. It is a reliable, hardy cereal and was probably known already before the appearance of agriculture as it is part of the indigenous flora and has been found wild at Mesolithic Franchthi. Although one cannot rely on myth, there is an association between the pre-Neolithic presence of barley and the exclusive use of barley in rituals for Demeter, the ancient Greek goddess of agriculture. Barley was considered as the first grain that appeared and was thus used in Demeter's celebrations for preparing the special ritual breads or was sprinkled over animals' heads during sacrifice (Valamoti et al. 2022a). Oral history, memories, and stories going back to the first encounter of the Mesolithic gatherer-hunters with agriculture might be at the core of later myths that narrate the spread of agriculture initiated by a female goddess (Demeter) and mediated by a mortal man turned hero, Triptolemos. Plant food ingredients used in rituals are often those that connect the community to its collective past and ancestral traditions. So, even when they have been replaced in daily meals by other ingredients, they continue to be used for rituals (e.g., Peña-Choccarro 1999 for examples from Asturias, Spain).

Similar microregional differences as those observed for cereals may be seen for other crops. The available record from prehistoric Greece, however, does not allow for such "high-resolution" insights at the moment. Ethnographic observations conducted on the island of Lemnos on the traditional cultivation of *Lathyrus ochrus* suggest that different plant species might have been consumed in almost neighboring villages: *Lathyrus ochrus* was eaten by the inhabitants of some villages, while *Lathyrus sativus* was eaten at others. I have seen no further explanation for this difference. Other differences in the selection of plants as food could be related to access to specific ingredients. Despite an overarching model of more or less egalitarian Neolithic farming based on reciprocity and diversification in plant food ingredients to secure subsistence (see Halstead 1981, 1994), subtle inequalities might have been present. For example, access to tended grapevines or even vineyards or access to cultivated fruit that would have added flavor, vitamins, and color. Modern observations on making fruit syrup from wild pears and grape (see Figures 5.12 and Figures 5.15 to 5.17) at the village of Kosmati reveal that wild pears were the only source for most people to make syrup, yet a wealthier inhabitant who owned a small vineyard could make a syrup using grapes. Acorn consumption offers yet another example of a plant food ingredient that can be perceived as an indication of famine and poverty during war (personal field notes, Kosmati, Grevena 2017) or as a sign of identity and noble origin: acorns were considered as the ancestral food of Arcadians, the inhabitants of mountainous central Peloponnese described by Herodotus as "acorn eaters" (Nielsen 1999:37).

Yet, culinary identities need not necessarily be expressed through differences in the choice of crops or choice of crops only. Even in areas where identical crops were grown, the ways to transform them into foodstuffs could have varied significantly, and through their culinary transformation into specific dishes they would have conveyed messages of belonging or being different at a collective or individual level. Even among groups using the same ingredients, each person involved in cooking could leave a personal mark in the way ingredients were turned into food, for example, in how fine the flour was ground, if it was roasted or not, with what it was mixed, what consistency it had, what fuel was used for cooking or baking a dish. Collective meals might have been the occasions where individual cooks could have had the chance to show their skills and through that define their position and reputation in the community, a process that might also have been ongoing on a household level through daily meals (for modern case studies exploring the role of gender, age, and culinary skill in daily and ritual meals in Mexico see, for example, Christie 2006, 2008; Hryciuk 2019). Identity could also be expressed on a more collective level, with communities or families being defined as "the tasty lentil producers" or as "einkorn eaters" or the "good cooks" of some special dish requiring special skills. Even when all members of a community share the same ingredient of choice or ritual prescription, identity could still have been expressed through the different textures and tastes achieved, the use of different seals on the prepared foodstufs, and other distinctive features such as decorative patterns and decorated containers (Figure 9.1, Figure 9.2, and Figure 9.3).

New Ingredients from Foreign Lands: Cultural Contacts, Identity, and Culinary Change during the Bronze Age

Crop species moving from western Asia via Greece to the Balkans and central Europe are eloquent indicators of networks between prehistoric communities of the region. Contacts between the Aegean and central Europe are

Figure 9.1. Seal from Thessaly, Middle Neolithic, Hara-Magoula Panagou. (Photograph © Ephorate of Antiquities of Larissa.)

Figure 9.2. Modern wooden bread seal from Mikri Mandineia, Peloponnese. (Soultana Maria Valamoti)

Figure 9.3. *Kollyva* (grain for the dead) in decorative patterns offered on All Souls Day, St. Eleftherios Church, Thessaloniki. (Photograph by Georgios-Vily Kapetanakis. Courtesy of PlantCult.)

evidenced at least since the Neolithic. *Spondylus* shells and shell objects traveled northward, starting their journey from their most likely places of production along the Aegean shores. By the time these marine shells and shell objects reached foreign lands of the north, they had perhaps acquired mythical dimensions, their owners demonstrating their involvement in networks with faraway places. In the world of Facebook and Instagram, it is perhaps difficult to imagine a world where what was manufactured at Late Neolithic Dimini or Dikili Tash, by the time it reached Denmark, had changed many hands, acquired special meanings, and its original story of manufacture had become a palimpsest of stories, the object possessing probably a totally different dynamic: the sea where the shells were harvested was nowhere near to be seen or felt, the skilled divers who culled them were unknown to the owners of the objects. These special artifacts probably traveled along the same pathways that allowed the spread of crops.

The Bronze Age in Greece witnesses increasing connectivity through exchange networks. From a culinary perspective, the Bronze Age demonstrates an impressive variability and complexity both in terms of ingredients and culinary facilities. A wider range of crop species is used, including plants with special properties due to their oil-rich seeds or their medicinal and/or hallucinogenic properties. Various lines of evidence suggest increased complexity and contacts during the third and second millennia BC, yet our archaeobotanical analyses of this complexity are still rather coarse-grained. Plant species with origins in distant places appear as new crops. Spelt wheat and opium poppy reach southeastern Europe from central Europe. *Lallemantia*, of a central Asian geographical origin, reaches the Aegean most likely from the east or northeast while millet, a later cereal of the second millennium BC, may have reached Greece along contact networks connecting southeastern Europe to central Asia and/or Anatolia (Valamoti 2007b; Valamoti 2016; Filipović et al. 2020). Other crops that appear in the Bronze Age for the first time are native to the Mediterranean. These include Celtic bean and coriander, the former cultivated as a crop in the Near East since the beginnings of farming, yet they fail to make it to the "package-basket" crossing from the east coasts of the Mediterranean across to southeastern Europe, for several millennia to come.

These new crops may well have acted as identity signifiers. Introducing an ingredient to a range of already established ingredients and recipes would not have been a simple and straightforward process. New ingredients may have been treated with caution or aversion, especially if they were dangerous at least to some, like Celtic bean, or unusual or unpleasant when cooked.

Fruit species such as the grapevine, the fig, and the olive appear as important elements of Late Bronze Age economies of the south. Thus, acquisition and access to their products, especially wine and olive oil, their use, and context of consumption might have entailed the existence of different levels of identity/identities. This might also have contributed toward the delineation of different groups consuming different foods and mind-altering substances such as wine. Andrew Sherratt (1991b) spoke of smoking and drinking cultures in prehistory, the former consuming opium, the latter, alcoholic drinks. Wine consumed over a wide region would have probably generated a sense of shared culinary tradition, yet the archaeobotanical record shows a much more complex picture. In prehistoric Greece, these two elements, which have been assumed to represent different worlds and identities in Sherratt's approach, coexist, as both opium poppy and wine were used during the Bronze Age.

At present we lack a systematic investigation of ingredients, equipment, and their distribution in space and time to be able to detect regions of specific plant food ingredients and practices. It would be very interesting to examine the distribution of braziers and brazier types across Greece, the east Mediterranean, and southeastern Europe to detect regional differences, idiosyncratic trends, and the available ingredients that might have been consumed. Does opium encounter wine on the Greek mainland, and if it does, where and when does this take place? Evidence for the actual mixing of the two comes from Bronze Age sites farther south, on the island of Cyprus (Collard 2012).

As discussed in Chapter 6, there is strong evidence that alcoholic drinks were prepared since the Neolithic in the north of Greece and during the Bronze Age in the south. In a region where grapes and grape juice are widely attested, wine appears to be the alcoholic drink par excellence, yet recent evidence indicates beer making in the Bronze Age. Did beer consumption reach Greece as the result of contacts with the Near East and/or Egypt? "Brewing beer in wine country" (Valamoti 2018), however, might have required something more than the mere contact with communities for which beer was the main alcoholic brew. It tastes different and has a different color and different seasonality regarding its production: beer can be easily produced on a regular, even daily basis; wine can only be produced once, during the grape harvest. Did people from these beer-consuming regions somehow reach this tip of southeastern Europe and settle there, bringing their drink and its recipe? The archaeobotanical finds of malt (Chapters 2 and 6) certainly raise this possibility, yet other lines of evidence are needed to lend further support to such a scenario. Indications for beer making around the end of the third millennium BC at two sites with a northerly distribution (Argissa and Archondiko) can be seen more as an exception to a long-lasting tradition of wine consumption. Although a comprehensive database of drinking cups and serving vessels is unavailable, the recent work of Jerry Rutter offers an interesting exploration of drinking sets and their variations in the Aegean (Rutter 2017). It is possible that different cups might have contained different drinks or corresponded to different contexts of consumption or consumers of different statuses. If we are not merely extrapolating to prehistory judging from the present, it is possible that changes in drinking apparatus might be suggestive of changes in the content that was drunk.

Two ceramic containers in particular (Figure 9.4) might be a hint for imported/introduced recipes for alcoholic drinks and drinking styles. So far, those known from Greece have been found in the Peloponnese, one at Lerna (Rutter 2008) and the other at P.O.T.A. Romanou (Rambach 2018). They resemble a similar vessel from Anatolia of a similar date (third millennium BC), and for this reason the vessel from Lerna has been associated with beer drinking (Rutter 2008). They also resemble vessels associated with drinking among Iron Age elites in the much later Hallstatt culture of central Europe (Dungel and Szombathy 1890:Table 2, Figure 9; Kromer 1958:Figure 77). The time span between the end of the third millennium BC in the Peloponnese and the first millennium BC in central Europe is many hundred years, and we lack any investigations between drinking sets and types of alcoholic drinks across the Near East, prehistoric Balkans, and central Europe that might allow a diachronic and spatial exploration of drinking customs. Was beer drunk in these special vessels that so strongly resemble each other despite the distance in space and time between them? Does their rarity signify special, ritual uses? If the vessel and a special drink, some form of beer, were indeed associated, is there a connection between the Anatolian find and those of Greece and Austria? If these vessels are indeed related to beer, is it possible that the idea and recipe reached central Europe gradually from the Near East across the Balkans? How long did it take and through what processes for the idea of beer making to travel north to central Europe? An alternative hypothesis might be that the idea of brewing a cereal-based alcoholic drink was an independent invention of those communities inhabiting the lands of central Europe, far from the east Mediterranean world where beer may have had already a long tradition in the region (see Heiss et al. 2020 for Neolithic beer in central Europe). Whether brewing and consuming beer in first millennium BC central Europe were introduced from the south along

Figure 9.4. Unique monumental three-spouted vessel ("sauceboat") for communal drinking and feasting from Early Bronze Age Greece P.O.T.A. Romanou. (From Rambach 2018, Figure 49a–b. Photograph courtesy of Jörg Rambach.)

contact networks already in existence since the Neolithic, or whether these were independent inventions using the same basic principle of fermentation of sprouted cereal grain, being practiced in different parts of the world, or both, remains to be seen.

The ways of consumption might also have varied, even without changing the basic ingredient. Wine, for example, might have been consumed pure or diluted in water as is the case in the ancient Greek world. This element indeed played the role of an identity signifier as those consuming it undiluted were considered to be barbarians (Nelson 2005; Valamoti et al. 2022a). Regulations regarding access to wine consumption may have also acted as a strong identity signifier: position during consumption and the use of different cups might have also highlighted distinctions among those who had access to wine.

Besides wine and other alcoholic drinks, oil plants could also have differentiated regions, some areas using the olive, others using alternative oil plants, such as opium poppy, mustard, and linseed, in regions marginal for olive cultivation (Jones and Valamoti 2005). The olive tree is part of the Greek wild flora, yet its mark on the archaeobotanical record of Greece occurs rather late in the Neolithic and is located in the southern parts of the country, mainly regions of southeastern Crete. Olive oil acquires great importance in the Mycenaean world, its production and trade being controlled by the palaces of the south, as can be seen in the Linear B tablets (Fappas 2008). It is a luxury product that only rarely seems to reach the shores of the north Aegean: the few finds of stirrup jars found in funerary contexts are rare indications of olive oil being traded to the north. The arrival of such a special product and access to it most likely signified a special status for those possessing it. Are we to perceive the inhabitants of northern Greece as "smelly barbarians" (Andreou et al. 2013) left outside the perfumed, sophisticated world of the south where olive oil and perfumed olive oil were produced and consumed in special rituals including rituals related to the dead? Animal fat is always an alternative, as well as a wide range of plant oils other than olive oil, in regions where olives do not grow (linseed, terebinth, *Lallemantia*). Yet, recent olive wood finds from the end of third millennium BC Skala Sotiros on the island of Thassos in the north Aegean (Maria Ntinou, personal communication; Theodosaki 2021) show that the northeastern Aegean may have been an area where olive was cultivated, earlier than assumed before (see Valamoti et al. 2018), making the olive pollen finds from Philippoi no longer stand out as dubious, isolated finds. Moreover, the recent olive wood finds from Skala Sotiros contribute further evidence toward discussions on island identities and maritime connectivity in the Aegean during the Bronze Age (e.g., Girella and Pavúk 2016 for a recent discussion). Thassos island, where the olive wood remains from Skala Sotiros have been found, forms part of the particular "small world" of the northeast Aegean where "incorporation, negotiation with and resistance to the outside world" can be detected, as recently argued by Girella and Pavúk (2016:17). This "outside world" comprised mainland Greece, northwest Anatolia, and the south Aegean world, and different levels of interaction recognized among the material cultural remains suggest complex processes being under way during the course of time (Girella and Pavúk 2016). "Migration, acculturation and hybridization" were probably under way in this process (Girella et al. 2016). Minoan Crete has certainly played an important role in this interaction as clearly evidenced by the impressive finds from the island of

Samothrace and the site of Mikro Vouni that indicate a "minoanization" of Samothrace (Matsas 1991, 1995; Matsas et al. 2011; Girella et al. 2016). As Minoan Crete seems to have played a crucial role in olive cultivation and oil production (e.g., Valamoti et al. 2018 for a review of the evidence) in the south Aegean, the olive charcoal finds from Skala Sotiros are indeed intriguing and invite more systematic archaeobotanical studies in the future as they can contribute important evidence toward the investigation of island connectivity during the Bronze Age in the Aegean. A new tree crop, the olive, might have emerged in areas of the northeastern Aegean partly at least due to this maritime connectivity during the Bronze Age. The wider absence of olive charcoal from coastal sites of the north Aegean Greek mainland (Valamoti et al. 2018) lends support to what Girella and Pavúk (2016:17) describe as a "small world" with its own processes in relation to the "outside" world. In this respect the appearance first of olive charcoal from the middle of the fourth millennium onward at Kumtepe and later in the third millennium BC of olive stones in Troy (Riehl and Marinova 2008) as well as the presence of olive at Early Bronze Age Poliochni on the island of Lemnos in the northeast Aegean (Arobba et al. 2010) invites a closer examination in the future of the role of this tree in the economies of the northeast Aegean and the possible cultural and/or environmental factors underlying olive tree management or even cultivation in this region.

It is along routes connecting communities since the onset of the Neolithic as well as new ones established during the Bronze Age that the Bronze Age additions to the crop repertoire reached Greece. Opium poppy seeds traveled south sometime in the Bronze Age or the end of the Neolithic and Celtic bean traveled north toward the end of the third millennium BC. If indeed opium poppy arrives in Greece toward the end of the Neolithic in the north, how, when, and in what context does it reach the south of Greece, Crete in particular, where it seems to hold a special place in rituals? Does it reach Egypt via communication networks of the east Mediterranean? In what context does it become a plant that is used in rituals for the living and the dead? Does it travel, crossing the small plains inhabited by Early Bronze Age farmers of mainland Greece? Does it reach the south of Greece along sea routes that connect northern Italy where opium poppy is also known from the Early Neolithic at the site of La Marmotta (Rotolli and Pessina 2007)? Despite the significance of such special plants that acquired great importance in healing and rituals, there are many gaps in our evidence that render the tracking of their journey to new territories where they thrived a real challenge.

Seeds of cultivated crops, however, do not travel alone, they travel with people who take them along because they want to cook a meal during their journey or because they wish to give them to their hosts or because they want to be able to cultivate them in their new homelands (see Hastorf 1998). There is still a lot of ground that needs to be covered before we are in position to explore these contact networks beyond hypothetical scenarios. The potential of plant food ingredients to reveal such contact networks has been neglected, yet, burned seeds correspond to foods and recipes that were exchanged and/or shared in ways that must have generated a sense of belonging to broader groups that shared similar ingredients and ways to cook them. When new ingredients appear, especially ingredients that were not in the "neighborhood" before reaching Greece, their introduction to new territory and their identity connotations become even more challenging to investigate. This is the case presented by *Lallemantia* and millet.

Crops along the "Silk Roads"? The Cases of *Lallemantia* and Millet

Lallemantia, a plant that appears in northern Greece in the third millennium BC, arrives at the same time as the "foreign metals" (*Fremdmetallen*), tin bronzes, that connect the north Aegean shores of mainland Greece with the northeast Aegean and networks through which tin bronzes reached northern Greece (Valamoti and Jones 2010). The lands where *Lallemantia* grew seem to overlap with Bronze Age tin sources (Valamoti and Jones 2010), yet there is inadequate evidence to track the locations where *Lallemantia* was grown before arriving in Europe, Greece first, and then slightly later Bulgaria and Serbia (Marinova and Valamoti 2014). The hypothesis that *Lallemantia* was introduced via networks responsible for bringing tin or tin bronzes to northern Greece remains to be further substantiated with solid evidence from the east and areas of the plant's natural distribution. Regarding the processes involved in the introduction of such a species, we can envisage traders giving some *Lallemantia* seeds as a medicine to some traveler from the Aegean, or, following Christine Hastorf's model for the chili pepper (Hastorf 1998), perhaps brides who traveled to western lands and foreign homes that became their own brought some seeds along to sow. Those Bronze Age communities that knew the properties of this plant and harvested it or grew it somewhere in central Asia and the Near East before it reached southeastern Europe remain elusive. Could people have moved westward bringing their own seeds along, settling among foreigners? The evidence for *Lallemantia* is as yet extremely flimsy from regions beyond southeastern Europe (Jones and Valamoti 2005; Valamoti and Jones 2010). Very few seeds of *Lallemantia* have been found at Adji Kui 1 in the Begash region of Turkmenistan (Spengler et al. 2018), and no substantial story can be built on a couple of *Lallemantia* seeds from central Asia before rich concentrations of the plant from different locations across central Asia and Anatolia or the north shores of the Black Sea come to light.

Millet is another "foreign" crop that reaches Greece from the east in the Bronze Age. The Greek finds are not directly dated except for the early finds from Skala Sotiros and Archondiko (Appendix B) which fall around 1500 BC. Those from Skala Sotiros correspond to dense, rich concentrations found in burned destruction layers dated to the early second millennium BC (Valamoti 2016). Early dates for directly dated millet seeds from other parts of Europe are not earlier than 1500 BC (Motuzeite-Matuzeviciute et al. 2013; Filipović et al. 2020), and the recent dates of the Greek finds place them among the earliest occurrences in Europe. The east–west exchange of cereal crops, millet from China and Old World cereals, wheat, and barley, seems to have been operating along the Hexi Corridor around the third millennium BC (Dong et al. 2018). Taking millet as a special case, it has been found in central Asia, in Begash (Spengler et al. 2018), but between these finds and southeastern Europe much space intervenes, with no evidence for millet. The oldest finds from Greece come from Skala Sotiros on Thassos island, in the north Aegean, a site where indications for elements of circum-Pontic influences are visible in the form of built-in *menhirs* (upright stones, in this case anthropomorphic stone stelae) that were used as building material in a late Early Bronze Age wall (Figure 9.5) (Koukouli-Chrysanthaki 2004). Therefore, it is legitimate to wonder whether Skala Sotiros was involved in contact networks with circum-Pontic cultures. Although not previously suggested, on the basis of these cultural features at Skala Sotiros and the presence of millet at the site, an alternative route, sailing the Black Sea, might have brought millet to southeastern Europe, rather than the overland path suggested in the literature. It is known from later traditions that the Black Sea was inhospitable, and it seems to have been beyond the reach of the Mycenaeans, the lords of the south in the second half of the second millennium BC (cf. Carpenter 1948; Tartaron 2013:136), at least directly, despite their apparent interest in these regions evidenced in toponyms in Linear B (Woudhuizen 2009). Thassos island, where Skala Sotiros is situated, was part of a northeast Aegean network that might have acted as an intermediary between south Aegean cultures and the Black Sea peoples. Weak artifactual evidence from the region of Macedonia suggests access to such networks as they point toward circum-Pontic regions, yet these contacts are still highly elusive (see Valamoti 2016). The role of the north Pontic regions in the spread of millet to Europe has recently been pointed out (Filipović et al. 2020), and in light of the recent dates from the Skala Sotiros millet, this appears to take place at the same time in the north Aegean. This might have a complex network that involved nomadic people inhabiting the regions

Figure 9.5. One of the menhirs used as building material from Early Bronze Age Skala Sotiros, which suggests possible contacts with circum-Pontic regions. (From Koukouli-Chrysanthaki 2004, Figure 30. Archive of the Ephorate of Antiquities of Kavala and Thassos. Photograph courtesy of Chaido Koukouli-Chrysanthaki.)

around the Black Sea coast, able to move fast on horseback, as well as maritime routes connecting the shores of the Black Sea to the northeast Aegean (see de Boer 2006–2007 for a critical discussion on contacts between the Aegean and the Black Sea during the Bronze Age).

As with *Lallemantia*, the journey of millet westward somehow evokes images of a prehistoric forerunner of the Silk Road, yet the paths along which it reached southeastern Europe still remain difficult to track down. These recent finds provide only a trigger to generate hypothetical scenarios about a central Asian corridor active as a network of contact routes that contributed to the spread of people, animals, and crops (see Valamoti 2016 for the arrival of millet and *Lallemantia* in southeastern Europe). As evidence from central and east Asia accumulates, it becomes clear that the anachronism of using the attractive reference to the concept of the "Silk Road" runs the risk of generating the false impression of a caravan route along which crops traveled from east to west Asia and Europe. It has, however, very convincingly been argued that cental Asian nomadic people of prehistory contributed to the creation of complex networks established along the movement paths of people and their animals across the steppes of central Asia during the third millennium BC (Frachetti et al. 2017). These nomadic networks, operating across the steppes, were responsible for the relatively quick transmission of crops and livestock both westward and eastward. Different corridors might have been operating as well as different timings in the adoption of new crops (Li 2021; Liu et al. 2019). It is probably along these routes and in the hands of nomadic pastoralists that crops such as *Lallemantia* and millet may have moved across the central Asian steppes westward to Europe and wheat to East Asia (e.g., Valamoti 2016; Dong et al. 2018).

The arrival of a new crop in a region probably requires more than mere contacts, and the scenario of the traveler with some foreign seeds in a linen bag might have worked only in some cases. Millet is a different cereal with a different growth cycle and culinary properties, lacking gluten and tasting different from wheat and barley. Therefore, its introduction must have been highly noticeable as something distinct; for this reason, it may have been little desirable as a culinary novelty by the local inhabitants of Greece. Indeed, it might have been a crop brought to northern mainland Greece sometime in the course of the third or second millennium BC by people originating from the circum-Pontic-Caspian steppes, with millet being part of their food crops, thus forming a clear culinary identity signifier. This possibility now becomes even stronger on the basis of new ancient DNA evidence that indicates gene input during the third millennium of Pontic-Caspian Steppe populations

(Clemente et al. 2021). Ancient Greek writers know millet as a crop, in many cases; however, it is associated with foreign people and its taste was not particularly appreciated (Valamoti et al. 2022a). In this context millet is rare in the south of Greece during the Bronze Age, something that is further supported by more recent evidence from central mainland Greece that suggests a limited, if any, consumption of millet among the Mycenaeans (see Valamoti 2016; Ntinou et al. 2022). Millet's presence, however, at the foothills of Mount Olympus, together with evidence for "fusion" culinary practices regarding cooking pot shapes (Dimoula et al. 2022b) underscores the complex interplay between cuisine and identity among communities consuming millet in the north of Greece and the Mycenaean world. Different levels of socioeconomic interaction might have brought millet into contact with Mycenaeans: slaves and/or brides from millet-eating areas, or, millet-eating locals with close interaction with the Mycenaeans guarding the northern limits of their world at Mount Olympus. At the other end of a "corridor" that indirectly might have connected southeastern Europe and east Asia during the third and second millennia BC, wheat spread eastward (Li 2021) but was not widely accepted in all parts of China (Liu et al. 2014): it was a foreign cereal, rich in gluten, with a taste and texture different from those of rice and millet, which the inhabitants of China had been used to since prehistoric times (Lu et al. 2019). It was only when grinding of wheat grain was adopted in the third century BC that this new ingredient could be successfully incorporated in Chinese cuisine (Lu et al. 2019). Yet, wheat found in ritual contexts, as offerings for the dead in Xinjang (Jiang et al. 2015), reveals the complexities involved in the adoption of a new crop, its profane or ritual use, and its significance for expressing identity and status.

Moving from a coarse-grained approach of plant food choices to more subtle choices within settlements or households or, choices for special occasions, the challenges posed to the archaeobotanical record to reveal such aspects of past lives in prehistoric Greece become obvious. This first synthesis attempted here raises a series of new questions. Does the existence of a Minoan and Mycenaean cuisine, established on the basis of cooking pots (Borgna and Levi 2017), correspond to different plant ingredients used for cooking? Is the later Mycenaean cuisine using the same ingredients continuing Minoan traditions, perhaps only "translated" differently regarding combinations of ingredients and different consistencies achieved through differences in cooking by boiling or simmering? Were griddles used for some form of pancake (Hruby 2017) that was widely consumed all over Greece judging from the griddle's widespread distribution, or were they a multifunctional gadget, used differently in the varied regional cuisines of different parts of Bronze Age Greece? Do new recipes arrive with new plant food ingredients, requiring their own, specific apparatus for their preparation? Did millet require the *pyraunos* for its conversion into a dish, and do these two elements of southeast European Bronze Age cuisine signify culinary variations between the north where millet is common and the south where it is hard to find in the archaeobotanical record? There seems to be a geographical limit of the *pyraunos*. Its southernmost occurrences are in the borderlands of the Mycenaen world and Late Bronze Age communities farther north (Dimoula et al. 2022b). It is possible that such a limit in a Bronze Age cooking facility coincides with millet, this new cereal ingredient.

Associating culinary variability across space and through time with changing identities requires adequate data for a high-resolution approach. Future syntheses of culinary identities of prehistoric Greece would benefit from an integrated approach whereby variability of plant food ingredients is considered together with processing equipment and context, spanning the Neolithic and the Bronze Age in different areas in Greece, southeastern Europe, and the east Mediterranean. Hopefully this exploration of the available archaeobotanical data from Greece will stimulate discovery of further connections in plant food ingredients, cooking techniques, and identity among prehistoric farmers of the region.

10

Eating the Past

Ancient Grains, Prehistoric Cuisine, and the Public

Plant foods have been used to sustain human communities from prehistory to the present, and tracking their history offers a valuable tool with which to understand the past. Starting backward from the present, food globalization has allowed us to taste fruit and vegetables during seasons they are not locally available or imported from the tropics, from landscapes and vegetation totally different from the local one. Spices and stimulants like tea, cocoa, and coffee are readily available in supermarkets, and hardly anybody realizes that their trade back in antiquity and more recent times triggered political struggles and merciless wars as well as the massacre of indigenous peoples (e.g., van der Veen and Morales 2015; Ghazanfar 2018). The discovery of new lands has enriched the ingredients available in Europe through the introduction and cultivation of plants like New World corn, tomatoes, peppers, potatoes, and beans and east Asian oranges, lemons, and rice. These foodstuffs have been so well incorporated into European and other Old World cuisines that many people consider them indigenous as they nowadays form part of daily meals. Yet, citrus fruits and rice (among other plant food ingredients) were introduced to Europe in the last centuries BC, initially via trade routes and/or by the "scientific team" of Alexander the Great, as exotica, and later, through the trading and horticultural activities in the contexts of the Roman Empire, introduced to orchards, European landscapes, and markets (Hughes 2003; Langgut 2017; Pagnoux et al. 2013; Pagnoux et al. 2016; Ferrero and Vidotto 2010). The national dish of Greece, bean soup made of *Phaseolus vulgaris,* is based on a New World introduction (Angioi et al. 2010; Crosby 2003). The only shared element with the prehistoric past is that it is a pulse seed. In this book, the prehistory of the plant foods of Greece reveals the core elements of cooking with plants in this part of the world, some resilient to the passage of time, others more susceptible to becoming marginalized.

The minuscule remains of plants preserved in archaeological deposits hold valuable information on ancient food. They are the scientific methodological tool for unlocking past plant-based culinary practices, as the previous chapters in this book have shown. In our inquiry into the origins of cereal cultivation or wine making, for example, it is the archaeobotanical data that hold the answers: changing proportions of tiny, fragile cereal chaff fragments reveal the long process of wild cereal cultivation that ultimately led to domestication, while the tedious measurement of grape pips may reveal morphologically cultivated grape varieties. These days we no longer need myths to explain how viticulture emerged in Greece or agriculture reached Europe, yet the archaeobotanical evidence collected to address these and other questions form the basis for explaining ancient crops to modern cooks and generates modern food mythologies. They are also the connecting thread that links small-scale food production and the marketing of foodstuffs that prevailed in prehistory but were later abandoned as needs changed through time (Figure 10.1).

An archaeobotany of prehistoric food, as presented in this book, aims to provide ways to understand past societies through plant food preparation and consumption. Parallel to the scientific work conducted in the context of various ancient food projects, through the outreach activities for which there is often a serious commitment, an arena for generating modern food consumption trends and identities emerges and poses various challenges to the archaeologists involved in this exercise. In the past twenty years, my encounters with the public through my archaeobotanical investigations of prehistoric food have generated a series of questions about the role of the archaeologist studying ancient food in the context of the modern world. This role touches on current trends of creating, safeguarding, and promoting intangible cultural heritage food traditions and becoming involved in lively debates about who owns traditional food and its past (e.g., Brulotte and Di Giovine 2014; Ducros 2015).

The theme explored in this chapter falls within a wider trend witnessed globally: food has become a very popular subject in the social media, television, film, and publications. Star chefs have emerged, generating a legendary status for their restaurants and personalities, some featuring in various media with their cookbooks (Matta 2019; Bergh 2016). Television cooking shows engage the public with gourmet terms and culinary approaches.

Figure 10.1. Einkorn galettes sold in Italy. (Courtesy of Andrea Brandolini.)

Food, with its power to evoke emotions and memories, has also featured in documentaries and films (Shugart 2008; Bower 2012).

In this wider "frenzy" about food, the archaeology of food has played its role, albeit unintentionally, in generating modern food fads and myths that acquire global dimensions in the "Brave New World" of social media, with food featuring in blogs or sites that offer recipes or promote specific dietary habits. Besides the medical or pharmacological literature that is based on clinical observations, modern cooking trends often claim their special benefits from being rooted in the prehistoric past, whether containing or excluding certain plant foods. Paleodiet, veganism, eating raw, eating according to blood type, gluten-free cereals and fake "gluten-free" cereals are promoted as beneficial in many ways, partly because of their roots in prehistory or antiquity (Hyman 2015; Hills and Erpenbeck 2018; Dixit et al. 2011; Leitzmann 2014; Bijlefeld and Zoumbaris 2015). In this general framework, an interesting trend has emerged toward the revival or reintroduction of cereals from prehistoric times such as einkorn in Provence, France, or in Brescia, Italy. Yet, many of these species were largely abandoned by past human societies for economic reasons. What have been the roles of archaeobotany and archaeobotanists in the promotion of some of these modern food trends, fads, and myths?

Archaeobotany, especially its culinary aspect, can operate within the established, institutionalized framework of public archaeology, through museum exhibitions and events, or conferences with special outreach activities or publications. It can also operate in various other ways and with a wide range of audiences/recipients.

At all levels and contexts, these outreach events and their nuanced features are inevitably marked by the interests, personality and communication skills, biases, and energy level of the researcher. They aim to explain the scientific results as well as the specific role of the archaeobotanist as an agent for communicating the findings, influenced to a large extent by any institutional position held. In this interaction between the archaeologist and the public, a dialectical relationship is established, potentially operating on many levels. My archaeobotanical cooking explorations and public outreach experience, which I share in this chapter, were largely defined by my first steps in archaeobotany.

My life experience of cooking for my loved ones and my students, thinking about prehistoric food, and trying to imagine what cooking plants in the Neolithic would have been like, is personal. My explorations of ancient food preparation, from sampling in the field, studying pressed grapes or sprouted barley grain, to preparing my own sourdough with grape juice for experimental breads (Figure 10.2) are closely interwoven with my everyday cooking. My experimental sourdough ended up by having a life of its own in our daily meals at home; after the first experimental breads, I started baking my own bread with the flour of barley, einkorn or spelt, creating a different loaf each time, equally tasty, heavy and really whole meal, not like the processed commercial type sold in my local bakeries or supermarkets. They were tasty to me but not to everybody: my mother refused to eat this "sort of bread," while a friend was delighted by its taste and smell. I could argue that the first level of outreach of my scientific adventures has been with my family. My daughter learned to love *trachanas* (fermented cracked grain with yogurt or milk, like frumenty) because of the experimental ones I made one summer during fieldwork in Crete and Mani (Valamoti and Anastasaki 2007). I later realized how good it was for her, when I researched its nutritional benefits in the literature during the preparation of another paper (Valamoti 2011b). I am sure that my grandma, who taught me how to cook bulgur *pilav* (a dish based on parboiled cracked wheat) among other things, would have looked down on my dark, heavy, ultra "rustic" breads that I prepare with great pride, puzzled by such an eccentric granddaughter trying to disinvent the fluffy, super-white, and light baguette-type bread that she loved so much. My culinary adventures, however, inspired by the archaeobotanical record, have involved wider audiences, during various events, and I have generated texts, talked to journalists, and given interviews to communicate the findings of my solitary research to a wider audience, beyond my family and friends.

The desire to communicate archaeobotanical finds beyond scientific publication started in 1988 when I did MSc and PhD research at the University of Sheffield as well as during my wanderings at excavations, by train or bus, carrying along my test sieves, plastic mesh, and volumetric buckets for the collection of archaeobotanical material. My specialization was somehow a reason for marginalization too. In the field of prehistoric Greek archaeology, I was an unusual, hybrid type. My archaeobotanical skills were seen as some kind of "secondary archaeology." I was the *sporou* ("seed girl") doing her "hobby," as the archaeobotanical work was mostly unpaid. This uncomfortable status, in combination with the endless hours of lonely endeavors with the stereomicroscope, sorting and identifying seeds in a laboratory, generated an urge to prove that my "seeds" could tell many stories, and interesting ones too, both for the archaeologists who study the past and for nonspecialists and nonarchaeologists, defining and justifying myself in my discipline and to a wider audience in my country (to paraphrase Bettina Arnold in Stoddart and Malone 2001:466).

Later, I elaborated further along this line of outreach as a faculty member of the Department of Archaeology at the Aristotle University of Thessaloniki. There I communicated the knowledge on plant foods that emerged from the datasets I was studying. From 2002 onward, archaeobotanical research was conducted systematically at the Aristotle University of Thessaloniki and taught at both graduate and postgraduate levels. This generated a group of actively involved researchers studying carbonized plant remains, including the remains of processed plant foods, thus placing the emphasis on prehistoric cooking practices. The results of this research have been published in scientific journals, and they were also disseminated to the wider public in ways that will be discussed herein.

Figure 10.2. Generating experimental breads (*top*) (photograph courtesy of Sandra Prévost-Dermarker) for PlantCult project, with grape juice sourdough (*bottom*), einkorn, and barley whole meal flour. (Soultana Maria Valamoti)

This interaction between the archaeobotanist and the public has been an ongoing process in my more than twenty years at the Aristotle University of Thessaloniki and the thirty years of my archaeobotanical research. Here, I critically reflect on this interaction, exploring the motives of the parties involved as well as the repercussions of reaching out to the public, communicating archaeological knowledge on ancient plant foods. Knowledge about ancient food preparation addressed the public, via the press, radio, and television, and meetings with farmers, small to medium enterprises active in the plant food sector, and NGOs focusing on safeguarding Greek landraces. Lectures as well as contributions to exhibitions in museums and other cultural events offered the opportunity to further communicate ancient plant foods to the wider public. This process can generate areas of ambiguity and lead to the creation of modern mythologies beyond the control of the archaeologist/archaeobotanist who started this.

Communicating Ancient Plant Foods to the Public

The archaeobotany of food has often featured in archaeological exhibitions focusing on various themes, ranging from the discards from daily life to special feasting events in a specific spatiotemporal context. Specific thematic topics presented in exhibitions include, for example, the very popular exhibition titled *Minoans and Mycenaeans, Flavours of Their Time* (Tzedakis and Martlew 1999), or as elements of past lives and specific lifestyles, for example, cooking in the Bronze Age at Stonehenge, United Kingdom or at Bad Buchau, Germany.

The media are attracted by archaeological science publications such as the one about the first noodles (Lu et al. 2005), the oldest bagel (Heiss et al. 2019), Bronze Age porridge (Valamoti et al. 2008), or Neolithic wine (Valamoti et al. 2007; McGovern et al. 2017), the "first bread" made by humans in the Near East (Arranz-Otaegui et al. 2018) or prehistoric beer in wine country (Valamoti 2018). "Booze" ("*Prost*" in the German issue) was the title on the front cover of the February 2017 issue of *National Geographic*, featuring an extensive article on alcoholic drink production and consumption since the Paleolithic (Curry 2017).

The actual place where the archaeobotanical information on plant foods departs from the academic sphere to reach out to the public may be a museum, a conference room, an exhibition hall, or a private place, such as, for example, a farmstead, a food shop, a field, a winery or the local distillery, a city square, a yard, an archaeological site, a village school, or even a private house. The transmission of the information may be instant (in a talk or during a visit to a museum) or repetitive (from printed matter distributed to visitors or by online material including articles and TV or radio shows). The relationship that is formed between the archaeobotanist, the medium, and the recipients can be very variable, with specific features and particularities and multiple levels of interpretation and perception. Individuals will inevitably receive the presented knowledge according to their own comprehension of the world and the past (e.g., Copeland 2004:134). The information can thus be understood in many different ways and can acquire its own dynamic, a life of its own. Prehistoric plant foods can be presented to the public through a simplified but robust version of the data by the scholar. For example, Aristotle University of Thessaloniki provides special publications for the dissemination of academic work to a wider audience. Such publications have also been embraced by other bodies, such as NGOs, which offer the public the scientific background of their activities. The texts might be entirely authored by the archaeologist or they can take the form of an interview with a journalist or a press release. From the original printed text to the live interviews on radio and TV or via participation in specialized productions such as documentaries, or participation in live events, the nature and the dynamic of the relationship established between the scientist and the public can be radically transformed, thus acquiring several facets that will be explored here. In all these activities, valuable assistance may be offered by a press office at the academic or research institution where the specialist is based. In countries other than Greece, this aspect of archaeology can be highly organized and managed (Stoddart and Malone 2001).

Apart from the output in various printed and digital media, direct contact with the public may operate on a small scale, with a limited outreach through the invitation/challenge of archaeology/archaeobotany students to communicate with their older relatives and to approach the old ways of plant food processing and potential uses of wild or marginalized plant species. Although this type of communication has an indirect, random nature and is initiated by teaching in an academic environment, so not strictly an element of public archaeology, this interaction then becomes public, because students will share the scientific information with their family who may be farmers or hotel owners or civil servants and teachers or even high-profile administrative staff of ministries, but with a country background. This is a two-way process, as scientific knowledge reaches individual families and their foodways, cooking traditions, and oral culinary histories.

In the case of recording the making of *xinochondros* (the Cretan *trachanas*) in the village of Kakodiki in western Crete, my student Stela Anastasaki learned how to make it after she had displayed this ingredient/food preparation during my class (Interdisciplinary Approaches in Archaeology) and after I had asked her to bring me the recipe. As she later confessed, this opportunity enriched the bond with her grandmother, who taught her how to prepare *xinochondros*. This type of interaction can result in scientific publications and conference presentations (Valamoti and Anastasaki 2007; Valamoti and Mimi 2016). This noninstitutional, small-scale public archaeology could have multiple dimensions, depending on the parties involved and their perceptions of what is edible and what is an unacceptable ingredient to consume. During my research in the village of Kosmati in Grevena looking into acorns as human food, some of the villagers when interviewed would deny their consumption of acorns, feeling embarrassed to admit that they ate food usually considered fit only for pigs, while others were reluctant to refer to it, or, alternatively, they would joke, when under the influence of the *tsipouro* (a kind of Greek grappa) that was being distilled at the same time, saying that with the Greek crisis people might again resort to acorns as food.

In this context of small-scale ethnographic fieldwork to record a plant-based recipe at risk of extinction, the interaction with the public requires the dissemination of the archaeobotanical information in a clear and simple way that will function as an explanation and justification for the researcher's interest in a specific regional way of cooking. This interaction can also operate between the archaeobotanist and a small-scale business owner, for example, specializing in acorn food products. This is the case in 2016 with Marcie Mayer, who produces various acorn products ranging from flour to pasta and cookies, on the island of Kea in the Cyclades. We had a fruitful exchange of information: Marcie shared her deep knowledge of processing acorns for food while I provided information on prehistoric acorn harvesting and consumption. An event whereby the archaeobotany of acorns could reach the interested public, residents, and tourists on the island was discussed but has not taken place. Yet, the seed for future collaborations and archaeobotanical outreach was sown.

In 2013, the interaction between archaeobotany and the public was explored for the first time in the form of a session in the context of the 16th Conference of the International Work Group for Paleoethnobotany (IWGP) in Thessaloniki (Valamoti and Bittmann 2015), titled "Archaeobotany and the Public." The participants presented various aspects of the use of archaeobotanical data for the formation of identities and commercial products and of foods that are promoted through their connection to archaeological remains. The participants included archaeobotanists (Stika et al. 2013; Valamoti 2013c), plant geneticists (Brandolini et al. 2013), food producers (Syndicat du Petit Epeautre de Haute-Provence), and NGOs (AEGILOPS).

Writing about Ancient Plant Foods for the Nonspecialist

Archaeobotanists rarely have the luxury of time to prepare texts for the general public. When they do, it is mostly in the context of special editions for events, exhibitions, or newspaper features. Examples from Greece come from special features published by the newspaper *I Kathimerini*. The feature showcased initiatives to organize conferences funded by the Piraeus Bank Group Cultural Foundation (former ETVA Cultural Foundation), the Fani Boutari Foundation, and the Biological Society of Cyprus. Archaeobotanical contributions focused on prehistoric grapevine growing and wine (Valamoti and Mangafa 1993) and on medicinal plants and aromatic herbs (Valamoti 1997a). For me, after a long hiatus in such dissemination activities because of the the challenge of combining a doctorate while caring for two babies, there was a sudden "boom" of outreach triggered by a spontaneous end-of-term event for an undergraduate course (Archaeological Data and Their Interpretation) that I taught in 2004–2005 at the Aristotle University of Thessaloniki. With my students, we explored archaeological approaches to foodways and cooking practices of past societies. To celebrate the end of the course, students prepared posters on various topics relating to the prehistoric food of northern Greece, presenting them in the archaeological community there, as part of our departmental Colloquia. The Colloquium concluded with a food-tasting event, where the served dishes were in part cooked with plant ingredients that were used in prehistoric Greece. In addition, the students were asked to bring what they considered to be a typical, plant-based delicacy from home, something they could not find in Thessaloniki where they studied, as well as recipes related to plant food ingredients of prehistoric Greece. This resulted in a colorful buffet that helped all participants visualize regional culinary diversity. It was presented in the newspaper *Ethnos*. This led to a much larger-scale event on prehistoric diet at the annual conference of *The Archaeological Work in Macedonia and Thrace* held in Thessaloniki in February 2005 (Figure 10.3). Sometimes, journalists may ask the archaeologist to prepare the main text on which they will base their article; this contributes to a more or less accurate transmission of the scientific information.

Διατροφή

Προϊστορικά... μαγειρέματα

Μπορεί οι κάτοικοι του Βορειοελλαδικού χώρου να προτιμούσαν το μονόκοκκο σιτάρι, όμως αυτό δεν καλλιεργείται πια στην Ελλάδα.

Μονόκοκκο σιτάρι, λαθούρι, κεχρί, καμελίνα και λαλλεμάντια. **Τι είναι όλα αυτά; Σπάνια δημητριακά, όσπρια και αρωματικά φυτά με θεραπευτικές ιδιότητες. Αν και δεν κατέχουν πλέον εξέχουσα θέση στο τραπέζι μας, αποτελούσαν τη βάση της διατροφικής πυραμίδας των προγόνων μας. Ας γνωρίσουμε τα μυστικά της προϊστορικής κουζίνας.**

Κείμενο: ΛΙΑΝΑ ΤΣΩΝΟΥ
Φωτογραφίες: Γιώργος Κωνσταντινίδης

Αναζητάτε πρωτότυπες συνταγές για να εμπλουτίσετε το διαιτολόγιό σας; Καιρός να δοκιμάσετε «προϊστορικά» εδέσματα ή ακόμη καλύτερα να πειραματιστείτε με τα αυθεντικά συστατικά που κατανάλωναν οι κάτοικοι της Βόρειας Ελλάδας έως και 7.000 χρόνια πριν.

Η έρευνα της τροφής φαίνεται πως έχει κάνει σημαντικά βήματα την τελευταία δεκαετία και μας δίνει, πλέον, τη δυνατότητα να μάθουμε τις διατροφικές συνήθειες στη Νεολιθική Εποχή και την Εποχή του Χαλκού. Ακόμη, μας φανερώνει τους τρόπους επεξεργασίας και κατανάλωσης των βασικών συστατικών εκείνης της περιόδου.

98 | ΑΡΜΟΝΙΑ

Figure 10.3. Armonia kai Zoi publication for wider public, "Archaeological Work in Macedonia and Thrace" conference presentation and event, Thessaloniki, 2005. (Soultana Maria Valamoti)

Specific subjects, such as the processing of cereals in the form of bulgur or *trachanas*, and foodstuffs intended for long-term storage (Valamoti et al. 2008b) or wine making (Valamoti et al. 2007), turned out to be very attractive for the mass media, on a national and international level. The realization that prehistoric people also used to drink wine or cooked the same grains that are known today, sometimes in ways similar to those of the present, seems to generate sincere and intense interest. In several cases, the scientific publication in peer reviewed international journals such as *Antiquity* or *Vegetation History and Archaeobotany* was accompanied by the presentation of the study results to the public. The interest of journalists in the work of the archaeobotanist made me think about titles for future scientific publication that would appeal to the public as well. Attractive titles attract further media attention. A good example is the use of the term "prehistoric fast foods" in the title of the scientific article to describe preprocessed cereals made by boiling them in water or milk and drying afterward (Valamoti 2011b). This term emerged through interaction with the media.

Efforts to present the right information are sometimes unsuccessful. In a very interesting article by Andrew Curry in *National Geographic* in February 2017, Curry offered a 9,000-year overview of alcohol consumption, titled "Our 9,000-Year Love Affair with Booze." This article covered the entire world and went back to Paleolithic times, in search of the roots of the preparation and consumption of alcohol by humans. The reader was offered exciting new evidence in a fascinating narrative, focusing primarily on cereal beer and grape wine from Paleolithic gatherer-hunters to the modern world. In this case, the interaction with the journalist, which involved an exchange of emails and a long telephone discussion, did not lead to the presentation of the earliest wine evidence in Europe, although the research on *Prehistoric Viticulture and Wine Making in Northern Greece* had been financially supported by the National Geographic Society. The recent publication of beer finds from Greece, however, made its way to various kinds of media, including national broadcasters, web pages, science for the public journals, as well as international media.

Outreach Activities: Presenting Ancient Food to the Public

A pot with some carbonized seeds in it or a grinding stone with grain often features in museum exhibitions of past human lives as related to food and the environment. This is a more controlled and mainstream case of public archaeology. An alternative approach to plant-people interactions in the past was offered by the project PaCe. It was organized by European institutions, under the direction of Anna Maria Mercuri (Mercuri 2009). This exhibition was displayed in several museums in Europe, including the Archaeological Museum of Thessaloniki in Greece, and was also made available on the web. In this way, a virtual visit is possible at any time. The archaeobotanical contribution from Greece in this project and exhibition focused on the grapevine and wine, which appealed to the public, especially as the oldest evidence so far for wine making in Europe comes from northern Greece and the Neolithic site of Dikili Tash (Valamoti et al. 2007).

In autumn 2010, the 12th Venice Biennale of Architecture hosted the exhibition *The Ark: Old Seeds for New Cultures*. The curator of the exhibition was Kesuo Sijima, and the Greek section was headed by Zissis Kotionis and Phoebe Giannisi. The Greek group focused on the dual dimensions of architecture, built and open-air spaces, and the landscape. In this context seeds were perceived as constituting, per the organizers' press release, the "organic departure point of the landscape, the morphogenetic software of its architecture, the core of a process of continuous transformation." The approach attempted to "bring the sites of culture back to contemporary architectural practices approaching the word 'culture' in both senses, that of civilization and that of agriculture." Archaeobotanical results from archaeological excavations were incorporated in the bilingual volume that accompanied the exhibition (Valamoti 2010b). Archaeobotany, unexpectedly, reached an audience that had visited an exhibition on architecture.

Cooking with plants for the public offers sensory experiences that form an alternative way to communicate prehistoric cuisine beyond scientific publications and museum exhibitions. The departmental Colloquium that took place at the end of the winter term in January 2005 to mark the completion of the course on interdisciplinary approaches to the archaeology of food was a big success. Many colleagues and archaeologists attended and had the chance to see the work of the students and taste the foods they brought as representative of their home villages. It was probably the first time in which food was brought to the archaeology scene of Thessaloniki. It also brought people together through food tasting in the present, inviting the consumers of today to reflect on consumption of the same ingredients in the prehistoric past. The success of this event and its positive overall atmosphere led to the organization of a larger event with a much greater outreach. With the collaboration of Polyxeni Veleni, then curator of archaeology at the Archaeological Museum of Thessaloniki, the student posters were presented at the annual Conference of the Archaeological Works in Macedonia and Thrace

(AEMTH; To Arxaiologiko Ergo sti Makedonia kai ti Thraki) together with my scientific talk, followed by another food tasting event. For this, Polyxeni Veleni and I prepared whole-grain breads, lentils, and *Lathyrus fava*, and offered them to the conference participants along with wine by the winemaker and former mayor of Thessaloniki, Giannis Boutaris. The event attracted the attention of the media, this time without any initiative from me and on an unprecedented scale.

This rather spontaneous event was the first step to several others involving the citizens of Thessaloniki. Archaeobotany, especially its culinary aspects, started to feature in various outreach events organized by the university. First, it was presented in the event "Mithoi sto Fos" (Myths to the Light) in 2005, at the Tellogleion Foundation on campus. This was an initiative of the Department of Archaeology to raise public awareness of its archaeological work. A wider outreach event was organized by the university and funded by the General Secretariat for Research and Technology with the title "Hermes: Research and Science Week." Work conducted at the university was presented in the context of the Thessaloniki International Fair, and archaeobotany was selected as one of the topics (Valamoti 2006b).

Through another initiative, research carried out at Aristotle University of Thessaloniki, the largest in Greece, was presented for the first time to the public, and the university was opened to the public on selected Sundays of the year. In this series of "open house" events called "AUTh on a Sunday," archaeobotany was one of the research fields that were presented to the public. A presentation, "From the Stereomicroscope to Prehistoric Cuisine," was made about plant foods used in Greek prehistory and ways of food preparation known from archaeobotanical evidence. In the gardens on campus, the participants experienced the processing of glume wheats by pounding, grinding, and sieving and cooking of bulgur using a replica of a Bronze Age hearth and *fava* (pulse seed stew) made of *Lathyrus sativus*. Participants were impressed that these foods that we consume today were most probably enjoyed thousands of years ago.

Participants, especially young children, learned about some similarities and differences between prehistoric plants and modern foods and saw the laborious and time-consuming steps involved in the preparation of plant food ingredients, glume wheats in particular, in prehistoric times. Milling grain into flour was hard work. In the last few years, archaeologists working in northern Greece and trained in archaeobotany have further developed this interaction with the public (e.g., Jenny Gatzogia, Dimitra Kotzachristou, and Charoula Stilianakou).

In Crete, Jerolyn Morrison has been communicating Bronze Age cuisine to the wider public by cooking the Minoan way through the "Minoan Tastes" company that she founded. An American archaeologist specializing in pottery, she used her archaeological and cooking skills developed during her doctoral studies on Crete to share her scientific knowledge on ancient cooking in a tangible way. She uses Minoan-style cooking pots, an assortment of herbs and onions, vegetables, legumes, and meat and fish, which were all available in Minoan times (Figure 10.4). Morrison's fascinating and highly original initiative to connect the archaeology of food with tourism and the public provided another opportunity for modern people to better understand how the ancients cooked.

This Greek experience is of course not unique, and many colleagues have had similar outreach experiences, communicating about ancient foodstuffs in different parts of Europe and elsewhere in the world. Over the last few years, for example, ancient brewing has been at the heart of events organized for the public involving producing beer, tasting it, and educating

Figure 10.4. Jerolyn Morrison presents Minoan cooking to His Royal Highness the Prince of Wales at an event held on Crete, organized by the British School at Athens and Branding Heritage at the Knossos Research Centre. (Photograph by Vassilis Goumas. Courtesy of the British School at Athens.)

people about its past. An example is archaeologist Bettina Arnold's work in the United States where the craft brewing industry has sparked an increased interest in the archaeological and historical evidence for the brewing of fermented drinks, especially beer. Based at the University of Wisconsin Milwaukee (UWM), "Brew City" in the late nineteenth and early twentieth centuries was the world's number one producer of beer. Today, Milwaukee (Brew City) has a booming craft beer scene. In response, UWM has developed an undergraduate certificate in the culture and science of fermentation and has organized educational events in collaboration with the Wisconsin Science Festival, Discovery World Museum, and Brew City MKE, a museum run by the Milwaukee Historical Society and dedicated to the history of brewing in Milwaukee. Participating units and departments at UWM, including Honors College, have created a Hortus Academicus (a university botanic garden) with a focus on plants, shrubs, and trees used in flavoring and producing alcoholic drinks. Arnold and Department of Anthropology doctoral student Josh Driscoll, whose thesis focuses on strategic feasting and the shelf life of Iron Age European beer (Arnold and Driscoll 2017), have used these plants in a recurring event that challenges local home brewers to devise unhopped drinks based on archaeological and historical evidence (Figure 10.5). They have partnered with brewers both large (Lakefront) and small (Gathering Place) to produce interpretations of ancient beer types that can be used as a way to educate the general public about archaeology and the cultural contexts in which these remains were found. Arnold also presented a program over two days at the J. Paul Getty Museum in Los Angeles in July 2017 featuring the "barbarian's beverage," or beer, at an event usually dedicated to classical archaeological and historical evidence for wine, known as "Bacchus Uncorked." Craft brewer Santa Monica Brew Works replicated and served a braggot (beer/honey/meadowsweet/mint drink) at the event that was based on the botanical remains from a sheet bronze cauldron that Arnold and University of Mississippi colleague Matthew L. Murray found in a burial dating to about 450 BC during their excavations in southwest Germany (Arnold and Murray 2016).

This involvement of the public in outreach events is not unique to archaeobotany or the archaeology of food but signifies a general trend in contemporary archaeology to connect with the public. This outreach of archaeobotany to the public is also partly related to funding and justifying the significance of archaeological or archaeobotanical research for modern societies.

Figure 10.5. Bettina Arnold holding glasses of the braggot that Santa Monica Brew Works brewed for an event held at the Getty Museum, Los Angeles, 2017. (Photograph by Lisa Guzzetta. Courtesy of Bettina Arnold.)

Archaeobotany through Images and the Digital Universe

Newscasts and documentaries have played important roles in communicating the culinary aspects of archaeobotanical research, as well as the relationships between people, environment, and food through time, to the public. In some cases, newscasts or programs present archaeobotanical work or a new archaeobotanical find, as happened recently with a scientific publication on beer brewing in prehistoric Greece (Valamoti 2018; Municipal Broadcasting Channel of Thessaloniki TV 100). For special documentaries, video footage is the main medium, occasionally accompanied by the narration of a script composed by the documentary producer(s). In other cases, archaeobotany and prehistoric food have made only a minor contribution to a documentary dealing with another topic, where they are used to highlight the dimension of plant foods through time, their origins in a distant place and in prehistory. An example is a special feature for the nongovernmental organization Peliti, which is active in the preservation of landraces and local breeds by modern farmers. The series *Alithina Senaria* (*Real Scenarios*) broadcast on the television channel ERT3 in April 2010 presented the results of archaeobotanical research connected to traditional plant food species (Figure 10.6). More

Figure 10.6. Soultana Maria Valamoti preparing for an interview for TV show *Alithina Senaria* during a Peliti Seed Exchange Festival, Paranesti Dramas, April 2009. (Photograph courtesy of Georgios-Vily Kapetanakis.)

specialized was the documentary by Katerina Patroni, again for PELITI, in May 2011, for the series *Kivotos* (*Ark*) broadcast on National Greek Television (NET).

The 10-minute documentary *Vivre avec la Terre* (*To Live with the Soil*) by the Canadian cinematographer Julie Perron was made during fieldwork collecting archaeobotanical material at the excavation of a Neolithic house in Dikili Tash. Perron remained for two weeks in the field, with her crew, participating as observers of the whole excavation work, interviewing me and the other contributors, and the lead archaeologists were remarkably positive and hospitable with her. Archaeology was not her basic interest; she worked on a project with Patrice Fortier, a romantic Canadian gardener and his efforts to safeguard endangered plant food species. The interaction with Perron was particularly interesting; she wanted to connect the prehistoric seeds with those of Fortier. The connections were apparently difficult to establish. The film sought to present an idealized image, however, of the Neolithic farmer as a gardener who works the earth peacefully in a society where reciprocity prevailed. I did not fully share this view.

An excellent example of public archaeobotany combined with traditional knowledge on the properties of plants was the BBC series titled *Wild Food* with Ray Mears and Gordon Hillman, presenting the potential wild plant foods of Britain prefarming to the wider public in a fascinating way. In February 2019, a three-episode documentary, a production of CNN Türk, directed by Sevinç Baloğlu, titled *Tohum* (*Seeds*), explored the spread of agriculture from the Near East to Europe, using archaeobotanical evidence to approach this journey of crop seeds across vast regions and through complex processes.

Archaeologists and archaeobotanists are often cautious in interpreting their data, so a certain degree of doubt and alternative interpretations is offered in academic publications. Although we stress the ambiguities of our evidence in such work and the different possible interpretations, this "hypothetical" element is usually filtered out by the intervention of a journalist or a documentary production company. When making a documentary, departing from the script and matching the text with video footage is a very special sphere of interaction both during the making of a film/documentary and afterward, during the presentation to the public. My recent collaboration with ANEMON Productions for the documentary *The Journey of Food* showed me how this interaction could be positive when there is mutual respect. Simultaneously, the sensitive balance of such a cooperation was revealed: my point of view as a specialist proved to be too hypothetical for the requirements of the producers. Thus, for the Neolithic wine of Dikili Tash, it was Jane Renfrew's footage that was included: she clearly stated that wine was produced from wild grapes at Dikili Tash. This was a much better option for communicating a clear message about the past, compared with my more hypothetical discourse offering alternative possible interpretations of an early management or cultivation of the vines, despite the morphologically wild type seeds that have been found at the site.

In the world of the web, new ways have been made available to make scientific results accessible for the public. This can sometimes act as a parallel universe; for example, information may be transmitted via social media, in which the researcher generating the results might not even have an account. Once scientific information is disseminated through the web, its retransmission in various places can be manipulated in many ways and can be used for various claims about the past. For example, the wine of fifth millennium BC Dikili Tash became the wine of people who lived centuries later, specifically the "Macedonian" people, and the archaeobotanical remains were arbitrarily claimed as a past owned by modern ethnic states. The food of the ancestors, in a vague sense of people who lived in a distant past, becomes the food of "our ancestors"; in this way, it can acquire a special meaning in expressing contemporary national identities.

Prehistoric Plant Foods, Current Socioeconomic Trends, and the Food Market

In this exercise of interacting with the public, one could ask whether the information provided by archaeobotany on past food preparation methods is used to make the present look inevitable by making the past look like a precedent for modern conditions (after Preucel and Mrozowski 2010:12). In the case of plant foods, perhaps the answer is no: prehistoric plant foods become part of modern movements toward either healthier or presumably healthier food habits. Rather than using the past to justify the present situation, archaeobotanical results are used by modern movements to introduce alternative lifestyles, diverging from a fast-food, starch-/sugar-rich diet of intensively produced food to choose alternative food habits involving ancient grains, organic cultivation, raw plant foods, vegan diets, and more. Thus, archaeobotany may be contributing, without this being its intention, toward creating images of the past that are being used to justify modern-day practices (see Preucel and Mrozowski 2010:4 for archaeology in general). Archaeobotanical research into ancient foods seems to have repercussions for present lifestyle trends that develop outside archaeology but that try to use archaeological finds for their legitimatization.

The foods that we eat today represent a palimpsest of foods from our distant pasts, transformed in the course of time through contacts, choices, and social changes. The food choices of prehistoric/ancient people seem to transcend modern boundaries, generating different maps of culinary identities from those that use modern nation-state borderlines. Archaeobotany can make a significant contribution to this realization. Many examples from southeastern Europe highlight the shared dishes that underline a common culinary identity. At the same time, their many variants can emphasize local identities even within the same ethnic entity (see Valamoti 2017). However, archaeobotanical finds have been (mis)used in the public sphere to claim modern national identities and continuity.

Along with the attention paid by the mass media, contemporary alternative communities and NGOs are interested in archaeobotany as a means for the documentation of continuity or discontinuity regarding the cultivation and promotion of certain plants in Greece. This connection with a distant past is sometimes, but not always, a necessary element of campaigns to safeguard local landraces of indigenous and foreign crops. This is the case of Peliti, a community that organizes every year the "Feast of Seeds and Seed Exchange" (Figure 10.7).

Figure 10.7. Seed Exchange Festival, Paranesti Dramas, April 2009. (Photograph courtesy of Georgios-Vily Kapetanakis.)

My participation in events organized by Peliti was particularly interesting, as I realized how the ancient plant remains that I studied could be perceived in many other ways by people who participated in this annual feast, who have their own images and perceptions of what an "ancient seed" is. Many of the participants, when questioned, revealed that they often have an idealized view of a prehistoric or ancient past when life is perceived as having been healthier and better than nowadays ("Hopeful and idealistic creations reflecting the contemporary search for a social utopia," Meskell 1995:74). As discussed in Chapter 2, indeed some of these ancient cereals are nutritionally advantageous, but how far back in time would a modern cook wish to go to prepare a meal? The ancient process, from preparing the field to weeding, harvesting, processing, to cooking, would be too laborious for any modern consumer to desire to return to.

The group AEGILOPS is interested in the preservation of biodiversity in cereals, especially in Greece, at an experimental/research level. They invited me to talk about prehistoric glume wheats based on archaeobotanical research. Thus, I had the chance to meet and establish contacts with farmers who cultivated glume wheats and learned from them that einkorn grows really well and that their main problem was in de-husking this grain. I tried to find a source of grain for the experiments needed in PlantCult. Learning about the experiences of these farmers in the fields and in marketing their products, such as einkorn wheat, as well as providing them with archaeobotanical knowledge on the antiquity of the glume wheats, was a process that contributed toward establishing a connection between archaeobotany and small local businesses active in the agro-food sector.

The past as a culinary present is interesting for the public, and this seems to be the case not only in Greece but also in other parts of Europe and elsewhere. In this framework, the hulled cereals are particularly important, and farmers and producers in Greece cultivate and trade them. Suddenly, emmer wheat gained "mythical" status, and under the brand name *zea* (one of its possible ancient Greek names) (see Valamoti et al. 2022a), it has become commercially available in various forms. It seems that there are many people who want to buy *zea*, which was introduced to the consumers as a "magical" wheat that nurtured ancient Greeks. The market somehow neglected the "humble" barley, which was the basis of daily meals and rituals in ancient Athens. On a more substantiated basis, *fava* from Santorini showed the cultural dimension of a crop with its origins in the Bronze Age, establishing the brand "Protected Designation of Origin" (PDO), with the approval of the European Union (Figure 10.8). The arguments for the designation of the

Figure 10.8. Varieties of packaged *fava* Santorinis and *Afkos* from Lemnos (bottom right package labeled Άφκος). The certification of the product as Protected Designation of Origin (PDO) for *fava* Santorinis relied partly on archaeobotanical evidence. (Soultana Maria Valamoti)

antiquity of the product on Santorini were based on archaeobotanical data (Sarpaki 1987a).

Archaeobotany can be further linked to the business world, as when archaeobotanical research for the design of a new product is funded by a private company. The expert secures research funding, the information is provided to the company, and the management of this information relies on mutual cooperation. I had a good experience working with Olyra-FlourMills Thrakis, but I can see some cases where information may be used incorrectly; the balance is sensitive. The official presentation/announcement of the cooperation leads to the need to consider whether the name of the university or institution can be used in launching a new product. Who is entitled to use the archaeological information and how, especially when it becomes part of the market economy? These steps in the connection of academic research with the business world create the need for a new framework setting a price that corresponds to the value of the generated archaeological information, if this is to be used for the promotion of a product in a market economy.

Whether this knowledge has a price and should be treated as a product is, of course, a matter open to debate.

Making archaeology relevant for the wider public and the connection with the taxpayer have been discussed for several decades now within archaeology (see Stoddart and Malone 2001). The connection of research funding with the dissemination of results to the public and connecting with small and medium enterprises active in the food sector is often a priority for European Union projects. For example, the project PLANTCULT: Investigating the food cultures of ancient Europe: an interdisciplinary investigation of plant ingredients, culinary transformations, and evolution through time (funded by European Research Council [ERC], Consolidator Grant Horizon, 2020 Research and Innovation Program, GA 682529) aims to make its results accessible not only to the scientific community but also to the public (Figure 10.9). After all it is taxpayers' money that essentially funds such projects. Thus, what for years has been more or less my own spontaneous interaction with the public, occasionally monitored by the university press office, in the context of my ERC project, has become something I need to

Figure 10.9. Presenting project PlantCult to the public at Aristotelous Square in Thessaloniki, October 2018. (Photograph © Yiannis Tsouflidis.)

always remember to do. Acknowledging the funding body requires better organization and awareness of my role in mediating our results to the public as well as networking with producers and bodies responsible for public events.

When the results of archaeobotanical research concern ancient food, the connection with the public seems to possess a special charm, and it becomes very powerful as the scientific knowledge that is presented to people can evoke their memories and senses, of special events and contexts of consumption. In this process people can associate themselves with the first wine or the first bread ever made on a subconscious level: it is food transcending time, generating perhaps a sense of continuity, of belonging, of being deeply connected with some very distant ancestors. Numerous archaeobotanical finds from Greek prehistoric sites correspond to food, such as the remains of bulgur (cracked wheat), or pressed grapes as indications of wine making, or split cotyledons of bitter vetch indicating some form of Bronze Age stew *fava*. They correspond to direct indications for special food preparations. Moreover, these preparations continue to have elements of daily and special meals even nowadays, thus making a direct connection between modern and ancient dishes or drinks. Ingredients can be equally fascinating, lentils, for example, being omnipresent from the Neolithic houses through to modern supermarket shelves and pressure cookers.

Does archaeobotany create past foodscapes that have never existed? The mere materiality of food remains that we are dealing with creates an inevitable link between the present and the past, in my opinion. The specific food remains at an archaeological site do indeed reflect snapshots of a culinary past. The sounds, the gestures that led to their preparation, the exact tastes and textures, and the senses their consumption generated in the specific contexts will inevitably be the product of our intervention as screenwriters of a past that may indeed exist more in our own perceptions rather than the past itself.

After 25 years of work in archaeobotany and a long journey from the mud of the water sieve to the recognition and responsibility of the research program PlantCult, I think that the contact with the public is one of the most interesting experiences, making me cautious about drawing hasty conclusions and suspicious about the distortion of data by the different points of view of journalists, food bloggers, or members of the public. In communicating our archaeobotanical results, we probably do not realize the effect our work has on the public or which aspects of our work interest the public (see Holtorf 2007:10). That soon changes once the archaeobotanist engages with the public, as I have found.

The information drawn from the laborious work of archaeobotany, I believe, concerns the present, as it reveals continuities and discontinuities in food choices, and it contributes to some critical reflection on food stereotypes and their manipulation, stereotypes that stem from arbitrary correlations, such as of food and place, as in, for example, wine, oil, and Greece or a glorious past and emmer wheat. The relationship between the archaeobotany of food and the public over all these years, in Greece and elsewhere, reveals a vivid interest of the people and has been shaped on many levels, from the experimental cooking and food tasting with students to the promotion of traditional products. The balance is delicate, but the experience is both fascinating and challenging. Our data gain a magical dimension, as they come from seeds kept in the ground for thousands of years, and, whether we like it or not, whether we pursue it or not, they are components of contemporary mythologies for products destined for body and mind. When these data leave the laboratory and our scientific publications, they acquire a life and autonomy of their own and contribute toward creating modern culinary mythologies.

11

Cooking with Plant Ingredients of Prehistoric Greece

Recipes

The culinary landscape of prehistoric Greece would have been significantly different from that of today, with our unprecedented variety of plant foods from all over the world available year-round. In the "globalized" Hellenistic world established by Alexander the Great and his successors and the even larger territory of the Roman Empire where foodstuffs were traded over long distances as far east as the shores of Malabar in India (Liu 1988), plants available, at least to some, would have offered a greater variety than that available in earlier historic periods, for example, the Classical and Archaic; yet, plant food products demonstrated a level of variability in colors, tastes, aromas, textures, and cooking behavior (e.g., their consistency when boiled) (Valamoti et al. 2022a).

This book presents an overview of the plant ingredients available to the prehistoric farmers of Greece, as evidenced in the archaeobotanical record. Here, a selection of recipes is offered as a guide to prepare plant-based dishes that might have been cooked in the Neolithic and in the Bronze Age. A wide range of cereals, pulses, fruit, and other plant ingredients have persisted, vanished, and reemerged from the culinary inventories of the different peoples who inhabited Greece 9,000 years ago. For example, some plants, such as *Lallemantia*, from which oil is extracted, are totally absent today from the Greek kitchen; others, such as wheat, are still present even if just an ingredient in ready-made frozen pizza dough. Some complex foodstuffs have persisted, including bulgur and *trachanas*, dried figs, wine, and split pulses.

The quantities provided in the recipes are approximate. When I cook, I rarely measure, unless I am making cakes. The dishes that I prepare are never identical, but this lack of consistency does not prevent them from being tasty. I often think of prehistoric people cooking, especially during recent baking adventures using my homemade sourdough. My bread is never the same from one batch to the other, and I know the sourdough is slightly different from the previous one: it is alive and evolves in a process of interaction between the ingredients, the cook's hands, temperatures, and natural yeasts.

The recipes here are meant to be executed in modern kitchens. This is certainly different from cooking in prehistoric times as was realized through experimentation with prehistoric replica ovens and hearths. For example, in a Neolithic or Bronze Age context one could not remove the food from the pot immediately after removal from fire. Experimental cooking of *fava* resulted in the leg of the tripod pot breaking off, probably due to the thermal shock of the vessel once its content was immediately removed from the pot. However, it would have probably worked fine had the contents been emptied gradually, suggesting that serving was probably done directly and in small batches from the cooking pot.

The ways in which the various plant ingredients were consumed in the Neolithic and the Bronze Age are to a great extent speculative. There is limited evidence on actual food remains, which correspond to mere fragments of past culinary practice. When these have been found, they offer a fragmentary recipe book, one with part of the ingredients missing, the instructions corresponding to just a few words. Take for example a fragment one might call "bread" or "porridge." We have difficulties distinguishing the two in the rare occasions when we do we find them archaeologically in a form that might correspond to a recipe. The recipes presented here are basic, leading to simple, hearty meals, and are based on my cooking, shaped over the years by recipes I have found in several cookbooks as well as advice by other women, especially my mother and maternal grandmother, as well as my father. I improvise and include ingredients unavailable to the prehistoric inhabitants of Greece. In a way, the recipes correspond to a "fusion" cuisine, connecting prehistory to the present.

Cereals

The recipes with cereals as the main component include one with whole-wheat kernels and parboiled, cracked wheat (bulgur) (for a more detailed account, see Valamoti 2011b).

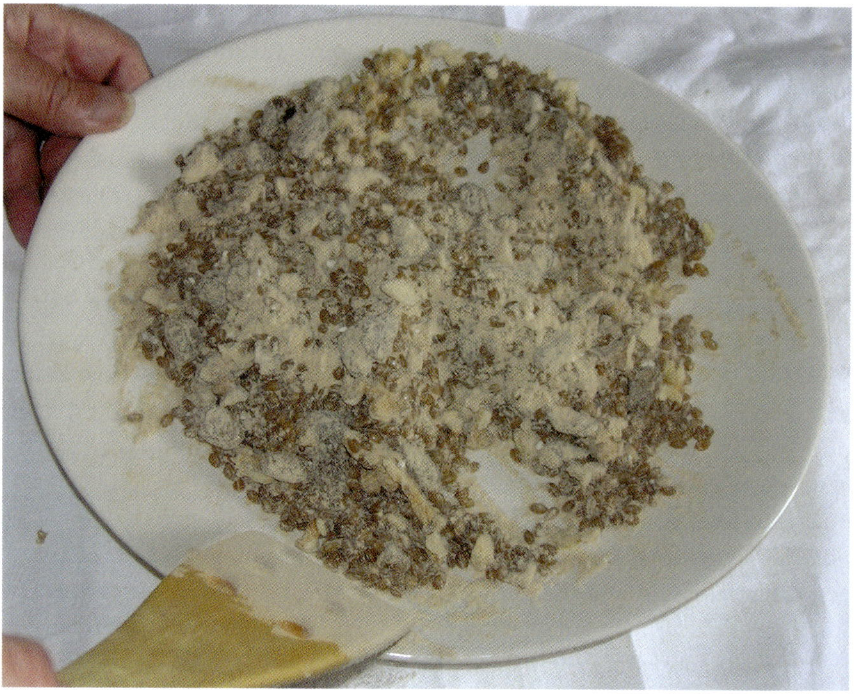

Figure 11.1. Mixing ingredients for *kollyva* preparation. (Soultana Maria Valamoti)

Einkorn or Emmer Kollyva (Figure 11.1)

In Greece today, *kollyva* is associated with the ritual of blessing in the church and memorializing of the dead and funerals and is only eaten in these contexts. This is my mother's recipe and is inspired by *kollyva* rather than a traditional version that has religious connotations. It lacks some basic ingredients used in traditional *kollyva* preparation, such as parsley, candy, and cumin seeds (for other *kollyva* recipes, see, for example, Psilakis and Psilaki 2001:316–321). It also omits the special decoration with icing sugar and the sign of the cross.

Ingredients (Serves 4 people)

1 cup de-husked einkorn grain, totally cleaned from chaff and weeds

3 cups water

½ cup einkorn flour

1/4 cup raisins
1/4 cup almonds, coarsely chopped
1 pomegranate, separated into individual *sarcotestas* (seeds)

Soak the einkorn grain overnight or longer placing it in a pot large enough to contain the grain and enough water to cover the grain by two fingers. Then boil the grain for approximately an hour, until the grain is crunchy in a pleasant way when bitten. Strain the grain, spread on a clean towel, and wrap it until all moisture has been absorbed. Meanwhile, roast the einkorn flour in a frying pan until it acquires a nice golden-brown color. Mix the grain, roasted flour, and remaining ingredients. Serve.

Einkorn kollyvozoumo

Kollyvozoumo is made at the same time as *kollyva* (see previous recipe). The name of this dish in our family was *aşure*, perhaps reflecting the origin of the recipe going back to my great-grandmother Marianthi, born in the Peran neighborhood of Constantinople, Turkey.

Ingredients
Liquid strained from boiled einkorn grain (see recipe above)
1 cup coarsely chopped almonds
2 spoonfuls sultanas (golden raisins)
1 cup roasted einkorn flour plus 4 spoonfuls
3 spoonfuls *petimezi* (boiled down grape syrup) (see recipe in Fruit section)

Keep the liquid of the boiled wheat from the kollyva recipe and reduce its quantity further. Add chopped almonds and 5 spoonfuls of *petimezi*. As the mixture thickens, add 1 cup of roasted einkorn flour to give the mixture a creamy consistency. Serve in individual bowls. Mix the 4 spoonfuls of roasted flour with the *petimezi* and form the mixture into small balls. Use them together with chopped almonds to garnish.

Bulgur with Nettles (Figure 11.2)

This recipe is very popular in Pontic Greek cuisine (e.g., Kiziridou 2002). Bulgur is an ingredient and also the result of a recipe that involves precooking through parboiling of the grain, then drying, and dehulling (Valamoti 2011b; Valamoti et al. 2021).

Ingredients
2 cups bulgur
1 onion finely chopped
2 bunches nettles
1 1/2 cup water
Some olive oil

Wash the nettles wearing gloves, as the leaves sting. I have heard stories, however, of Pontic grandmothers who did not wear gloves. Lightly fry the onion until golden brown, then add the nettle leaves, constantly stirring over medium heat, for 3 minutes. Then add the bulgur and water. Cover and simmer until bulgur is soft and nettles are cooked. You may need to add a little bit of water. Nettles can be replaced by other seasonal greens such as young poppy shoots or wild asparagus.

Pulses

Pulses are essential for communities that rely on plants for the vast majority of their meals. They are very important elements of traditional Greek cuisine. Here, two very common dishes are presented: a soup and a mush known under the generic name *fava*.

Tania's Lentil Soup (Figure 11.3)

Ingredients
2 cups small lentils, sorted and rinsed to remove debris
water for soaking and cooking
bay leaves
oregano
1 large onion
3 cloves garlic
salt
2 tsp. vinegar (optional)

Place the lentils in a bowl full of water and leave to soak overnight. Even though my mother tells me that lentils are the only pulse that do not need soaking before cooking, I refuse to listen. I find that soaking reduces cooking time, and the lentils add a thickness to the soup. The next day, drain them, and place them in a pan with adequate water, approximately 6 cups. Add the onion whole, or chopped finely if you prefer, 3 cloves of garlic, and the

Figure 11.2. Bulgur and nettles can be combined in a delicious dish. (Soultana Maria Valamoti)

Figure 11.3. Lentils come in different varieties with different seed sizes and colors. (Soultana Maria Valamoti)

bay leaves. Bring to boil, then reduce the heat and simmer for approximately 2 hours, checking regularly to see that the water is adequate and tasting to determine whether the lentils are soft. Depending on the field where the lentils grew, or the variety, or both, the soup may be ready in 1 hour or perhaps longer than 2 hours. Some lentil types are known as "quick boiling" (*vrastera*), whereas others may turn out to be too difficult to make a decent soup. So, it is advisable to try different brands and select the tastier and quicker-to-make version. At 15 minutes before completing the cooking, add oregano and the (optional) vinegar.

Basic Fava (Figure 11.4)

On the island of Lemnos as well as in the various recipe books I have read, *fava* is cooked until the cotyledons (seed-bearing leaves) are soft enough, the mixture is then placed in a food processor. Here, instead, it is cooked until it reaches a smooth, creamy consistency through constant stirring during the last 5 to 10 minutes of cooking. Also, a whole onion and a carrot can be added to the *fava* cotyledons during boiling and then incorporated in the mixture with the food processor. However, I prefer the unadulterated taste of *fava* and enjoy the long, slow cooking and occasional stirring of the mixture.

Figure 11.4. *Lathyrus sativus* fava. (Soultana Maria Valamoti)

Ingredients
2 cups split *Lathyrus* seeds, clean of weeds and chaff, or other pulses, de-husked and split in cotyledons
Water
Salt

Take the clean pulse seeds and split them with a hand mill. Winnow them and sieve to remove the testa surrounding the seeds. Hand clean the resulting split seeds. Bring to boil, discard the water, then place in ample water and cook, regularly removing the scum; simmer for 2 to 3 hours, or until it turns into a mush, stirring regularly. Add salt toward the end of the cooking process.

You can garnish in different ways. Spread the hot *fava* in a shallow bowl. Make a "well" in the center, fill it with olive oil, oregano, and finely sliced onion. Season with salt and oregano. You can also add capers. Alternatively, you can serve it with caramelized onions and salted sardines.

Fruit

Cornelian Cherry Sweet Cream (Figure 11.5)

This recipe is inspired by ingredients available in prehistoric Greece and is adapted from a cookbook on Pontic Greeks (Kiziridou 2002:345). The recipe originates from the region of Kromni and is called "kissel." Instead of sugar and corn flour in the original recipe, I use *petimezi* and barley flour; I have omitted lemon.

Ingredients
2 lb. Cornelian cherries
1/2 cup water
1 cup *petimezi*
2 cups barley flour, finely ground
chopped almonds

Rinse the fruit. Simmer in water until the fruit liquifies. Remove from heat and strain the fruit so that the pulp is separated from the stones. Place the pulp in the pan and add *petimezi*. Stiring slowly, add the barley flour until it thickens. If needed, add more flour. Place the mixture in serving bowls and allow to cool.

Figure 11.5. Cornelian cherries (*Cornus mas*) used for the preparation of kissel. (Soultana Maria Valamoti)

Dried Fruit Soup

Ingredients

1/2 lb. dried fruits such as blackberries, elderberries, wild pears, figs

1/2 cup *petimezi* (more or less depending on desired sweetness)

4 and 1/4 cups water

Let the fruit soak overnight. Slowly bring to boil and simmer until the fruit is tender. The resulting fruit "syrup" can be mixed with einkorn flour over low heat to produce a sweet dessert with a consistency of thick custard.

Appendix A: Tables

TABLE I.1. GREEK NEOLITHIC AND BRONZE AGE SITES INVESTIGATED ARCHAEOBOTANICALLY, WITH REFERENCES.

		Site	Bibliography	Sampling method	
				total	random
Thrace	1	Makri	Valamoti 2004	x	
Macedonia	2	Limenaria	Megaloudi 2006		
Macedonia	3	Agios Ioannis Loukas	Megaloudi 2006		
Macedonia	4	Dikili Tash	Kokkidou 2018; Renfrew 2003; Valamoti 2004; 2015a	x (Valamoti)	
Macedonia	5	Arkadikos	Valamoti 2004	x	
Macedonia	6	Sitagroi	Renfrew 2003	x	
Macedonia	7	Photolivos	Logothetis 1970; Megaloudi 2006		
Macedonia	8	Dimitra	Foster 1997; Renfrew 1997	x	
Macedonia	9	Promachon-Topolnitsa	Valamoti 2007c	x	
Macedonia	10	Koroneia	Paraskevopoulou 2019	x	
Macedonia	11	Vasilika Kyparissi	Lathiras 2020		x
Macedonia	12	Agios Mamas	Becker, Kroll 2008		
Macedonia	13	Olynthus	Mylonas 1929; Vickery 1936		
Macedonia	14	Mesimeriani Toumba	Valamoti 2002b		
Macedonia	15	Thermi B	Valamoti 1992		
Macedonia	16	Toumba Thessalonikis	Andreou, Kotsakis 1997; Karathanou, Valamoti 2016; Kotsachristou 2009	x	
Macedonia	17	Stavroupoli	Margaritis 2002	x	
Macedonia	18	Assiros	Jones 1981; 1987; Jones et al. 1986	x	
Macedonia	19	Kastanas	Kroll 1983		
Macedonia	20	Agios Athanasios	Moniaki 2009		
Macedonia	21	Giannitsa B	Valamoti 1995		
Macedonia	22	Archondiko	Petridou 2016; Telioridou 2014; Valamoti 1997b; 1999	x	
Macedonia	23	Mandalo	Valamoti, Jones 2003	x	
Macedonia	24	Sosandra	Valamoti 2015b		
Macedonia	25	Apsalos	Valamoti 2006a	x	
Macedonia	26	Avgi	Margaritis 2007	x	
Macedonia	27	Dispilio	Mangafa 2002; Margaritis 2011	x	
Macedonia	28	Angelochori	Stefani et al. 2006; Valamoti 2010a		
Macedonia	29	Nea Nikomedeia	van Zeist, Bottema 1971		
Macedonia	30	Paliambela	Kotzamani, Livarda 2018	x	
Macedonia	31	Makrygialos	Valamoti 2004	x	
Macedonia	32	Kleitos	Stylianakou 2013	x	
Macedonia	33	Fyllotsairi	Valamoti 2011a	x	
Macedonia	34	Toumba Kremastis Koiladas	Karathanou 2009	x	
Macedonia	35	Vassilara Rahi	Stylianakou, Valamoti 2013; Valamoti et al. 2018		
Macedonia	36	Servia	Housley 2000; Hubbard 1979		
Macedonia	37	Revenia	Kotzamani 2009; Kotzamani, Livarda 2018	x	
Thessaly	38	Theopetra	Mangafa 2000	x	
Thessaly	38	Theopetra	Kotzamani 2009	x	
Thessaly	39	Prodromos	Halstead, Jones 1980		
Thessaly	40	Tsani Magoula	Vickery 1936		

		Retrieval method		Sieve aperture							
subjective / justified	collection of visible seeds	hand picking	flotation	200µm	250µm	300µm	500µm	0.75mm	1.5mm	1.65mm	2mm
			x			x					
			x				x				
			x				x				
	x (Renfrew)	x (Renfrew)	x (Valamoti)			x					
			x			x					
			x						x		
			x					x			
			x			x					
			x			x					
	x		x			x					
			x			x					
	x	x									
x			x			x					
x			x				cloth				
			x			x					
			x		x						
			x			x					
x			x			x					
x	x		x			x					
x			x			x					
			x			x					
			x			x					
x			x			x					
			x			x					
			x		x						
			x			x					
x			x			x					
	x	x									
			x			x					
			x			x					
			x			x					
			x			x					
		x	x			x					
x			x			x					
x		x	x		x						
			x			x					
			x		x						
			x			x					
x		x									
		x									

		Site	Bibliography	Sampling method	
				total	random
Thessaly	41	Achilleion	Renfrew 1989		
Thessaly	42	Koutroulou Magoula	Koromila et al. 2017; 2018		
Thessaly	43	Platia Magoula Zarkou	Jones, Halstead 1993b		
Thessaly	44	Soufli Magoula	Renfrew 1966		
Thessaly	45	Marmariani	Vickery 1936		
Thessaly	46	Otzaki Magoula	Kroll 1981; 1983		
Thessaly	47	Rachmani	Renfrew 1966		
Thessaly	48	Arapi Magoula	Kroll 1981; 1983		
Thessaly	49	Gendiki	Renfrew 1966		
Thessaly	50	Argissa	Hopf 1962b; Kroll 1981; 1983; Renfrew 1966		
Thessaly	51	Sesklo	Kroll 1981; 1983; Renfrew 1966; Tsountas 1908		
Thessaly	52	Visviki Magoula	Renfrew 1966		
Thessaly	53	Dimini	Kroll 1979; 1981; Tsountas 1908		
Thessaly	53	Dimini	Karathanou, Valamoti 2013a	x	
Thessaly	54	Iolkos	Renfrew 1966		
Thessaly	55	Pefkakia	Kroll 1981; 1983		
Thessaly	56	Pyrassos	Renfrew 1966		
N. Aegean	57	Cave of Cyclops	Sarpaki 2011		x
N. Aegean	58	Poliochni	Arobba et al. 2010	x	
Cyclades	59	Kephala	Renfrew 1977		
Cyclades	60	Ftelia	Megaloudi, Marinval 2002		x
Cyclades	61	Saliagos	Renfrew 1968		
Cyclades	62	Dhaskalio	Margaritis 2013a	x	
Cyclades	63	Phylakopi	Renfrew 1982		
Cyclades	64	Markiani	Renfrew 2006		
Cyclades	65	Akrotiri	Sarpaki 1987a; 1992b		
Cyclades	66	Therasia	Vickery 1936		
Aegean	67	Kataphygadi Cave	Trantalidou et al. 2019	x	
Aegean	68	Kalythies	Halstead, Jones 1987		
Aegean	68	Kalythies	Sarpaki 1987b		x
Central Greece	69	Agia Paraskevi	Gkotsinas et al. 2014	x	
Central Greece	70	Kynos	Karathanou, Valamoti 2013a; 2016		
Central Greece	71	Kalapodi	Kroll 1993		
Central Greece	72	Mitrou	Karathanou, Valamoti 2013a	x	
Central Greece	73	Toumba Balomenou	Sarpaki 1995		
Central Greece	74	Orchomenos	Vickery 1936		
Central Greece	75	Gla	Jones 1995		
Central Greece	76	Sarakenos	Samspon, Megaloudi 2006		
Central Greece	77	Delphi (Pilier des Rhodiens)	Luce, Marinval 2008		
Central Greece	78	Kirrha	Margaritis 2014		
Central Greece	79	Thebes	Jones, Halstead 1993a		
Central Greece	80	Skoteini Tharounia	Mangafa 1993		
Central Greece	81	Agora	Hopf 1971		
Central Greece	82	Agios Kosmas	Mylonas 1959		
Peloponnese	83	Kassaneva	Colaianni et al. 2012	x	
Peloponnese	84	Agia Sotira	Kotzamani, Livarda 2017	x	
Peloponnese	85	Tsoungiza	Hansen 1988; 2000		
Peloponnese	85	Tsoungiza	Forste 2012, Hansen, Allen 2011, Dabney et al. 2020	x	

		Retrieval method		Sieve aperture							
subjective / justified	collection of visible seeds	hand picking	flotation	200µm	250µm	300µm	500µm	0.75mm	1.5mm	1.65mm	2mm
x			x								
x			x			x					
	x		x			x					
	x	x									
x			x	x							
	x	x									
x			x	x							
	x	x									
x (Kroll)	x	x	x (Kroll)	x (Kroll)							
	x	x	x (Kroll)	x (Kroll)							
	x	x									
x (Kroll)	x (Tsountas)	x (Tsountas)	x (Kroll)	x (Kroll)							
	x	x									
x			x	x							
	x	x									
			dry sieve								
			x								x
	x	x									
			x				x				
	x	x									
			x		x						
			x								
			x								
x	x	x	x								
		x									
			x			x					
			x			x					
x			x								
			x	x							
			x								
x			x			x					
		x									
x	x	x									
			x	x							
	x	x									
x			x								
		x									
		x									
			x								
			x			x					
			x								
			x (+ dry sieve)		x						

		Site	Bibliography	Sampling method	
				total	random
Peloponnese	85	Tsoungiza	Allen, Forste 2020	x	
Peloponnese	86	Barnavos	Wright et al. 2008	x	
Peloponnese	87	Mycenae	Wace 1921		
Peloponnese	88	Aspis	Philippa-Touchais et al. 2014	x	
Peloponnese	89	Midea	Margaritis et al. 2014; Shay et al. 1998	x	
Peloponnese	90	Tiryns	Kroll 1982		
Peloponnese	90	Tiryns	Vetters et al. 2016		
Peloponnese	91	Lerna	Hopf 1961; 1962a		
Peloponnese	92	Synoro	Willerding 1973		
Peloponnese	93	Iria	Willerding 1973		
Peloponnese	94	Franchthi	Hansen 1991	x	
Peloponnese	95	Mt. Lykaion	Margaritis 2014		
Peloponnese	96	Kouphovouno	Megaloudi 2006; Vaiglova et al. 2014		
Peloponnese	97	Agios Vasileios	Karathanou, Valamoti 2013a; 2013b; 2016	x	
Peloponnese	98	Nichoria	Shay, Shay 1978		
Peloponnese	99	P.O.T.A. Romanou	Valamoti et al. 2020	x	
Peloponnese	100	Alepotrypa	Margaritis 2018		
Crete	101	Palaikastro	Sarpaki 1989; 2007a; 2019		
Crete	102	Ourania	Kopaka 2011; Sarpaki 2009b		
Crete	103	Livari	Margaritis 2014		
Crete	104	Kephala Petras	Livarda, Kotzamani 2013	x	
Crete	105	Petras	Margaritis 2017	x	
Crete	106	Papadiokampos	Brogan et al. 2013	x	
Crete	107	Mochlos	Sarpaki, Bending 2004		
Crete	107	Mochlos	Bending 2011		
Crete	108	Pseira	Jones, Smith 2009	x	
Crete	109	Chrysokamino	Jones, Schofield 2006		
Crete	110	Vronda	Flint-Hamilton 2016		
Crete	111	Gournia	Watrous et al. 2015		
Crete	112	Priniatikos Pyrgos	Molloy et al. 2014	x	
Crete	113	Aphrodite's Kephali	Margaritis 2013c	x	
Crete	114	Myrtos	Renfrew 1972		
Crete	115	Sissi	Livarda, Kotzamani 2013	x	
Crete	116	Malia	Sarpaki 2007b		
Crete	117	Myrtos-Pyrgos	Cadogan 1978; Livarda, Kotzamani 2013		
Crete	118	Nirou Chani	Evans 1900-1901		
Crete	119	Knossos	Evans 1968; Jones 1984		
Crete	119	Knossos	Efstratiou et al. 2004; Sarpaki 2013		
Crete	119	Knossos, Villa Dionysus	Livarda 2012		
Crete	120	Zominthos	Livarda, Kotzamani 2013; 2020		
Crete	121	Kommos	Shay et al. 1995		
Crete	121	Agia Triada	Follieri 1986		
Crete	123	Monastiraki	Fiorentino, Solinas 2006; Sarpaki, Kanta 2011		
Crete	124	Chamalevri-Tzambakas	Sarpaki 1999a		
Crete	125	Kasteli	Sarpaki 2016	x	
Ionian	126	Drakaina	Sarpaki 2009c		

		Retrieval method		Sieve aperture							
subjective / justified	collection of visible seeds	hand picking	flotation	200µm	250µm	300µm	500µm	0.75mm	1.5mm	1.65mm	2mm
		x	x		x						
			x								
			x								
		x	x								
x			x	x							
			x (+ water sieve)			x					
		x									
			x								
			x								
			x								
			x							x	
			x			x					
			x								
x		x	x (+ water sieve)		x						
			x			x					
			x								
		x	x								
			x		x						
x			x			x					
			x			x					
			x (+ york-type separation machine)				x				
		x									
			x			x					
		x									
	x (Jones)	x (Evans)									
			x								
x			x			x					
x			x			x					
		x									
x			x								
			x		x						
x			x (+ water sieve)		x						
			x								

TABLE 1.1

	Site	*T. monococcum*	*T. dicoccum*	*T. timopheevii*	Free-threshing wheat
Thessaly	Achilleion	x	x		
Thessaly	Argissa	x	x		
Macedonia	Fyllotsairi	x	x[a]	x	
Peloponnese	Franchthi		x		
Macedonia	Giannitsa B	x	x		x
Crete	Knossos	x	x		x[a]
Thessaly	Koutroulou Magoula	x	x		
Macedonia	Nea Nikomedeia	x[a]	x[a]		
Thessaly	Otzaki Magoula	x	x		x
Macedonia	Paliambela	x	x	x	x
Thessaly	Prodromos	x[a]	x[a]		
Macedonia	Revenia	x	x		
Thessaly	Sesklo	x	x		x
Macedonia	Sosandra	x			
Thessaly	Soufli Magoula	x	x		
Central Greece	Toumba Balomenou	x[a]			

Table 1.1. Cereal and Pulse Crop Species Identified at Early Neolithic Sites of Greece.
[a]Presence of species in dense, pure concentrations established on the basis of a minimum of 100 quantifiable items in a single sample.

Hordeum sp.	*Avena* sp.	*Cicer arietinum*	*Lathyrus sativus/cicera*	*Lens* sp.	*Pisum* sp.	*Vicia ervilia*	*Vicia faba*
x	x			x			
x	x			x	x	x	
x				x	x		
x	x			x			
x				x			
x	x			x	x		x
x				x	x		
x[a]				x[a]	x	x	
x[a]	x	x		x	x	x	
x			x	x		x	
x			x[a]	x	x[a]		
x	x		Lathyrus sp.	x		x	
x	x			x		x	
x			x	x[a]			
x				x	x		
x	x			x	x	x	

TABLE 2.1. CEREAL CROP SPECIES IDENTIFIED FROM GREEK NEOLITHIC SITES.

Phase	Site	*Triticum monococcum*	*T. monococcum* grains	*T. monococcum* glume bases	*T. dicoccum*	*T. dicoccum* grains	*T. dicoccum glume* bases	*T. spelta* grains
AN[b]	Achilleion					x		
	Argissa	x			x			
	Gendiki		x			x		
	Knossos		x	x		x	x	
	Sesklo	x			x			
	Soufli Magoula				x			
EN[c]	Achilleion		x			x		
	Argissa		x			x		
	Fyllotsairi		x	x		x[a]	x	
	Franchthi					x		
	Giannitsa B		x	x		x	x	
	Knossos		x			x		
	Koutroulou Magoula	x			x			
	Nea Nikomedeia		x[a]			x[a]	x	
	Otzaki Magoula	x			x			
	Paliambela		x	x		x	x	
	Prodromos		x[a]			x[a]	x	
	Revenia		x	x		x	x	
	Sesklo	x			x			
	Sosandra		x					
	Soufli Magoula		x			x		
	Toumba Balomenou		x	x[a]				
AN/MN[d]	Cave of Cyclops							
MN	Achilleion		x	x		x	x	
	Apsalos		x	x[a]		x	x	
	Dispilio		x	x		x	x	x
	Knossos	x						
	Limenaria		x	x		x		
	Otzaki Magoula	x			x			
	Photolivos	x			x			
	Platia Magoula Zarkou		x			x		
	Servia		x[a]	x[a]		x	x[a]	
	Sesklo	x			x			
	Sitagroi		x[a]			x	x	
	Toumba Balomenou			x	?			
	Tsoungiza		x					

T. timopheevii glume bases	Free-threshing wheat	Free-threshing wheat grains	Free-threshing wheat rachis inter-nodes	*Triticum* sp.	*Hordeum* sp.	*Hordeum* sp. rachis internodes	*Avena* sp.	*Bromus* sp.	*Panicum miliaceum*	*Panicum* sp.
							x			
					x				x?	
					x					
		x[a]	x	x	x	x				
					x		x			
					x		x	x		
					x		x			
x					x					
					x		x			
		x	x		x					
	x[a]	x		glume base	x		x	x		
					x			x		
					x[a]					
	x				x[a]		x			
x		x	x		x	x		x		
					x					
					x	x	x	x		
	x				x		x	x		
					x					
					x					
				x	x	?	x			x
					x					
					x		x			
x[a]					x	x				
		x	x		x	x				
				x	x					phytoliths
		x		x	x			x	x	
					x		x		x	
					x					
					x					
				x	x[a]	x				
					x		x			
		x			x		x	x		
				x	x					
					x					

Phase	Site	*Triticum monococcum*	*T. monococcum* grains	*T. monococcum* glume bases	*T. dicoccum*	*T. dicoccum* grains	*T. dicoccum* glume bases	*T. spelta* grains
MN/LN[e]	Avgi		x	x		x	x	
	Dispilio		x	x		x		x
	Kouphovouno	x			x			
LN	Arapi Magoula	x			x			?
	Arkadikos		x[a]	x[a]		x[a]	x[a]	
	Dikili Tash		x[a]	x[a]		x[a]	x[a]	
	Dimini	x			x			
	Dimitra		x			x		
	Drakaina	x			x			x ?
	Kalythies	? impression			? impression			
	Kleitos		x	x		x[a]	x[a]	
	Koroneia		x	x		x	x	
	Kremasti Koiladas		x	x[a]		x	x	
	Vasilika Kyparissi		x	x[a]		x[a]	x	
	Makri		x[a]	x[a]		x	x[a]	
	Makrygialos		x[a]	x[a]		x[a]	x[a]	
	Olynthus							
	Pefkakia	x			x			
	Platia Magoula Zarkou		x			x		
	Pyrassos	x			x[a]			
	Saliagos		?			x		
	Sarakenos		x[a]	x[a]		x		
	Servia		x	x		x[a]	x	
	Sesklo	x			x			
	Sitagroi		x[a]	x		x	x	
	Skoteini Tharounia		x			x		
	Visviki Magoula	x						
LN/FN[f]	Agios Ioannis Loukas	x			x			
	Agora							
	Alepotrypa		x			x		x
	Dispilio		x	x		x		x
	Photolivos							
FN	Alepotrypa		x	x		x		x
	Dikili Tash		x[a]	x		x		
	Ftelia				?			
	Kephala							

T. timopheevii glume bases	Free-threshing wheat	Free-threshing wheat grains	Free-threshing wheat rachis inter-nodes	*Triticum* sp.	*Hordeum* sp.	*Hordeum* sp. rachis internodes	*Avena* sp.	*Bromus* sp.	*Panicum miliaceum*	*Panicum* sp.
		x			x					
					x			x		
		x			x					
					x		x			
x[a]		x	x		x[a]	x				
					x[a]		x	x	x	
					x[a]					
	x			x	x		x			
	x				x		x			
	? impression			x (& glume impression)	x					
x			x		x	x				
x					x					
x		x	x		x	x		x		
x[a]		x	x		x[a]	x	x[a] (sativa)	x		
x[a]		x	x[a]	x	x[a]	x[a]	x	x[a]		
x[a]			x	x	x[a]	x[a]	x	x		
				x & impression					x	
	x				x					
					x		x			
					x[a]					
					x[a]					
	x				x[a]					
				x	x[a]		x	x		
					x		x			
		x			x					
		x		x	x		x			
				x					x	
					impression					
					x					
		x			x		x			
				x	x					
		x			x					
					x[a]		x	x	x	
				x	x					
					x					

Phase	Site	*Triticum monococcum*	*T. monococcum* grains	*T. monococcum* glume bases	*T. dicoccum*	*T. dicoccum* grains	*T. dicoccum* glume bases	*T. spelta* grains
	Kleitos		x[a]	x		x		
	Mandalo		x[a]	x[a]		x[a]	x[a]	
	Pefkakia	x			x			
	Rachmani							
	Sesklo					x		
	Sitagroi		x[a]	x		x	x	
	Stavroupoli		x	x		x	x	x
	Tsoungiza		x			x		
MN/FN	Franchthi		x	x		x	x	
FN/EBA[g]	Lerna							
	Sesklo					x		
	Vassilara Rahi		x	x		x	x	
	Theopetra		x	x		x[a]	x	x
	Thermi B		x	x		x	x	

[a]Presence of species in dense, pure concentrations established on the basis of a minimum of 100 quantifiable items in a single sample.
[b]AN = Aceramic Neolithic (early seventh millennium BC)
[c]EN = Early Neolithic (7000–5800 BC)
[d]MN = Middle Neolithic (5800–5300 BC)
[e]LN = Late Neolithic (5300–4500 BC)
[f]FN = Final Neolithic (4500–3500 BC)
[g]EBA = Early Bronze Age (3500–2100 BC)

T. timopheevii glume bases	Free-threshing wheat	Free-threshing wheat grains	Free-threshing wheat rachis inter-nodes	*Triticum* sp.	*Hordeum* sp.	*Hordeum* sp. rachis internodes	*Avena* sp.	*Bromus* sp.	*Panicum miliaceum*	*Panicum* sp.
x					x[a]	x				
x			x		x		x	x	x	
		x			x		x			
	?				x					
					x[a]		x			
		x	x		x					
					x					
		x			x	x	x			
					x					
					x					
x		x			x	x	x	x	x	
		x	x	x	x	x				

TABLE 2.2. CEREAL CROP SPECIES IDENTIFIED FROM GREEK BRONZE AGE SITES.

Phase	Site	*T. monococcum* grains	*T. monococcum* glume bases	*T. dicoccum* grains	*T. dicoccum* glume bases	*T. spelta* grains	*T. spelta* glume bases
EBA[b]	Agios Athanasios	x	x	x[a]	x		
	Archondiko	x[a]	x[a]	x[a]	x[a]	x[a]	x[a]
	Argissa	x	x	x			
	Chrysokamino						
	Dhaskalio		x	x	x		
	Kassaneva			x			
	Kastanas	x[a]	x[a]	x[a]	x	x	
	Kasteli						
	Lerna	x		x			
	Mandalo	x[a]	x[a]	x[a]	x[a]		
	Markiani						
	Mesimeriani Toumba	x		x			
	Myrtos						
	Pefkakia	x		x			
	Petras						
	Platia Magoula Zarkou			x[a]	x		
	Poliochni	x	x				
	Priniatikos Pyrgos						
	Servia	x		x			
	Sitagroi	x	x[a]	x	x		
	Skoteini Tharounia	x	x	x	x	x	
	Tiryns				x		
	Agios Mamas	x	x	x	x		x
	Agora						
	Argissa	x		x		x	
	Aspis						
	Chamalevri-Tzambakas				x		
	Kasteli		x				
	Kommos						
	Kouphovouno			x			
	Lerna			x			
	Monastiraki	x		x			
	Nichoria						
	Orchomenos			x			
	Pefkakia	x		x		x	
	Toumba Thessalonikis	x	x[a]	x	x	x	
	Tsani Magoula						

T. timopheevii glume bases	Free-threshing wheat grains	Free-threshing wheat rachis internodes	*Triticum* sp.	*Hordeum* sp. grains	*Hordeum* sp. rachis internodes	*Triticum*/ *Hordeum*	*Panicum milliaceum*
				x[a]			
x	x			x[a]	x		x
				x			
			x (glume)	x	x		
				x			
	x			x[a]	x		x
			x (glume bases)				
				x[a]			
	x			x[a]			
				impression			
				x			
			x (impression)	x (impression)	x (impression)		
				x			
				x			
				x[a]			
	x			x			
			x	x	x		
				x			
	x	x		x	x[a]		x
	x	x		x			
				x			
				x	x		x
				x			
	x			x[a]			x
				x			
			x				
				x			
						x	
				x			
				x			
	x			x	x		
			x			x	
			x	x			
	x			x			x
x[a]	x	x		x[a]	x		
			x	x			

Phase	Site	*T. monococcum* grains	*T. monococcum* glume bases	*T. dicoccum* grains	*T. dicoccum* glume bases	*T. spelta* grains	*T. spelta* glume bases
	Malia			x			
	Myrtos-Pyrgos						
	Ourania	x		x	x		
LBA	Agia Triada			x		x	
	Agios Mamas	x	x	x	x	x	x
	Agios Vasileios	x	x				
	Agora						
	Akrotiri	x		x			
	Angelochori	x	x	x	x	x	x
	Archondiko	x[a]	x[a]	x[a]	x[a]	x[a]	x
	Assiros	x	x	x	x	x	x
	Agia Sotira				x		
	Barnavos						
	Delphi (Pilier des Rhodiens)	x		x			
	Dimini						
	Dimitra	x		x[a]			
	Glas	x[a]					
	Iolkos			x			
	Iria			x			
	Kalapodi			x			
	Kastanas	x[a]	x	x[a]	x[a]	x	x
	Kasteli		x	x	x		
	Kataphygadi Cave	x		x			
	Knossos	x		x[a]	x		
	Kommos						
	Kynos						
	Marmariani			x			
	Mesimeriani Toumba						
	Midea	x					
	Mochlos			x			
	Mt. Lykaion						
	Mycenae						
	Nichoria						
	Orchomenos						
	Palaikastro	x					
	Palaikastro Building 1		x	x	x		
	Phylakopi						
	Therasia						
	Tiryns	x	x	x	x		x

T. timopheevii glume bases	Free-threshing wheat grains	Free-threshing wheat rachis internodes	*Triticum* sp.	*Hordeum* sp. grains	*Hordeum* sp. rachis internodes	*Triticum*/ *Hordeum*	*Panicum miliaceum*
			x	x			
				x[a]			
	x[a]			x[a]			
	x			x			
	x	x		x	x		x
	x	x		x	x		
				x			
				x			
x				x[a]			x
x[a]	x[a]	x		x[a]			x[a]
x	x			x			x
			x	x			
						x	
	x			x			
			x	x			
	x			x			
				x			
				x			
	x			x			
	x	x		x	x		x[a]
	x		x	x	x	x	x
	x			x			
	x[a]			x[a]			
						x	
				x	x		
							x
				x			
	x			x			
			x	x			
				x			
	x		x	x		x	
			x				
			x	x			
			x	x			
			x	x	x		
			x	x			
				x			
	x		x	x			x

Phase	Site	*T. monococcum* grains	*T. monococcum* glume bases	*T. dicoccum* grains	*T. dicoccum* glume bases	*T. spelta* grains	*T. spelta* glume bases
	Toumba Thessalonikis	x	x[a]	x	x[a]	x	
	Tsoungiza	x		x			
	Zominthos			x			

[a]Presence of species in dense, pure concentrations established on the basis of a minimum of 100 quantifiable items in a single sample.
[b]EBA = Early Bronze Age (3500–2100 BC)
[c]MBA = Middle Bronze Age (2100–1700 BC)
[d]LBA = Late Bronze Age (1700–1200 BC)

T. timopheevii glume bases	Free-threshing wheat grains	Free-threshing wheat rachis internodes	*Triticum* sp.	*Hordeum* sp. grains	*Hordeum* sp. rachis internodes	*Triticum/ Hordeum*	*Panicum milliaceum*
x	x	x		x	x		x
				x			
				x			x

TABLE 3.1. PULSE CROP SPECIES IDENTIFIED FROM GREEK NEOLITHIC SITES.

Phase	**Site**	***Cicer arietinum***	***Lathyrus sativus/cicera***	***Lens* sp.**	***Pisum* sp.**
AN[b]	Argissa			x	
	Gendiki			x	x
	Knossos			x[a]	x
	Sesklo			x	x
	Soufli Magoula			x	
EN[c]	Achilleion			x	
	Argissa			x	x
	Fyllotsairi			x	x
	Franchthi			x	
	Giannitsa B			x	
	Knossos			x	x
	Koutroulou Magoula			x	x
	Nea Nikomedeia			x[a]	x
	Otzaki Magoula	x		x	x
	Paliambela		x	x	
	Prodromos		x[a]	x	x[a]
	Revenia		x (Lathyrus sp.)	x	
	Sesklo			x	
	Sosandra		x	x[a]	
	Soufli Magoula			x	x
	Toumba Balomenou		x	x	x
MN[d]	Acilleion				x
	Apsalos		x	x	x
	Dikili Tash			x	
	Dispilio		x	x	x
	Knossos			x	x
	Limenaria		x	x	
	Otzaki Magoula			x	x
	Photolivos			x	
	Platia Magoula Zarkou			x	
	Servia		x[a]	x[a]	
	Sesklo				
	Sitagroi			x	x
	Toumba Balomenou		x	x	x
	Tsoungiza			x	
MN/LN[e]	Avgi		x	x	
	Dispilio			x	
	Kouphovouno		x	x	x
LN	Arapi Magoula			x	
	Arkadikos		x[a]	x[a]	x[a]
	Dikili Tash		x[a]	x[a]	x
	Dimini	x	x	x	x

Vicia ervilia	*Vicia faba*	*Vicia sativa*	*Vicia* sp.	*Lathyrus clymenum*	*Lupinus* sp.
			x		
x					
x					
	x				
x					
x					
x					
x					
x					
x					
x					
x					
x					
x[a]					
	x				
x					
x			x		
x					
x					
x					
	x				
x[a]					
x	x				

Phase	Site	*Cicer arietinum*	*Lathyrus sativus/cicera*	*Lens* sp.	*Pisum* sp.
	Dimitra		x	x	
	Drakaina		x	x	x
	Kalythies			x	
	Kleitos		x	x	
	Koroneia		x	x	
	Kremasti Koiladas		x[a]	x	
	Vasilika Kyparissi		x	x	
	Makri		x	x[a]	x[a]
	Makrygialos		x	x[a]	x
	Pefkakia			x	
	Platia Magoula Zarkou			x	
	Pyrassos				x
	Servia		x[a]	x[a]	x
	Sesklo			x	
	Sitagroi	x		x	x
	Skoteini Tharounia		x		x
	Stavroupoli			x	x
LN/FN[f]	Agios Ioannis Loukas			x	
	Agora				impression
	Alepotrypa			x	x
	Dispilio			x	
	Photolivos			x	
FN	Alepotrypa		x	x	x
	Dikili Tash		x[a]	x[a]	x
	Ftelia		x[a]	x	
	Kephala		x[a]		
	Kleitos		x[a]	x	
	Mandalo		x	x[a]	x
	Pefkakia		x	x	x
	Rachmani			x	x
	Sitagroi		x	x	x
	Tsoungiza		x	x	
MN/FN	Franchthi		x	x	x
FN/EBA[g]	Lerna			x	x
	Vassilara Rahi		x	x	
	Theopetra		x	x	x
	Thermi B		x	x	x

[a]Presence of species in dense, pure concentrations established on the basis of a minimum of 100 quantifiable items in a single sample.
[b]AN = Aceramic Neolithic (seventh millennium BC)
[c]EN = Early Neolithic (7000–5800 BC)
[d]MN = Middle Neolithic (5800–5300 BC)
[e]LN = Late Neolithic (5300–4500 BC)
[f]FN = Final Neolithic (4500–3500 BC)
[g]EBA = Early Bronze Age (3500–2100 BC)

Vicia ervilia	***Vicia faba***	***Vicia sativa***	***Vicia* sp.**	***Lathyrus clymenum***	***Lupinus* sp.**
x[a]					
				x ?	x ?
? impression					
x					
x					
x					
x					
x[a]	x				
x	x				
x					
x[a]					
x					
			x		
	x				
x					
x					
x[a]					
x[a]					
x[a]		x			
x					
x					
	x				
x					
x	x				

TABLE 3.2. PULSE CROP SPECIES IDENTIFIED FROM GREEK BRONZE AGE SITES.

Phase	Site	*Cicer arietinum*	*Lathyrus sativus/cicera*	*Lens* sp.
EBA[b]	Agios Athanasios		x[a]	x
	Archondiko		x[a]	x[a]
	Argissa			x
	Dhaskalio		x	x
	Kastanas	x	x	x
	Lerna		x	x
	Limenaria			x
	Mandalo		x[a]	x[a]
	Markiani			
	Mesimeriani Toumba		x	x
	Pefkakia			x
	Platia Magoula Zarkou			
	Poliochni			x
	Priniatikos Pyrgos		x	x[a]
	Servia		x	x
	Sitagroi			x
	Skoteini Tharounia	x	x	
	Tiryns			x
	Tsoungiza	x	x	x
EBA/MBA[c]	Kasteli			x
MBA	Agia Paraskevi		x	x
	Agios Mamas		x	x
	Akrotiri			
	Argissa		x	x
	Aspis		x	x
	Chamalevri-Tzambakas			x
	Kasteli			x
	Kirrha		x	
	Lerna		x	x
	Monastiraki	x	x	x
	Orchomenos		x	
	Pefkakia		x	x
	Pseira		x	
	Toumba Thessalonikis		x	x
	Tsoungiza			x
MBA/LBA[d]	Kasteli		x[a]	x
	Malia		x	x
	Myrtos-Pyrgos			
	Ourania		x[a]	x[a]
	Sissi		x[a]	
	Tsoungiza			x

Pisum sp.	*Vicia ervilia*	*Vicia faba*	*Vicia* sp.	*Lathyrus clymenum*	*Lathyrus ochrus*	*Lupinus* sp.
	x[a]					
x	x	x				
	x					
x	x			x		
x	x[a]	x				
x	x	x[a]				
	x[a]	x[a]				
	x			x		
	x	x				
	x					
	x[a]					
		x	x			
	x	x		x		
	x[a]	x[a]				
x		x[a]				
	x	x				
x	x	x	x			
	x					
	x	x				
						x
x	x	x				
x						
		x				
x		x				
		x[a]				
x	x	x	x	x	x	
	x	x				
	x	x				
x	x	x				
	x					
x	x	x[a]		x	x	
x		x				
	x[a]					
x		x[a]				

Phase	Site	*Cicer arietinum*	*Lathyrus sativus/cicera*	*Lens* sp.
LBA	Agia Triada		x	x
	Agios Mamas	x	x	x
	Agios Vasileios		x	x
	Akrotiri		x	x
	Angelochori		x	x
	Archondiko			x
	Assiros			x
	Delphi (Pilier des Rhodiens)			x
	Dimitra			x
	Iolkos			
	Iria			
	Kalapodi	x		x
	Kastanas		x	x
	Kasteli		x	x
	Kataphygadi Cave		x	x
	Knossos		x[a]	x
	Knossos Villa Dionysus			
	Kynos			
	Mesimeriani Toumba			
	Midea	x	x	x
	Mitrou			
	Mochlos		x	x
	Mycenae			
	Nirou Chani			
	Orchomenos		x	x
	Palaikastro		x	
	Palaikastro Building 1		x	x
	Phylakopi			
	Sissi		x[a]	
	Thebes		x	
	Therasia			
	Tiryns	x	x	x
	Toumba Thessalonikis		x	x
	Tsoungiza		x	x
	Vronda		x	x
	Zominthos			x

[a]Presence of species in dense, pure concentrations established on the basis of a minimum of 100 quantifiable items in a single sample.
[b]EBA = Early Bronze Age (3500–2100 BC)
[c]MBA = Middle Bronze Age (2100–1700 BC)
[d]LBA = Late Bronze Age (1700–1200 BC)

Pisum sp.	*Vicia ervilia*	*Vicia faba*	*Vicia* sp.	*Lathyrus clymenum*	*Lathyrus ochrus*	*Lupinus* sp.
x	x					
x	x	x				
	x					
x				x		
x	x					
	x[a]					
	x					
	x					
	x					
x		x[a]				
		x				
	x					
x	x[a]	x				
		x				
x	x	x				
x	x	x[a]		x	x[a]	
x						
		x				
	x					
x	x	x	x			
	x					
	x	x				
			x			
x	x	x				
x	x	x				
		x				
x		x				
			x			
		x				
			x			
x	x	x				
x	x	x				
x	x	x				
x	x					
x			x			

TABLE 4.1. OIL-YELDING SPECIES IDENTIFIED FROM GREEK NEOLITHIC SITES.

Phase	Site	*Camelina sativa*	*Lallemantia* sp.	*Linum usitatissimum*	*Linum* sp.	*Olea europea*
AN[b]	Gendiki					
	Sesklo					
	Soufli Magoula					
EN[c]	Achilleion					
	Fyllotsairi					
	Franchthi					
	Giannitsa B					
	Knossos			x	x	
	Otzaki Magoula			x		
	Paliambela				x	
	Revenia			x		
	Sesklo					
	Toumba Balomenou					
EN/MN[d]	Cave of Cyclops					
MN	Achilleion					
	Apsalos			x		
	Dispilio			x		
	Limenaria					
	Servia				x	
	Sesklo				x	
	Sitagroi					
	Toumba Balomenou					
	Tsoungiza					
MN/LN[e]	Avgi					
	Kouphovouno			x		
LN	Arapi Magoula			x		
	Arkadikos			x[a]		
	Cave of Cyclops					
	Dikili Tash			x		
	Dimini					
	Dimitra					
	Kleitos		x?			
	Koroneia					
	Kremasti Koiladas		x?	x		
	Vasilika Kyparissi					
	Makri			x		
	Makrygialos			x[a]		
	Pefkakia			x		
	Sarakenos					

Olea sp.	*Papaver* cf. *somniferum*	*Pistacia terebinthus*	*Pistacia atlantica*	*Pistacia vera*	*Pistacia* sp.
			x		
			x		
1?					
					x
					x
					x
		x	x		
		x			
			x		
		x			
		x			
					x
					x
		x			
		x			
			x		
			x		
					x (cf. P. *terebinthus/lentiscus*)
					x (P. *lentiscus*)
		x			
		x			
		x			
					x
x					
		x			x
		x			
		x			
		x			
		x			
	x	x			
		x			

Phase	Site	*Camelina sativa*	*Lallemantia* sp.	*Linum usitatissimum*	*Linum* sp.	*Olea europea*
	Sesklo				x	
	Sitagroi					
	Skoteini Tharounia					
	Stavroupoli					
LN/FN[f]	Sesklo					
FN	Dikili Tash			x	x	
	Kephala Petras					x
	Mandalo			x		
	Pefkakia	x		x		
	Sitagroi					
MN/FN	Franchthi					
	Theopetra					
	Thermi B					

[a]Presence of species in dense, pure concentrations established on the basis of a minimum of 100 quantifiable items in a single sample.
[b]AN = Aceramic Neolithic (seventh millennium BC)
[c]EN = Early Neolithic (7000–5800 BC)
[d]MN = Middle Neolithic (5800–5300 BC)
[e]LN = Late Neolithic (5300–4500 BC)
[f]FN = Final Neolithic (4500–3500 BC)

Olea sp.	*Papaver* cf. *somniferum*	*Pistacia terebinthus*	*Pistacia atlantica*	*Pistacia vera*	*Pistacia* sp.
			x		
x					
		x			
				x	
	x	x			
				x	
			x		
					x (cf. P. lentiscus)
x		x			
					x

TABLE 4.2. OIL-YELDING SPECIES IDENTIFIED FROM GREEK BRONZE AGE SITES.

Phase	Site	*Camelina sativa*	*Lallemantia*	*Linum usitatissimum*
EBA[b]	Agios Athanasios			x
	Aphrodite's Kephali			
	Archondiko		x[a]	x[a]
	Argissa			x
	Dhaskalio			
	Kassaneva			
	Kastanas	x		x[a]
	Kasteli			
	Lerna			x[a] (sp.)
	Livari			
	Mandalo		x[a]	x
	Markiani			
	Myrtos			
	Pefkakia			x
	Petras			x
	Poliochni			
	Priniatikos Pyrgos			
	Tiryns			
	Tsoungiza			
EBA/MBA[c]	Kasteli			
MBA	Agia Paraskevi			
	Agios Mamas	x		x
	Argissa			x
	Aspis			
	Chamalevri-Tzambakas			
	Kasteli			
	Kommos			
	Kouphovouno			
	Monastiraki			
	Nichoria			
	Pefkakia			x
	Toumba Thessalonikis			
	Tsoungiza			
MBA/LBA[d]	Kasteli			x
	Malia			
	Ourania			
	Tsoungiza			

Olea europea	*Olea* sp.	*Papaver somniferum*	*Pistacia terebinthus*	*Pistacia vera*	*Pistacia* sp.	*Sinapis arvensis*
x						
		x		x		
x						
x						
		x				
	x					
	x					
		x				
x						
x						
	x					
x						
x						
x						
x					x	
	x					
					x	
		x				
				x		
	x					
x			x		x	
	x					
x						
	x					
x						
x						
				x		
		x				
x						
x					x	
x						
x						

Phase	Site	*Camelina sativa*	*Lallemantia*	*Linum usitatissimum*
LBA	Agios Mamas	x	x	x
	Agios Vasileios			
	Akrotiri			
	Angelochori		x	
	Archondiko		x	x
	Assiros	x		x (sp.)
	Agia Sotira			
	Barnavos			
	Delphi (Pilier des Rhodiens)			
	Dimini			
	Dimitra			
	Iolkos			
	Kalapodi	x		
	Kastanas	x[a]		x
	Kasteli			x
	Kataphygadi Cave			
	Knossos Villa Dionysus			
	Kommos			
	Kynos			
	Midea			
	Mochlos			
	Nichoria			
	Palaikastro			
	Palaikastro Building 1			
	Papadiokampos			
	Pseira			
	Tiryns			x
	Toumba Thessalolnikis			
	Tsoungiza			x
	Vronda			

[a]Presence of species in dense, pure concentrations established on the basis of a minimum of 100 quantifiable items in a single sample.
[b]EBA = Early Bronze Age (3500–2100 BC)
[c]MBA = Middle Bronze Age (2100–1700 BC)
[d]LBA = Late Bronze Age (1700–1200 BC)

Olea europea	***Olea* sp.**	***Papaver somniferum***	***Pistacia terebinthus***	***Pistacia vera***	***Pistacia* sp.**	***Sinapis arvensis***
		x				
	x				x	
x						x
		x[a]				
		x				
x						
x						
x						
x						
			x		x	
x						
x						
		x				
x	x					
x						
x						
x						
x						
x						
x					x	
x						
x						
x					x (cf. *P. lentiscus*)	
	x					
x						
x		x				
	x					
x					x	
x				x		

TABLE 5.1. FRUIT SPECIES IDENTIFIED FROM GREEK NEOLITHIC SITES.

Phase	Site	*Amygdalus* sp.	*Arbutus* sp.	*Capparis* sp.	*Celtis* sp.	*Cornus mas*	*Ficus carica* seeds	*F. carica* fruits	*Juniperus* spp.	*Malus* sp.
AN[b]	Argissa						x			
	Gendiki									
	Knossos	x					x			
	Sesklo						x			
EN[c]	Achilleion									
	Argissa						x			
	Fyllotsairi					x				
	Franchthi	x								
	Giannitsa B					x				
	Knossos	x					x	x		
	Koutroulou Magoula					x	x			
	Nea Nikomedeia					x				
	Otzaki Magoula						x			
	Paliambela					x	x	x		
	Prodromos					x				
	Revenia					x	x	x		
	Sesklo						x			
	Toumba Balomenou						x			
EN/MN[d]	Cave of Cyclops									
MN	Achilleion									x
	Apsalos						x			
	Dispilio					x	x			
	Knossos	x		x			x	x		
	Limenaria						x		x	
	Otzaki Magoula						x			
	Servia	x			x (cf. *C. australis*)	x				
	Sesklo						x			
	Sitagroi	x				x				
	Toumba Balomenou						x			
	Tsoungiza						x			
MN/LN[e]	Avgi					x	x			
	Dispilio									
	Kouphovouno						x			
LN	Arapi Magoula						x			

Malus/ Pyrus/ Sorbus	*Prunus cerasus*	*Prunus spinosa*	*Prunus* sp.	*Punica* sp.	*Pyrus* sp.	*Querqus* sp.	*Rosa* sp.	*Rubus fruticosus*	*Rubus idaeus*	*Rubus* sp.	*Sambucus* sp.
											x (*S. ebulus*)
						x				x	
						x					
											x (*S. ebulus*)
										x	x
											x
		x				x					
x (*Malus/ Sorbus*)			x		x			x			x
						x					
x (*Malus/ Sorbus*)			x							x	x
			x	x		x			x	x	
						x					
								x			x
								x			x (*S. ebulus*)
			x (cf. *P. ceracifera/ institia*)		x			x			x (*S. nigra*)
										x	
										x	
											x (*S. ebulus*)

Phase	Site	*Amygdalus* sp.	*Arbutus* sp.	*Capparis* sp.	*Celtis* sp.	*Cornus mas*	*Ficus carica* seeds	*F. carica* fruits	*Juniperus* spp.	*Malus* sp.
	Arkadikos					x				
	Cave of Cyclops									
	Dikili Tash		x			x	x[a]	x		
	Dimini	x					x			
	Dimitra					x				
	Drakaina	x					x	x		
	Kalythies						x			
	Kleitos					x	x			
	Knossos	x						x		
	Koroneia					x				
	Kremasti Koiladas					x	x			
	Vasilika Kyparissi					x	x[a]	x[a]		
	Lerna		x				x			
	Makri			x		x	x[a]			
	Makrygialos					x	x[a]			
	Olynthus						x			
	Pefkakia	x					x			
	Sarakenos						x[a]			
	Servia					x				
	Sitagroi	x				x	x[a]			
	Skoteini Tharounia						x			
	Sesklo						x			
	Stavroupoli					x	x	x		
LN/FN[f]	Agios Ioannis Loukas						x			
	Alepotrypa	x					x			
	Dispilio									
	Sesklo									
FN	Alepotrypa	x					x			
	Dikili Tash					x	x	x		
	Kleitos									
	Mandalo					x	x			
	Pefkakia	x					x			
	Rachmani							x		
	Sitagroi	x				x				
	Tsoungiza						x			

Malus/ Pyrus/ Sorbus	***Prunus cerasus***	***Prunus spinosa***	***Prunus* sp.**	***Punica* sp.**	***Pyrus* sp.**	***Querqus* sp.**	***Rosa* sp.**	***Rubus fruticosus***	***Rubus idaeus***	***Rubus* sp.**	***Sambucus* sp.**
								x			x
					x	x				x	x (*S. ebulus*)
					x						
											x (*S. nigra*)
								x		x	x
							x			x	x
x								x			
			x			x	x	x			x (*S. ebulus*)
x						x		x			x
			x (cf. *P. ceracifera/ institia* & cf. *P. mahaleb*)		x	x		x			
x											
						x					
								x			x
					x						
						x					
					x[a] (P. amygdaliformis)	x		x			x (*S. nigra, S.ebulus*)
								x			x (*S. ebulus*)
x						x		x[a]			x (*S. ebulus*)
						x					
						x[a]					
						x					

Phase	Site	*Amygdalus* sp.	*Arbutus* sp.	*Capparis* sp.	*Celtis* sp.	*Cornus mas*	*Ficus carica* seeds	*F. carica* fruits	*Juniperus* spp.	*Malus* sp.
MN/FN	Franchthi	x								
FN/EBA[g]	Lerna		x				x			
	Vassilara Rahi					x				
	Theopetra			x (C. spinosa)	x	x	x	x	x	
	Thermi B									

[a]Presence of species in dense, pure concentrations established on the basis of a minimum of 100 quantifiable items in a single sample.
[b]AN = Aceramic Neolithic (seventh millennium BC)
[c]EN = Early Neolithic (7000–5800 BC)
[d]MN = Middle Neolithic (5800–5300 BC)
[e]LN = Late Neolithic (5300–4500 BC)
[f]FN = Final Neolithic (4500–3500 BC)
[g]EBA = Early Bronze Age (3500–2100 BC)

Malus/ Pyrus/ Sorbus	*Prunus cerasus*	*Prunus spinosa*	*Prunus* sp.	*Punica* sp.	*Pyrus* sp.	*Querqus* sp.	*Rosa* sp.	*Rubus fruticosus*	*Rubus idaeus*	*Rubus* sp.	*Sambucus* sp.
					x						
						x					
x (Malus/ Pyrus)			x		x (P. amygdaliformis)	x	x	x		x	x
								x			

TABLE 5.2. FRUIT SPECIES IDENTIFIED FROM GREEK BRONZE AGE SITES.

Phase	Site	*Prunus dulcis/ Amygdalus* sp.	*Cornus mas*	*Corylus* sp.	*Cucumis melo* L.	*Ficus carica*	*Fragaria vesca*
EBA[b]	Agios Athanasios					x	
	Aphrodite's Kephali	x				x	
	Archondiko		x			x	
	Argissa					x	
	Dhaskalio	x				x (sp.)	
	Kassaneva					x	
	Kastanas		x			x	
	Kasteli					x	
	Lerna					x	
	Limenaria					x	
	Mandalo		x			x	
	Mesimeriani Toumba						
	Myrtos						
	Pefkakia					x	
	Petras						
	Poliochni					x[a]	
	Priniatikos Pyrgos					x	
	P.O.T.A. Romanou						
	Servia		x				
	Sitagroi			x		x	
	Synoro						
	Tiryns					x	
	Tsoungiza	x			x (sp.)	x[a]	
EBA/MBA[c]	Kasteli	x					
	Petras					x	
MBA	Agia Paraskevi					x	
	Agios Mamas					x	
	Agora						
	Argissa					x	
	Aspis					x	
	Chamalevri-Tzambakas	x				x	
	Kasteli	x				x	
	Kommos	x					
	Kouphovouno						
	Lerna						
	Monastiraki	x				x	
	Nichoria						
	Orchomenos						
	Pefkalia					x	
	Platia Magoula Zarkou						
	Pseira						
	Toumba Thessalonikis		x			x	
	Tsoungiza					x	

Juniper sp.	*Malus* sp.	*Prunus* sp.	*Pyrus* sp.	*Quercus* sp.	*Rubus fruticosus*	*Rubus* sp.	*Sambucus ebulus*	*Sambucus* sp.	*Vitis vinifera*
							x		x
				x	x		x		x
			x				x		x
									x
									x
			x	x[a]	x		x		x
									x
		x		x					x[a]
				x[a]	x		x		x
									x
									x
			x				x		x
									x
									x
									x
									x
									x
x				x[a]					x
									x
									x
			x (type)	x		x			x
									x
						x			x
		x		x			x		x
									x
				x			x		x
									x
			x					x	x
									x
				x					
									x
				x					x (imprint)
									x
		x		x					x
									x
				x			x		x
				x					
									x
				x					x
									x[a]

Phase	Site	*Prunus dulcis/ Amygdalus* sp.	*Cornus mas*	*Corylus* sp.	*Cucumis melo* L.	*Ficus carica*	*Fragaria vesca*
MBA/LBA[d]	Kasteli	x				x	
	Malia	x				x	
	Tsoungiza					x	
LBA	Agia Triada					x	
	Agios Mamas		x			x	
	Agios Vasileios					x	
	Akrotiri					x	
	Angelochori					x	
	Archondiko		x			x	
	Assiros					x	
	Agia Sotira					x	
	Barnavos						
	Delphi (Pilier des Rhodiens)					x	
	Dimitra						
	Gournia					x	
	Iolkos						
	Iria					x	
	Kalapodi					x	
	Kastanas					x	x
	Kasteli	x				x	
	Kataphygadi Cave					x[a]	
	Knossos					x	
	Knossos Villa Dionysus					x	
	Kommos	x					
	Mesimeriani Toumba						
	Midea					x	
	Mochlos	x				x	
	Mycenae					x	
	Nichoria						
	Orchomenos						
	Palaikastro	x					
	Palaikastro Building 1	x				x	
	Papadiokampos						
	Phylakopi						
	Pseira	x *Prunus dulcis* shell fragment					
	Tiryns	x			x	x	
	Toumba Thessalonikis		x			x	
	Tsoungiza				x	x	
	Vronda	x				x	
	Zominthos					x	

[a]Presence of species in dense, pure concentrations established on the basis of a minimum of 100 quantifiable items in a single sample.
[b]EBA = Early Bronze Age (3500–2100 BC)
[c]MBA = Middle Bronze Age (2100–1700 BC)
[d]LBA = Late Bronze Age (1700–1200 BC)

Juniper sp.	*Malus* sp.	*Prunus* sp.	*Pyrus* sp.	*Quercus* sp.	*Rubus fruticosus*	*Rubus* sp.	*Sambucus ebulus*	*Sambucus* sp.	*Vitis vinifera*
	x								x
				x		x			x
									x
									x
							x		x
					x				x
					x				x
									x
									x
									x
									x
									x
				x					x
				x	x		x		x[a]
			x			x			x
		x							x
									x
				x					x
									x
								x	x
									x
									x
		x		x					x
									x
		x							x
			x						x
									x
									x
							x		x
				x		x			x
									x

TABLE 5.3. GRAPEVINE FINDS FROM GREEK NEOLITHIC SITES.

Phase	Site	*V. vinifera* pips	*V. vivifera* pedicels	Pressings
AN[b]	Argissa	x		
EN[c]	Achilleion	x		
	Giannitsa B	x		
	Knossos	x		
	Paliambela	x		
	Revenia	x		
	Sesklo	x		
	Toumba Balomenou	x	x	
MN[d]	Apsalos	x		
	Dispilio	x		
	Knossos	x		
	Linemaria	x		
	Photolivos	x		
	Servia	x		
	Sesklo	x		
	Tsoungiza	x		
MN/LN[e]	Avgi	x		
LN	Arapi Magoula	x		
	Dikili Tash	x		
	Dimini	x		
	Dimitra	x	x	
	Koroneia	x		
	Kremasti Koiladas	x		
	Vasilika Kyparissi	x		
	Makri	x		
	Makrygialos	x		
	Pefkakia	x		
	Servia	x		
	Sesklo	x		
	Sitagroi	x		
	Stavroupoli	x		
LN/FN[f]	Dispilio	x		
	Photolivos	x		
FN	Dikili Tash	x[a]	x	x[a]
	Mandalo	x	x	

Phase	Site	*V. vinifera* pips	*V. vivifera* pedicels	Pressings
	Pefkakia	x		
	Sitagroi	x		
	Vassilara Rahi	x		
	Theopetra	x		
	Thermi B	x		
MN/FN	Franchthi	x		

[a]Presence of species in dense, pure concentrations established on the basis of a minimum of 100 quantifiable items in a single sample.
[b]AN = Aceramic Neolithic (seventh millennium BC)
[c]EN = Early Neolithic (7000–5800 BC)
[d]MN = Middle Neolithic (5800–5300 BC)
[e]LN = Late Neolithic (5300–4500 BC)
[f]FN = Final Neolithic (4500–3500 BC)

TABLE 5.4. GRAPEVINE FINDS FROM GREEK BRONZE AGE SITES.

Phase	Site	*Vitis vinifera* pips	Pedicels	Pressings
EBA[b]	Agios Athanasios	x		
	Agios Kosmas	x		
	Aphrodite's Kephali	x		
	Archondiko	x	x	
	Argissa	x		
	Dhaskalio	x	x	
	Kassaneva	x		
	Kastanas	x		
	Kasteli	x		
	Lerna	x[a]		
	Mandalo	x		
	Mesimeriani Toumba	x		
	Myrtos	x	x	x
	Pefkakia	x		
	Petras	x		
	Poliochni	x		
	Priniatikos Pyrgos	x	x	
	P.O.T.A. Romanou	x		
	Servia	x		
	Sitagroi	x	x	
	Synoro	x		
	Tiryns	x		
	Tsoungiza	x	x	
	Vassilara Rahi	x		
EBA/MBA[c]	Petras	x		
MBA	Agia Paraskevi	x		
	Agios Mamas	x		
	Agora	x		
	Argissa	x		
	Aspis	x		
	Chamalevri-Tzambakas	x		
	Kasteli	x		
	Kouphovouno	x		
	Monastiraki	x	x	x
	Nichoria	x		
	Orchomenos	x		
	Pefkakia	x		
	Pseira	x		
	Toumba Thessalonikis	x		

Phase	Site	*Vitis vinifera* pips	Pedicels	Pressings
	Tsoungiza	x[a]	x	
MBA/LBA[d]	Kasteli	x		
	Malia	x		
LBA	Agios Mamas	x		
	Agios Vasileios	x		
	Akrotiri	x		
	Angelochori	x		
	Archondiko	x	x	
	Assiros	x		
	Agia Sotira	x	x	
	Barnavos	x		
	Delphi (Pilier des Rhodiens)	x		
	Dimitra	x		
	Gournia	x		x
	Iolkos	x		
	Kastanas	x[a]		
	Kasteli	x		
	Kataphygadi Cave	x		
	Knossos	x		
	Kommos	x		
	Mesimeriani Toumba	x		
	Midea	x		
	Mochlos	x	x	
	Mycenae	x		
	Nichoria	x		
	Orchomenos	x		
	Palaikastro	x		
	Palaikastro Building 1	x		
	Papadiokampos	x		
	Phylakopi	x		
	Tiryns	x		
	Toumba Thessalonikis	x		x
	Tsoungiza	x	x	
	Vronda	x	x	
	Zominthos	x	x	

[a]Presence of species in dense, pure concentrations established on the basis of a minimum of 100 quantifiable items in a single sample.
[b]EBA = Early Bronze Age (3500–2100 BC)
[c]MBA = Middle Bronze Age (2100–1700 BC)
[d]LBA = Late Bronze Age (1700–1200 BC)

TABLE 7.1. LIST OF SPECIES WITH MEDICINAL/AROMATIC PROPERTIES FROM NORTHERN GREEK SITES.

Species	Part of plant used	Properties
Quercus spp.	bark	astringent, antiseptic
Ficus caricaa	fruit	laxative, vulnerary (roasted fruit)
Urtica spp.	whole plant	diuretic, stimulant, astringent
Polygonum aviculare	whole plant	diuretic, against kidney disorders
Rumex acetosella	whole plant	antidysenteric, antiscorbutic, for stomach ailments
Stellaria spp.	whole plant	demulcent, refrigerant
Agrostemma githago		narcotic, toxic
Ranunculus spp.		poisonous
Papaver somniferuma	capsules, flower	narcotic, anodyne, expectorant, astringent, diaphoretic
Fumaria spp.	whole plant	tonic, diuretic, diaphoretic
Glaucium spp.		laxative, eye ulcer treatment
Camelina sativa	oil	emollient
Brassica nigra/rapa	seeds, oil	stimulant, diuretic, emetic
Capparis spp.		ear infections
Reseda cf. *luteola*	oil	hallucinogenic
Aphanes spp.		used in past medical recipes
Rubus fruticosus agg.[a]	root, leaves	astringent, tonic for diarrhea, dysentery
Rosa spp.	fruit	astringent, refrigerant, pectoral
Potentilla spp.		hemostatic, for diarrhea, dysentery, tonic, astringent, tonic
Fragaria vesca	leaves, fruit, roots	antirheumatic, diuretic, hepatic, astringent
Prunus cf. *spinosa*		antispasmodic, diuretic, diaphoretic, palliative, blood purifier
Prunus amygdalus	fruit, oil	antitussive, treatment kidney disorders
Vicia ervilia	seeds	treatment of skin conditions, poisonous
Trifolium spp.	flowers	palliative, antispasmodic
Melilotus spp.	whole plant	aromatic, emollient, carminative, insect repellent, flavor enhancer for cheese
Medicago spp.[a]	whole plant	weight increase
Linum usitatisimum	seeds, oil	emulcent, pectoral
Pistacia terebinthusa		aromatic resin
Vitis vinifera	fruit, leaves, juice	astringent, hemostatic, strengthening
Malva spp.		antiphlogistic, expectorant
Thymelea spp.		insect repellent
Hypericum spp.[a]	whole plant, flowers	vulnerary, sciatica, snake bites
Cornus mas		bowel disorders, treatment of cholera and diarrhea
Coriandrum sativuma	fruit and leaves	aromatic, stimulant, carminative
Anagalis cf. *arvensis*		poisonous, medicinal use in the past
Galium spp.	whole plant	laxative, diuretic, tonic
Heliotropium spp.		scorpion bites treatment, contraceptive

Verbena officinalis	whole plant	emetic, diaphoretic, astringent, antispasmodic, tonic
Ajuca spp.	whole plant	astringent, aromatic
Teucrium spp.	whole plant	antipyretic, diaphoretic, diuretic, stimulant
Hyoscyamus spp.	leaves, flowers	poisonous, sedative, palliative, pupil dilation
Physalis alkenkengi		diuretic, refrigerant, aperient, treatment for colic, hydrops
Solanum nigrum		poisonous
Plantago lanceolata type		antispasmodic, diuretic, vulnerary, gastroprotective, laxative
Sambucus ebulus	fruit, leaves	laxative, juice used for wine coloring, antirheumatic
Sambucus nigra	fruit, leaves, flowers, bark	purgative, for inhalations, insecticide
Centaurea spp.	flowers	tonic, stimulant, emmenagogue
Allium spp.	bulb	antiseptic, diuretic, stimulant
Hordeum vulgare	seeds	antipyretic, strengthening, treatment of intestinal disorders
Avena spp.	seeds	stimulant, antispasmodic
Lallemantia sp. *(Lallemantia royleana)*	seeds	fever, common cold, expectorant, coughing, anti-thirst
Lallemantia sp. (Lallemantia iberica)	seeds	reconstitute, stimulant, diuretic, expectorant

Note: Species/genera that appear often and/or in significant quantities in the archaeobotanical samples.
Sourced from Amiri et al. 2012, Bauman 1993, Brockhampton Reference 1996, Charistos 1988, Eleftherochorinos 1989, Heldreich 1909, Huxley και Taylor 1989, Naghibi et al. 2005, Panagiotakopulu et al. 1995, Pesexidis 1981, Polunin 1969 and 1987, Polunin and Huxley 1987, Scarborough 1991, Schönfelder and Schönfelder 1990, Sfikas 1983 and 1995.

TABLE 7.2. PLANT SPECIES/GENERA ATTRIBUTED TO LINEAR B IDEOGRAMS.

Linear B word	Common name	Latin name(s)	
ka-na-ko	Safflower	*Carthamus tinctorius*	a
ka-na-ko e-ru-ta-ra	Safflower	*Carthamus tinctorius* L.	b
ka-na-ko re-u-ka	Safflower seeds	*Carthamus tinctorius* L.	b
ka-ra-ko	*Pennyroyal*	*Mentha pulegium* L.	b
ku-mi-no	Cumin	*Cuminum cyminum*	a, b
ku-mi-no	Cumin	*Cuminum cyminum* L.	b
mi-ta	Mint	*Mentha* spp.	a
mi-ta	*Mint*	*Mentha viridis* L.	b
sa-sa-ma	Sesame	*Sesamum indicum*	a, b
ka-da-mi-ja	Cardamom	*Cardamomum* spp.	a
ka-da-mi-ja	Kardamia/Cress	*Nasturtium officinale* L. or *Lepidium sativum* L.	b
ko-no or *ko-i-no*	Ginger grass	*Cymbopogon schoenanthus*	a, b
ko-no or *ko-i-no*	Terebinth	*Pistacia terebinthus* or *P. lentiscus*	a
ki-ta-no	Terebith	*Pistacia terebinthus* L. or *P. atlantica* L.	b
ki-ta-no	Rockrose	*Cistus creticus* L.	a
KAPO	Cinnamon	*Cinnamomum* cf. *cassia*	a
se-ri-no	Celery	*Apium graveolens*	a, b
wi-ri-za	Iris root (?)	*Iris germanica, Iris* spp.	a
wi-ri-za	Iris root	*Iris germanica* L. or *I. cretica* L.	b
ma-ra-tu-wo	Fennel	*Foeniculum vulgare* Mill.	a, b
ku-pa-ro	*Cyperus*	*Cyperus* spp., *C. esculentus, C. rotundus* L., *C. longus* L.	a
ku-pa-ro	*Cyperus* grass	*Cyperus rotundus* L. or *C. esculentus* L.	b
ko-ri-ja-do-no/ ko-ri-ja-da-na/ ko-ri-a-da-na	Coriander	*Coriandrum sativum*	a, b
po-ni-ki-jo	Alkanet	*Alkanna tinctoria*	a
po-ni-ki-jo	Madder	*Rubia tinctorum/Anchusa tinctoria*	a
CROC	Saffron crocus	*Crocus* spp., *C. cartwrightianus, C. sativus*	a
ko-ro-ki-no	Saffron	*Crocus sativus* ssp., *C. oreocreticus* L. or *C. cartwrightiensis* L.	b
wo-do-we	Rose	*Rosa* spp. (e.g., *R. canina, R. gallica* subsp. *centifolia*)	a
wo-do-we	Rose	*Rosa centifolia* L.	b
pa-ko-we	Sage	*Salvia* spp.	a
pa-ko-we	Sage	*Salvia tribola* L. fils	b
e-ti-we	Henna	*Lawsonia inermis* L./*L. inermis* var. *alba*	a
ep-i-ka	Tree-mallow or ebiskos	*Malva silvestris* or *Lavatera arborea* L.	b
MU	Myrrh	*Balsamodendron* spp./*Commiphora*	a
su-za	Figs	*Ficus carica* L.	b
ri-ta	Flax	*Linum usitatissimum* L	b
ki-da-ro	Onions	*Allium cepa* L.	b

Note: Not all species/genera are accurately identified nor is there concensus on their interpretation.
[a]After Sarpaki 2001b
[b]After Arnott 2014

Appendix B: AMS Dates

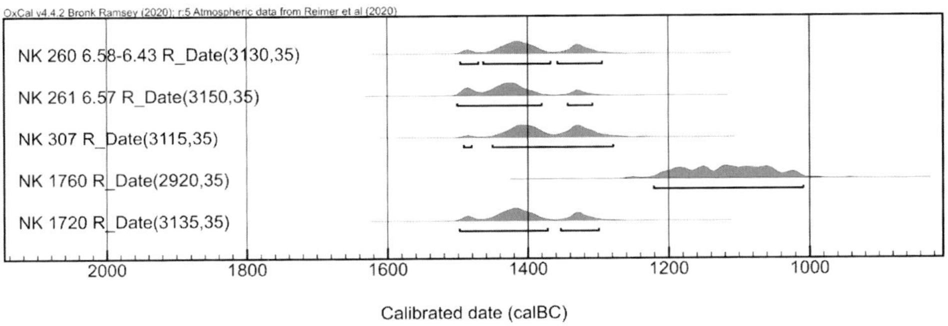

14C dates of millet grains from Skala Sotiros (individual flotation sample numbers: NK260, NK261, NK307) and Archondiko (individual flotation sample numbers NK1720, NK1760); 14C ages were calibrated against the curve INTCAL20 (Reimer et al. 2020), using the program OxCal ver 4.2 (Bronk Ramsey 2009; Bronk Ramsey, Lee 2013)

References Cited

Abanoudis, Kostas

2007 Τον καιρό που πουλούσαμε τσίκουδα [The Time We Used to Sell Tsikoudia]. www.vasileoniko.gr/index.php?mid=3&smid=12.

Abbas, Mazher, Tahir Mehmood, Arshed Bashir, Muneeb Zafar, and Aneela Afzal

2012 Economics of *Lallemantia royleana* (Tukham-E-Balangoo) Production in Low Intensity Cropping Zone of the Punjab, Pakistan. *Pakistan Journal of Agricultural Research* 25(2):110–119.

Abbo, Shahal, and Avi Gopher

2017 Near Eastern Plant Domestication: A History of Thought. *Trends in Plant Science* 22(6):491–511.

2020 Plant Domestication in the Neolithic Near East: The Humans-Plants Liaison. *Quaternary Science Reviews* 242:106412.

Abbo Shahal, Simcha Lev-Yadun, and Avi Gopher

2011 Origin of Near Eastern Plant Domestication: Homage to Claude Lévi-Strauss and "La Pensée Sauvage." *Genetic Resources and Crop Evolution* 58(2):175–179.

2012 Plant Domestication and Crop Evolution in the Near East: On Events and Processes. *Critical Reviews in Plant Sciences* 31(3):241–257.

Abbo Shahal, Zvi Peleg, Simcha Lev-Yadun, and Avi Gopher

2021 Does the Proportion of Shattering vs. Non-Shattering Cereal Remains in Archeobotanical Assemblages Reflect Near Eastern Neolithic Arable Fields? *Review of Palaeobotany and Palynology* 284:104339

Abbo, Shahal, Ruth Pinhasi van-Oss, Avi Gopher, Yehoshua Saranga, Itai Ofner, and Zvi Peleg

2014 Plant Domestication versus Crop Evolution: A Conceptual Framework for Cereals and Grain Legumes. *Trends in Plant Science* 19(6):351–360.

Abdalla, Michael

1990 Bulgur—An Important Wheat Product in the Cuisine of Contemporary Assyrians in the Middle East. In *Staple Foods: Proceedings of the Oxford Symposium on Food and Cookery 1989*, edited by Harlan Walker, pp. 27–37. Prospect Books, London.

Abdolahi, Mina, and Saeideh Maleki Farahani

2019 Seed Quality, Water Use Efficiency and Eco Physiological Characteristics of Lallemantia (*Lallemantia* sp.) Species as Effected by Soil Moisture Content. *Acta agriculturae Slovenica* 113(2):307–320.

Abrams, Leon. H.

1978 A Diachronic Preview of Wheat in Hominid Nutrition. *Journal of Applied Nutrition* 30:41–55.

Adams, Jenny L.
2008 Beyond the Broken. In *New Approaches to Old Stones: Recent Studies of Ground Stone Artefacts*, edited by Yorke M. Rowan and Jennie R. Ebeling, pp. 213–229. Equinox, London.
Adam-Veleni, Polyxeni, and Maria Mangafa
1996 Αρχαίο Ελαιοτριβείο στα Βρασνά Νομού Θεσσαλονίκης [Ancient Olive Press at Vrasna, Thessaloniki]. In *Ελιά και Λάδι (Olive and Oil). Proceedings of the 4th Three-Day Working Meeting on Olive and Oil, Kalamata, 7–13 May 1993*, pp. 92–104. ETVA Cultural and Technological Foundation, Athens.
Aijmer, Göran
1968 A Structural Approach to Chinese Ancestor Worship. *Bijdragen tot de Taal-, Land- en Volkenkunde* 124(1):91–98.
Akeret, Örni
2005 Plant Remains from a Bell Beaker Site in Switzerland, and the Beginnings of *Triticum spelta* (Spelt) Cultivation in Europe. *Vegetation History and Archaeobotany* 14(4):279–286.
Alagesaboopathi, Chinnappan
2012 Ethnobotanical Studies on Useful Plants of Sirumalai Hills of Eastern Ghats, Dindigul District of Tamilnadu, Southern India. *International Journal of Biosciences* 2(2):77–84.
Albala, Ken
2013 *Pancake: A Global History* (e-book). Originaly published 2008, Reaktion Books, London.
Albayaty, Nahida
2011 The Most Medicinal Plants Used in Iraq: Traditional Knowledge. *Advances in Environmental Biology* 5(2):401–406.
Alberti, Maria Emanuela
2016 Cooking Wares. In *Petras, Siteia I: A Minoan Palatial Settlement in Eastern Crete: Excavation of Houses I.1 and I.2*, edited by Metaxia Tsipopoulou, pp. 61–92. Prehistory Monographs 53. INSTAP Academic, Philadelphia.
Alcock, Susan E., Terence N. D' Altroy, Kathleen D. Morrison, and Carla M. Sinopoli (editors)
2001 *Empires: Perspectives from Archaeology and History*. Cambridge University Press, Cambridge.
Allen, Harriet D.
1990 A Postglacial Record from the Kopais Basin, Greece. In *Man's Role in the Shaping of the Eastern Mediterranean Landscape: Proceedings of the INQUA/BAI Symposium on the Impact of Ancient Man on the Landscape of the Eastern Mediterranean Region and the Near East, Groningen, Netherlands, 6–9 March 1989*, edited by Sytze Bottema, Gertie Entjes-Nieborg, and Willem van Zeist, pp. 173–182. Balkema, Rotterdam.
Allen, Susan E., and Kathleen M. Forste
2020 Archaeobotanical Remains. In *The Mycenaean Settlement on Tsoungiza Hill*, edited by James C. Wright and Mary K. Dabney, pp. 1029–1062. Nemea Valley Archaeological Project III. The American School of Classical Studies at Athens, Princeton.
Alonso, Natàlia
2019 A First Approach to Women, Tools, and Operational Sequences in Traditional

Manual Cereal Grinding. *Archaeological and Anthropological Sciences* 11(8):4307–4324.

Al-Snafi, Ali Esmail
2019 Medical Benefit of *Lallemantia iberica*—A Review. *To Chemistry Journal* 3:128–133.

Altmeyer, Peter
2018 Altmeyers Enzyklopädie, s.v. Weinsäure syn. *Acidum tartaricum*.

Amiri, Mohammad Sadegh, Parham Jabbarzadeh, and Mahdi Akhondi
2012 An Ethnobotanical Survey of Medicinal Plants Used by Indigenous People in Zangelanlo District, Northeast Iran. *Journal of Medicinal Plants Research* 6(5):749–753.

Ammerman, Albert J., and Luigi Luca Cavalli-Sforza
1971 Measuring the Rate of Spread of Early Farming in Europe. *Man* 6(4):674–688.
1973 Population Model for the Diffusion of Early Farming in Europe. In *The Explanation of Culture Change: Models in Prehistory. Proceedings of a Meeting of the Research Seminar in Archaeology and Related Subjects Held at the University of Sheffield*, edited by Colin Renfrew, pp. 343–357. Duckworth, London.
1984 *The Neolithic Transition and the Genetics of Population in Europe*. Princeton University Press, Princeton.

Amouretti, Marie-Claire
1986 *Le pain et l'huile dans la Grèce Antique: De l'araire au moulin*. Centre de Recherche d'Histoire Ancienne 67. Annales Littéraires de l'Université de Besançon 328. Les Belles Lettres, Paris.

Andreadaki-Vlazaki, Maria
1999 The Production of Aromatic and Pharmaceutical Oils in Minoan Crete: The Case of Chamalevri. In *Minoans and Mycenaeans: Flavours of Their Time. Exhibition Catalogue, National Archaeological Museum of Athens, 12 July–27 November 1999*, edited by Yannis Tzedakis and Holley Martlew, pp. 48–49. Kapon, Athens.

Andreou, Stelios, Michael Fotiadis, and Kostas Kotsakis
1996 Review of Aegean Prehistory V: The Neolithic and Bronze Age of Northern Greece. *American Journal of Archaeology* 100(3):537–597.

Andreou, Stelios, Carl Heron, Glynis Jones, Evangelia Kiriatzi, Kiki Psaraki, Maria Roumpou, and Soultana Maria Valamoti
2013 Smelly Barbarians or Perfumed Natives? An Investigation of Oil and Ointment in Use in Late Bronze Age Northern Greece. In *Diet, Economy, and Society in the Ancient Greek World: Towards a Better Integration of Archaeology and Science. Proceedings of the International Conference Held at the Netherlands Institute at Athens on 22–24 March 2010*, edited by Sofia Voutsaki and Soultana Maria Valamoti, pp. 173–185. Pharos Supplement 1. Peeters, Leuven.

Andreou, Stelios, and Kostas Kotsakis
1997 Η προϊστορική Τούμπα Θεσσαλονίκης: Παλιά και νέα ερωτήματα [The Prehistoric Mound (Toumba) of Thessaloniki: Old and New Questions]. *Proceedings of the Archaeological Work in Macedonia and Thrace (AEMTH)* 10A, 1996:369–387.

Andreou, Stelios, and Kyriaki Psaraki
2007 Tradition and Innovation in the Bronze Age Pottery of the Thessaloniki Toumba: Food and Drink Consumption and "Tableware" Ceramics. In *The Struma/Strymon River Valley in Prehistory: Proceedings of the International Symposium STRYMON PRAEHISTORIKUS, Kjustendil, Blagoevgrad (Bulgaria) and*

Serres—Amphipolis (Greece), 27/9–1/10/2004, edited by Henrieta Todorova, Mark Stefanovich, and Georgi Ivanov, pp. 397–420. In the Steps of James Harvey Gaul, Vol. 2. Gerda Henkel Stiftung, Sofia, Bulgaria.

Angel, J. Lawrence
1973 Human Skeletons from Grave Circles at Mycenae. In *Ο Ταφικός Κύκλος Β των Μυκηνών* [Grave Circle B at Mycenae], edited by Georg E. Mylonas, pp. 379–397. Monographs 73. The Library of the Archaeological Society at Athens.

Angioi, Simonetta A., Rau Domenico, Attene Giovanna, Nanni Laura, Bellucci Elisa, Logozzo Giuseppina, Negri Valeria, Spangoletti Zeuli Pierluigi, and Papa Roberto
2010 Beans in Europe: Origin and Structure of the European Landraces of *Phaseolus vulgaris* L. *Theoretical and Applied Genetics* 121(5):829–843.

Antolín, Ferran
2016 *Local, Intensive, and Diverse? Early Farmers and Plant Economy in the North-east of the Iberian Peninsula (5500–2300 BC)*. Advances in Archaeobotany 2. Barkhuis, Groningen.

Antolín, Ferran, Marian Berihuete-Azorín, and Oriol López-Bultó (editors)
2016 *Archaeobotany of Wild Plant Use: Approaches to the Exploitation of Wild Plant Resources in the Past and Its Social Implications*. Quaternary International 404(A). Elsevier, Oxford.

Antolín, Ferran, and Stefanie Jacomet
2015 Wild Fruit Use among Early Farmers in the Neolithic (5400–2300 cal BC) in the North-East of the Iberian Peninsula: An Intensive Practice? *Vegetation History and Archaeobotany* 24(1):19–33.

Appadurai, Arjun
1981 Gastropolitics in Hindu South Asia. *American Ethnologist* 8(3):494–511.

Aranguren, Biancamaria, Roberto Becattini, Marta Mariotti Lippi, and Anna Revedin
2007 Grinding Flour in Upper Palaeolithic Europe (25000 Years BP). *Antiquity* 81(314):845–855.

Arıtuluk, Zekiye Ceren, and Nurten Ezer
2012 Halk Arasında Diyabete Karşı Kullanılan Bitkiler (Türkiye)–II. *Hacettepe Üniversitesi Eczacılık Fakültesi Dergisi* 32(2):179–208.

Arnold, Bettina
1999 Drinking the Feast: Alcohol and the Legitimation of Power in Celtic Europe. *Cambridge Archaeological Journal* 9(1):71–93.

Arnold, Bettina, and Joshua Driscoll
2017 Keltischer Trank aus Brew City, USA. *Archäologie in Deutschland* 2017(2):64–65.

Arnold, Bettina, and Matthew L. Murray
2016 Two Hallstatt Burial Mounds in the Hohmichele Group in the "Speckhau." In *The Heuneburg and the Early Iron Age Princely Seats: First Towns North of the Alps*, edited by Dirk Krausse, Manuel Fernández-Götz, Leif Hansen, and Inga Kretschmer, pp. 121–127. Archaeolingua, Budapest.

Arnott, Robert
2014 Healers and Medicines in the Mycenaean Greek Texts. In *Medicine and Healing in the Ancient Mediterranean World*, edited by Demetrios Michaelides, pp. 44–53. Oxbow, Oxford.

Arobba, Daniele, Rosanna Caramiello, and Antonella Traverso
2010 Risultati preliminari di analisi archeobotaniche su campioni provenienti da Poliochni (Lemnos). *Annuario della Scuola Archeologica Italiana di Atene Atene e delle missioni italiane in oriente* 86, 2008:139–151.

Arranz-Otaegui, Amaia, Lara González-Carretero, Monica N. Ramsey, Dorian Q. Fuller, and Tobias Richter
2018 Archaeobotanical Evidence Reveals the Origins of Bread 14400 Years Ago in Northeastern Jordan. *Proceedings of the National Academy of Sciences of the United States of America (PNAS)* 115(31):7925–7930.

Arranz-Otaegui, Amaia, Juan José Ibáñez, and Lydia Zapata
2016 Hunter-Gatherer Plant Use in Southwest Asia: The Path to Agriculture. In *Wild Harvest: Plants in the Hominin and Pre-Agrarian Human Worlds*, edited by Karen Hardy and Lucy Kubiak-Martens, pp. 91–110. Studying Scientific Archaeology 2. Oxbow, Oxford.

Arroyo-García, Rosa Adela, Leticia Ruiz-Garcia, Laurence Bolling, Rafael Ocete, M. A. López, Claire Arnold, Ali Ergül, Gökhan Söylemezoğlu, Halil Ibrahim Uzun, F. Leyva Cabello, Javier Ibáñez, Mallikarjuna K. Aradhya, Atanas Atanassov, Ivan Atanassov, Susan Bálint, José L. Cenis, Luigi Costantini, S. Gorislavets, Maria Stella Grando, Benjamin Y. Klein, Patrick E. McGovern, Didier Merdinoglu, I. I. Pejic, Frédérique Pelsy, Nikolas I. Primikirios, V. Risovannaya, Kalliopi A. Roubelakis-Angelakis, Hager Snoussi, Petraq Sotiri, Shubhada A. Tamhankar, Patrice This, Leonid P. Troshin, José María Malpica, François Lefort, and José Miguel Martínez-Zapater
2006 Multiple Origins of Cultivated Grapevine (*Vitis vinifera* L. ssp. *sativa*) Based on Chloroplast DNA Polymorphisms. *Molecular Ecology* 15(12):3707–3714.

Arthur, John W.
2003 Brewing Beer: Status, Wealth, and Ceramic Use Alteration among the Gamo of South-Western Ethiopia. *World Archaeology* 34(3):516–528.
2014 Beer through the Ages: The Role of Beer in Shaping Our Past and Current Worlds. *Archaeology Now* 6(2):1–11.

Artzy, Michal
1994 Incense, Camels, and Collared Rim Jars: Desert Trade Routes and Maritime Outlets in the Second Millennium. *Oxford Journal of Archaeology* 13(2):121–147.

Asif, Muhammad, Lloyd W. Rooney, Rashida Ali, and Mian N. Riaz
2013 Application and Opportunities of Pulses in Food System: A Review. *Critical Reviews in Food Science and Nutrition* 53(11):1168–1179.

Asouti, Eleni
2003 Wood Charcoal from Santorini (Thera): New Evidence for Climate, Vegetation, and Timber Imports in the Aegean Bronze Age. *Antiquity* 77(297):471–484.

Asouti, Eleni, and Dorian Q. Fuller
2012 From Foraging to Farming in the Southern Levant: The Development of Epipalaeolithic and Pre-Pottery Neolithic Plant Management Strategies. *Vegetation History and Archaeobotany* 21(2):149–162.
2013 A Contextual Approach to the Emergence of Agriculture in Southwest Asia: Reconstructing Early Neolithic Plant-Food Production. *Current Anthropology* 54(3):299–345.

Asouti, Eleni, Maria Ntinou, and Ceren Kabukcu
2018 The Impact of Environmental Change on Palaeolithic and Mesolithic Plant Use and the Transition to Agriculture at Franchthi Cave, Greece. *PLoS ONE* 13(11):e0207805.

Athanasiadis, Nikolaos, Achilleas Gerasimidis, and Sampson Panagiotidis
1996 Η ελιά στα διαγράμματα γύρης και η σημασία της από ιστορικοαρχαιολογική άποψη. In *Ελιά και Λάδι* [Olive and Oil]. *Proceedings of the 4th Three-Day Working Meeting on Olive and Oil, Kalamata, 7–13 May 1993*, pp. 78–91. ETVA Cultural and Technological Foundation, Athens.
1998 Η άμπελος στα διαγράμματα γύρης της Βόρειας Ελλάδας και η σημασία της από ιστορικοαρχαιολογική άποψη [The Grapevine in Pollen Diagrams from Northern Greece and Its Historical-Archaeological Significance]. In *Αμπελοοινική ιστορία στο χώρο της Μακεδονίας και της Θράκης* [The Viticultural History of Macedonia and Thrace]. *Proceedings of the Three-Day Working Meeting on the Viticultural History of Macedonia and Thrace, Naoussa, 17–19 September 1993*, pp. 170–178. ETVA Cultural and Technological Foundation, Athens.

Attia, Elshafaey A. E., Elena Marinova, Ahmed G. Fahmy, and Masahiro Baba
2018 Archaeobotanical Studies from Hierakonpolis: Evidence for Food Processing during the Predynastic Period in Egypt. In *Plants and People in the African Past: Progress in African Archaeobotany*, edited by Anna Maria Mercuri, Catherine D' Andrea, Rita Fornaciari, and Alexa Höhn, pp. 76–89. Springer, Cham, Switzerland.

Aubaile-Sallenave, Françoise
1994 Al-Kishk: The Past and Present of a Complex Culinary Practice. In *A Taste of Thyme: Culinary Cultures of the Middle East*, edited by Sami Zubaida and Richard Tapper, pp. 104–139. I. B. Tauris, London.

Avitsur, Shmuel
1977 The Way to Bread: The Example of the Land of Israel. *Tools and Tillage* 2(4):228–241.

Ayerdi, Miren, Amaya Echazarreta-Gallego, Sara de Francisco-Rodríguez, Hugo H. Hernández, and Izaskun Sarasketa-Gartzia
2016 Acorn Cake during the Holocene: Experimental Reconstruction of Its Preparation in the Western Pyrenees, Iberia. *Vegetation History and Archaeobotany* 25(5):443–457.

Aykroyd, Wallace Ruddell, Joyce Doughty, and Ann F. Walker
1982 *Legumes in Human Nutrition*. FAO Food and Nutrition Paper 20. Food and Agriculture Organization of the United Nations, Rome.

Ayora-Díaz, Steffan-Igor
2012 *Foodscapes, Foodfields, and Identities in Yucatán*. CEDLA Latin America Studies series 99. Berghahn, New York.

Azcan, Nezihe, B. Ozturk-Kalender, and Mehmet Kara
2004 Investigation of Turkish Poppy Seeds and Seed Oils. *Chemistry of Natural Compounds* 40(4):370–372.

Badr, Amr, Karl Müller, Ralf Schäfer-Pregl, Ha El Rabey, Sigi Effgen, H. H. Ibrahim, Carlo Pozzi, Wolfgang Rohde, and Francesco Salamini
2000 On the Origin and Domestication History of Barley (*Hordeum vulgare*). *Molecular Biology and Evolution* 17(4):499–510.

Bacvarov, Krum

2007 Jar Burials as Early Settlement Markers in Southeast European Neolithic. In *A Short Walk through the Balkans: The First Farmers of the Carpathian Basin and Adjacent Regions. Proceedings of the Conference held at the Institute of Archaeology UCL on June 20th–22nd, 2005*, edited by Michela Spataro and Paolo Biagi, pp. 189–205. Società per la preistoria e protostoria della regione Friuli-Venezia Gulia, Trieste.

Badgujar, Shamkant B., Vainav V. Patel, Atmaram H. Bandivdekar, and Raghunath T. Mahajan

2014 Traditional Uses, Phytochemistry, and Pharmacology of *Ficus carica*: A Review. *Pharmaceutical Biology* 52(11):1487–1503.

Baik, Byung-Kee, and Steven E. Ullrich

2008 Barley for Food: Characteristics, Improvement, and Renewed Interest. *Journal of Cereal Science* 48(2):233–242.

Bakels, Corrie

1996 Fruits and Seeds from the Linearbandkeramik Settlement at Meindling, Germany, with Special Reference to *Papaver somniferum*. *Analecta Praehistorica Leidensia* 25:55–68.

2005 Crops Produced in the Southern Netherlands and Northern France during the Early Medieval Period: A Comparison. *Vegetation History and Archaeobotany* 14(4):394–399.

Balls, Edward K.

1962 *Early Uses of California Plants*. University of California Press, Berkeley.

Barker, Graeme

1985 *Prehistoric Farming in Europe*. Cambridge University Press, Cambridge.

Barley, Nigel

1984 Placing the West African Potter. In *Earthenware in Asia and Africa. A colloquy held 21–23 June 1982*, edited by John Picton, pp. 93–105. Colloquies on Art and Archaeology in Asia 12. Percival David Foundation of Chinese Art/School of Oriental and African Studies/University of London, London.

Barnard, Hans, Alek N. Dooley, Gregory Areshian, Boris Gasparyan, and Kym F. Faull

2011 Chemical Evidence for Wine Production around 4000 BCE in the Late Chalcolithic Near Eastern Highlands. *Journal of Archaeological Science* 38(5):977–984.

Başaran, Uğur, Ozlem Onal Asci, Hanife Mut, Zeki Acar and Ilknur Ayan

2011 Some Quality Traits and Neurotoxin β-N-oxalyl-L-α,β-diaminopropionic Acid (β-ODAP) Contents of *Lathyrus* sp. Cultivated in Turkey. *African Journal of Biotechnology* 10(20):4072–4080.

Batsatsashvili, Ketevan, Naiba Mehdiyeva, George Fayvush, Zaal Kikvidze, Manana Khutsishvili, Inesa Maisaia, Shalva Sikharulidze, David Tchelidze, Alla Aleksanyan, Valida Alizade, Narel Y. Paniagua Zambrana, and Rainer W. Bussmann

2016 *Cornus mas* L. Cornaceae. In *Ethnobotany of the Caucasus*, edited by Rainer W Bussmann, pp. 1–7. European Ethnobotany. Springer, Cham, Switzerland.

Bauer-Petrovska, Biljana

2012 Historical Review of Medicinal Plants' Usage. *Pharmacognosy Reviews* 6(11):1–5.

Baumann, Hellmut

1993 *Η ελληνική χλωρίδα στο μύθο, στην τέχνη, στη λογοτεχνία* [The Greek Plant World in Myth, Art and Literature]. The Hellenic Society for the Protection of Nature, Athens.

Bayram, Mustafa

2000 Bulgur around the World. *Cereal Foods World* 45(2):80–82.

2008 An Analysis of Scorched Immature Wheat: Frekeh. *Cereal Foods World* 53(3):134–138.

Becker, Cornelia, and Helmut Kroll

2008 *Das Prähistorische Olynth: Ausgrabungen in der Toumba Agios Mamas, 1994–1996. Ernährung und Rohstoffnutzung im Wandel.* Prähistorische Archäologie in Südosteuropa 22. Marie Leidorf, Rahden/Westfalen.

Bekiaris, Tasos

2020 Ground Stone Technology in Context: Consumption of Grinding Tools and Social Practice at Neolithic Avgi, NW Greece. *Journal of Lithic Studies* 7(3).

Bekiaris, Tasos, Danai Chondrou, Ismini Ninou, and Soultana Maria Valamoti

2020 Food-Processing Ground Stone Tools in the Greek Neolithic and Bronze Age: A Synthesis of the Published Data. *Journal of Greek Archaeology* 5:135–195.

Belfer-Cohen, Anna, and A. Nigel Goring-Morris

2011 Becoming Farmers: The Inside Story. *Current Anthropology* 52, Supplement 4:S209–S220.

Bendall, Lisa M.

1998–1999 A Time for Offerings: Dedications of Perfumed Oil at Pylian Festivals. In *A-NA-QO-TA. Studies presented to J. T. Killen*, edited by John Bennet and Jan Driessen, pp. 1–9. Minos 33–34. University of Salamanca.

2007 *Economics of Religion in the Mycenaean World: Resources Dedicated to Religion in the Mycenaean Palace Economy*. Monograph 67. Oxford University School of Archaeology.

Bender, Barbara

1978 Gatherer-Hunter to Farmer: A Social Perspective. *World Archaeology* 10(2):204–222.

Bending, Joanna

2011 Archaeobotanical Assemblages. In *Mochlos IIC: Period IV. The Mycenaean Settlement and Cemetery: The Human Remains and Other Finds*, edited by Jeffrey S. Soles and Costis Davaras, pp. 133–140. Prehistory Monographs 32. INSTAP Academic, Philadelphia.

Benedikt, Rudolf

2007 [1895] *Chemical Analysis of Oils, Fats, Waxes, and of the Commercial Products Derived Therefrom*. Macmillan, London. 2007 reprint. Read Books.

Bennet, John

2016 The Work of the British School at Athens, 2015–2016. *Archaeological Reports* 62:7–22.

Ben-Shlomo, David, Eleni Nodarou, and Jeremy B. Rutter

2011 Transport Stirrup Jars from the Southern Levant: New Light on Commodity Exchange in the Eastern Mediterranean. *American Journal of Archaeology* 115(3):329–353.

Bentley, R. Alexander, Rüdiger Krause, T. Douglas Price, and Bruno Kaufmann

2003 Human Mobility at the Early Neolithic Settlement of Vaihingen, Germany: Evidence from Strontium Isotope Analysis. *Archaeometry* 45(3):471–486.

Bergh, Justin L.

2016 *Hungry for More: Anthony Bourdain and the Cultural Valorization of Chefs and Cooks*. PhD dissertation, University of Minnesota.

Berihuete-Azorín, Marian, Hans-Peter Stika, Moritz Hallama, and Soultana Maria Valamoti
2020 Distinguishing Ripe Spelt from Processed Green Spelt (Grünkern) Grains: Methodological Aspects and the Case of Early La Tène Hochdorf (Vaihingen an der Enz, Germany). *Journal of Archaeological Science* 118:105143.

Bettinger, Robert L., Ripan Malhi, and Helen McCarthy
1997 Central Place Models of Acorn and Mussel Processing. *Journal of Archaeological Science* 24(10):887–899.

Bierl, Anton
2012 Dionysos' Epiphany in Performance. The God of Ecstatic Cry, Noise, Song, Music, and Choral Dance. *Electra 2: Dionysus: Myth, Cult, Ritual*:1–12.

Bijlefeld, Marjolijn, and Sharon K. Zoumbaris
2015 *Encyclopedia of Diet Fads: Understanding Science and Society*. 2nd ed. Greenwood, Santa Barbara.

Bjerver, Kjell, John Jonsson, Anna Nilsson, Jan Schuberth, and Jutta Schuberth
1982 Morphine Intake from Poppy Seed Food. *Journal of Pharmacy and Pharmacology* 34(12):798–801.

Blake, Emma
2008 The Mycenaeans in Italy: A Minimalist Position. *Papers of the British School at Rome* 76:1–34.

Blatter, Robert H. E., Stefanie Jacomet, and Angela Schlumbaum
2004 About the Origin of European Spelt (*Triticum spelta* L.): Allelic Differentiation of the HMW Glutenin B1-1 and A1-2 Subunit Genes. *Theoretical and Applied Genetics* 108(2):360–367.

Blitzer, Harriet
2014 Preliminary Notes on Olive Domestication and Cultivation in the Prehistoric Aegean. In *Physis: L'environnement naturel et la relation homme-milieu dans le monde égéen protohistorique. Actes de la 14e rencontre égéenne internationale, Paris, Institut National d'Histoire de l'Art (INHA), 11–14 décembre 2012*, edited by Gilles Touchais, Robert Laffineur, and Françoise Rougemont, pp. 239–243. Aegaeum 37. Peeters, Leuven/Liège.

Blouet, Brian
1986 Development of the Settlement Pattern. In *Excavations at Sitagroi: A Prehistoric Village in Northeast Greece, Vol. 1*, edited by Colin Renfrew, Marija Alseikaitė Gimbutas and Ernestine S. Elster, pp. 133–143. Monumenta Archaelogica 13. Institute of Archaeology, University of California, Los Angeles.

Boardman, Sheila, and Glynis Jones
1990 Experiments on the Effects of Charring on Cereal Plant Components. *Journal of Archaeological Science* 17(1):1–11.

Boeyens, Jan C. A., Maria van der Ryst, Francois P. Coetzee, Maryna Steyn, and Marius Loots
2009 From Uterus to Jar: The Significance of an Infant Pot Burial from Melora Saddle, an Early Nineteenth-Century African Farmer Site on the Waterberg Plateau. *South African Humanities* 21(1):213–238.

Bofill, Maria, Danai Chondrou, Antoni Palomo, Hara Procopiou, and Soultana Maria Valamoti
2020 Processing Plants for Food: Experimental Grinding within the ERC-Project PlantCult. *Journal of Lithic Studies* 7(3).

Bogaard, Amy, Dragana Filipović, Andrew Fairbairn, Laura Green, Elizabeth Stroud, Dorian Q. Fuller, and Michael Charles

2017 Agricultural Innovation and Resilience in a Long-Lived Early Farming Community: The 1,500-Year Sequence at Neolithic to Early Chalcolithic Çatalhöyük, Central Anatolia. *Anatolian Studies* 67:1–28.

Bogaard, Amy, Rebecca Fraser, Tim H. E. Heaton, Michael Wallace, Petra Vaiglova, Michael Charles, Glynis Jones, Richard P. Evershed, Amy K. Styring, Niels H. Andersen, Rose-Marie Arbogast, László Bartosiewicz, Armelle Gardeisen, Marie Kanstrup, Ursula Maier, Elena Marinova, Lazar Ninov, Marguerita Schäfer, and Elisabeth Stephan

2013 Crop Manuring and Intensive Land Management by Europe's First Farmers. *Proceedings of the National Academy of Sciences of the United States of America (PNAS)* 110(31):12589–12594.

Borghi, Basilio, Renato Castagna, Maria Corbellini, Manfred Heun, and Francesco Salamini

1996 Breadmaking Quality of Einkorn Wheat (*Triticum monococcum* ssp. *monococcum*). *Cereal Chemistry* 73(2):208–214.

Borgna, Elisabetta

1997 Kitchen-Ware from LM IIIC Phaistos: Cooking, Tradition, and Ritual Activities in Late Bronze Age Cretan Society. *Studi Micenei ed Egeo-Anatolici* 39(2):189–217.

2012 From Minoan Crete to Mycenaean Greece and Beyond: The Dissemination of Ritual Practices and Their Material Correlates in Ceremonial Architecture. In *Materiality and Social Practice: Transformative Capacities of Intercultural Encounters,* edited by Joseph Maran and Philipp Stockhammer, pp. 137–151. Oxbow, Oxford.

Borgna, Elisabeta, and Sarah T. Levi

2017 Cooking Vessels and Cooking Installations in the Mediterranean Bronze Age: A Comparative Evaluation of Household Practices in LMIIIC Crete and LBA Italy. In *From Cooking Vessels to Cultural Practices in the Late Bronze Age Aegean,* edited by Julie Hruby and Debra Trusty, pp. 116–126. Oxbow, Oxford.

Bottema, Sytze

1974 *Late Quaternary Vegetation History of Northwestern Greece.* PhD dissertation, University of Groningen.

1979 Pollen Analytical Investigations in Thessaly (Greece). *Palaeohistoria* 21:19–40.

1982 Palynological Investigations in Greece with Special Reference to Pollen as an Indicator of Human Activity. *Palaeohistoria* 24:256–289.

Bottema Sytze, and Anaya Sarpaki

2003 Environmental Change in Crete: A 9000-Year Record of Holocene Vegetation History and the Effect of the Santorini Eruption. *The Holocene* 13(5):733–749.

Bottema, Sytze, and Hendrik Woldring

1990 Anthropogenic Indicators in the Pollen Record of the Eastern Mediterranean. In *Man's Role in the Shaping of the Eastern Mediterranean Landscape. Proceedings of the INQUA/BAI Symposium on the Impact of Ancient Man on the Landscape of the Eastern Mediterranean Region and the Near East, Groningen, Netherlands, 6–9 March 1989,* edited by Sytze Bottema, Gertie Entjes-Nieborg, and Willem van Zeist, pp. 231–264. Balkema, Rotterdam.

Bottéro, Jean
2004 *The Oldest Cuisine in the World: Cooking in Mesopotamia*. University of Chicago Press, Chicago.

Bottino, Aldo Paolo
2014 The Trees of Laertes: An Epic Environment of Nóstos. Available through www.chs.harvard.edu/wp-content/uploads/2020/11/The-Trees-of-Laertes.pdf.

Bouby, Laurent
2014 *Agriculture dans le bassin du Rhône du Bronze Final à l'antiquité: Agrobiodiversité, économie, cultures*. Archives d'écologie préhistorique, Toulouse.

Bouby, Laurent, Vincent Bonhomme, Sarah Ivorra, Thierry Pastor, Núria Rovira, Margaux Tillier, Clémence Pagnoux, and Jean-Frédéric Terral
2018 Back from Burn Out: Are Experimentally Charred Grapevine Pips too Distorted to Be Characterized Using Morphometrics? *Archaeological and Anthropological Sciences* 10(4):943–954.

Bouby, Laurent, Isabel Figueiral, Anne Bouchette, Núria Rovira, Sarah Ivorra, Thierry Lacombe, Thierry Pastor, Sandrine Picq, Philippe Marinval, and Jean-Frédéric Terral
2013 Bioarchaeological Insights into the Process of Domestication of Grapevine (*Vitis vinifera* L.) during Roman Times in Southern France. *PLoS One* 8(5):e63195.

Boulotis, Christos
1996 Το λάδι στους μυκηναϊκούς χρόνους [Olive Oil during Mycenaean Times]. In *Ελιά και Λάδι* [Olive and Oil]. *Proceedings of the 4th Three-Day Working Meeting on Olive and Oil, Kalamata, 7–13 May 1993*, pp. 19–58. ETVA Cultural and Technological Foundation, Athens.

Bourdieu, Pierre
1977 *Outline of a Theory of Practice* [Esquisse d'une theorie de la pratique: Précédé de Trois études d'ethnologie kabyle]. Translated by Richard Nice. Cambridge Studies in Social Anthropology 16. Cambridge University Press.

Bowe, Patrick
2009 The Sacred Groves of Ancient Greece. *Studies in the History of Gardens and Designed Landscapes* 29(4):235–245.

Bower, Anne L.
2012 Watching Food: The Production of Food, Film, and Values. In *Reel Food: Essays on Food and Film*, edited by Anne L. Bower, pp. 1–13 (e-book). Originally published 2004, Routledge, London.

Bown, Deni
1995 *The Royal Horticultural Society Encyclopedia of Herbs and Their Uses*. Dorling Kindersley, London.

Bozi, Soula
1997 *Καππαδοκία, Ιωνία, Πόντος: Γεύσεις και παραδόσεις* [Kappadokia, Ionia, Pontos: Tastes and Traditions]. Asterismos, Athens.

Braadbaart, Freek, Elena Marinova, and Anaya Sarpaki
2016 Charred Olive Stones: Experimental and Archaeological Evidence for Recognizing Olive Processing Residues Used as Fuel. *Vegetation History and Archaeobotany* 25(5):415–430.

Braidwood, Robert J.
1953 Symposium: Did Man Once Live by Beer Alone? *American Anthropologist* 55(4):515–517.

Brain, Charles K., and Andrew Sillent
1988 Evidence from the Swartkrans Cave for the Earliest Use of Fire. *Nature* 336:464–466.

Brandolini, Andrea, Alyssa Hidalgo, Norberto Pogna, and Riccardo Geminati
2013 MonICA—A Model Project for Einkorn Reintroduction. In *Book of Abstracts of 16th Conference of the International Work Group for Palaeoethnobotany, Thessaloniki 17–21 June 2013,* edited by Soultana Maria Valamoti, pp. 221–213. Aristotle University of Thessaloniki.

Brandolini, Andrea, M. Marturini, Luca Plizzari, J. C. Hidalgo, Carlo Pompei, and Alyssa Hidalgo
2008 Chemical and Technological Properties of *Triticum monococcum, Triticum turgidum,* and *Triticum aestivum. Technica Molitoria International* 59(9/A):85–93.

Braun, Thomas
1995 Barley Cakes and Emmer Bread. In *Food in Antiquity: Studies in Ancient Society and Culture,* edited by John Wilkins, David Harvey, and Michael J. Dobson, pp. 25–37. University of Exeter Press, Exeter.

Breslow, Jan L.
2006 n-3 Fatty Acids and Cardiovascular Disease. *American Journal of Clinical Nutrition* 83(6), Supplement S1477–S1482.

Brockhampton Reference Series
1996 *Guide to Herbal Remedies.* Brockhampton Press, London.

Brogan, Thomas M., Chrysa Sofianou, Jerolyn E. Morrison, Dimitra Mylona, and Evi Margaritis
2013 Living off the Fruits of the Sea: New Evidence for Dining at Papadiokampos, Crete. In *Diet, Economy, and Society in the Ancient Greek World: Towards a Better Integration of Archaeology and Science. Proceedings of the International Conference Held at the Netherlands Institute at Athens on 22–24 March 2010,* edited by Sofia Voutsaki and Soultana Maria Valamoti, pp. 123–132. Pharos Supplement 1. Peeters, Leuven.

Bronk Ramsey, Christopher
2009 Bayesian Analysis of Radiocarbon Dates. *Radiocarbon* 51(1):337–360.

Bronk Ramsey, Christopher, and Sharen Lee
2013 Recent and Planned Developments of the Program OxCal. *Radiocarbon* 55(2–3):720–730.

Brown, Jack
2005 *Oil Crop Potential for Biodiesel Production: Summary of Three Years of Spring Mustard Research—Methodologies, Results, and Recommendations (2000–2003).* National Renewable Energy Laboratory, Golden, Colorado.

Brown, Linda
2001 Feasting on the Periphery: The Production of Ritual Feasting and Village Festivals at the Cerén Site, El Salvador. In *Feasts: Archaeological and Ethnographic Perspectives on Food, Politics, and Power,* edited by Michael Dietler and Brian Hayden, pp. 368–390. Smithsonian Institution Press, Washington, DC.

Brown, Terence A., Robin G. Allaby, Robert Sallares, and Glynis Jones
1998 Ancient DNA in Charred Wheats: Taxonomic Identification of Mixed and Single Grains. *Ancient Biomolecules* 2(2):185–193.

Brulotte, Ronda L., and Michael A. Di Giovine (editors)
2014 *Edible Identities: Food as Cultural Heritage.* Ashgate, Farnham, Surrey.
Brumberg-Kraus, Jonathan, and Betsey Dexter Dyer
2011 Cultures and Cultures: Fermented Foods as Culinary "Shibboleths." In *Cured, Smoked, and Fermented. Proceedings of the Oxford Symposium on Food and Cookery 2010,* edited by Helen Saberi, pp. 56–65. Prospect Books, Totnes.
Brun, Jean-Pierre
2000 The Production of Perfumes in Antiquity: The Cases of Delos and Paestum. *American Journal of Archaeology* 104(2):277–308.
Buddha, Paul D., Catherine Dreka, Eric S. Knight, and Michael L. Smith
1996 Gas Chromatographic/Mass Spectrometric Detection of Narcotine, Papaverine, and Thebaine in Seeds of *Papaver somniferum. Planta Medica* 62(6):544–547.
Bulut, Gizem Emre, and Ertan Tuzlacı
2008 Bozcaada Yöresinin Halk İlacı Olarak Kullanılan Bitkile [Uses of Medicines on Bozcaada]. In *Bozcaada Değerleri Sempozyumu, 25–26 Ağustos 2008,* edited by Ali Akdemir, Osman Demircan, Selehattin Yılmaz, Turan Takaoğlu, and Evren Erginal, pp. 165–172. Aynalı Pazar, Çanakkale.
Bunimovitz, Shlomo
1996 Minoan-Mycenaean Olive Oil Production and Trade: A Review of Current Research. In *Olive Oil in Antiquity: Israel and Neighbouring Countries from the Neolithic to the Early Arab Period,* edited by David Eitam and Michael Heltzer, pp. 49–53. History of the Ancient Near East Studies, Vol. VII. Sargon, Padova.
Bunimovitz, Shlomo, and Zvi Lederman
2016 Opium or Oil? Late Bronze Age Cypriot Base Ring Juglets and International Trade Revisited. *Antiquity* 90(354):1552–1561.
Burkhart, Louise
1997 Mexican Women on the Home Front: Housework and Religion in Aztec Mexico. In *Indian Women of Early Mexico,* edited by Susan Schroeder, Stephanie Wood, and Robert Haskett, pp. 25–54. University of Oklahoma Press, Norman.
Butler, Ann, Zelealem Tesfay, Catherine D'Andrea, and Diane Lyons
1999 The Ethnobotany of *Lathyrus sativus* L. in the Highlands of Ethiopia. In *The Exploitation of Plant Resources in Ancient Africa. Proceedings of the 2nd International Workshop on Archaeobotany in Northern Africa, Held June 23–25, 1997, in Leicester, United Kingdom,* edited by Marijke van der Veen, pp. 123–136. Kluwer Academic/Plenum, New York.
Butterworth, Peter J., Peter R. Ellis, and Michele Wollstonecroft
2016 Why Protein Is Not Enough: The Roles of Plants and Plant Processing in Delivering the Dietary Requirements of Modern and Early Homo. In *Wild Harvest: Plants in the Hominin and Pre-Agrarian Human Worlds,* edited by Karen Hardy and Lucy Kubiak-Martens, pp. 31–54. Studying Scientific Archaeology 2. Oxbow, Oxford.
Buxó, Ramon
2007 Crop Evolution: New Evidence from the Neolithic of West Mediterranean Europe. In *The Origins and Spread of Domestic Plants in Southwest Asia and Europe,* edited by Sue Colledge and James Conolly, pp. 155–172. Left Coast Press, Walnut Creek.
2008 The Agricultural Consequences of Colonial Contacts on the Iberian

Peninsula in the First Millennium B.C. *Vegetation History and Archaeobotany* 17(1):145–154.

Buxó, Ramon, and Raquel Pique-Huerta

2008 *Arqueobotánica: Los usos de las plantas en la península Ibérica.* Ariel, Barcelona.

Cadogan, Gerald

1978 Pyrgos, Crete, 1970–7. *Archaeological Reports* 24:70–84.

Campbell, Clayton G.

1997 *Grass Pea:* Lathyrus sativus *L. Promoting the Conservation and Use of Underutilized and Neglected Crops, Vol. 18.* Institute of Plant Genetics and Crop Plant Research, Gatersleben/International Plant Genetic Resources Institute, Rome.

Cañellas-Boltà, Nuria, Santiago Riera-Mora, Hector A. Orengo, Alexandra Livarda, and Carl Knappett

2018 Human Management and Landscape Changes at Palaikastro (Eastern Crete) from the Late Neolithic to the Early Minoan Period. *Quaternary Science Reviews* 183:59–75.

Cantuel, Jean, Frédéric Mercier, and Virginie Thomas

2010 Les équidés dans le monde égéen à l'âge du Bronze: Approche archéologique et iconographique. In *Histoire d'équidés: Des textes, des images et des os. Actes du colloque organisé par l'UMR5140 du CNRS, Montpellier, 13–14 mars 2008,* edited by Armelle Gardeisen, Emmanuelle Furet, and Nicolas Boulbes, pp. 157–175. Monographies d'Archéologie Méditerranéenne, Hors-Série 4. ADAL, Lattes.

Capparelli, Aylen, Soultana Maria Valamoti, and Michèle M. Wollstonecroft

2011 After the Harvest: Investigating the Role of Food Processing in Past Human Societies. *Archaeological and Anthropological Sciences* 3(1):1–5.

Cappers, René, Reinder Neef, Renée Bekker, Federica Fantone, and Yasar Okur

2016 *Digital Atlas of Traditional Agricultural Practices and Food Processing, Vol. 1–3.* Groningen Archaeological Studies 30. Barkhuis and Groningen University Library, Groningen.

Caracuta, Valentina, Jacob Vardi, Ytzhak Paz, and Elisabetta Boaretto

2017 Farming Legumes in the Pre-Pottery Neolithic: New Discoveries from the Site of Ahihud (Israel). *PLoS ONE* 12(5):e0177859.

Caramia, Giuseppe, Alessandro Gori, Enrico Valli, and Lorenzo Cerretani

2012 Virgin Olive Oil in Preventive Medicine: From Legend to Epigenetics. *European Journal of Lipid Science and Technology* 114(4):375–388.

Carpenter, Rhys

1948 The Greek Penetration of the Black Sea. *American Journal of Archaeology* 52(1):1–10.

Carrigan, Matthew A., Oleg Uryasev, Carole B. Frye, Blair L. Eckman, Candace R. Myers, Thomas D. Hurley, and Steven A. Benner

2015 Hominids Adapted to Metabolize Ethanol Long Before Human-Directed Fermentation. *Proceedings of the National Academy of Sciences of the United States of America (PNAS)* 112(2):458–463.

Cartwright, Caroline R.

2003 Grapes or Raisins? An Early Bronze Age Larder under the Microscope. *Antiquity* 77(296):345–348.

Castle, Sarah E.

1993 Intra-Household Differentials in Women's Status: Household Function and

Focus as Determinants of Children's Illness Management and Care in Rural Mali. *Health Transition Review* 3(2):137–157.

Cauvin, Jacques
1994 *Naissance des divinités, naissance de l'agriculture: La révolution des symboles au Néolithique.* CNRS, Paris.

Cavanagh, William, and Christopher Mee
1998 *A Private Place: Death in Prehistoric Greece.* Studies in Mediterranean Archaeology 125. Paul Åströms, Jonsered.

Chamber's Encyclopedia
1959 New Edition, Vol. 10. G. Newnes, London.

Charistos, Pavlos
1988 *Αρωματικά και φαρμακευτικά φυτά του Νομού Κοζάνης (Aromatic and Medical Plants of Kozani Prefecture).* Society of Arts and Cultere, Kozani.

Chauhan, M. P., Sadhna Singh, and Ashok Kumar Singh
2009 Post Harvest Uses of Linseed. *Journal of Human Ecology* 28(3):217–219.

Chlouveraki, Stephania, Philip P. Betancourt, Costis Davaras, Heidi M. C. Dierckx, Susan C. Ferrence, Jane Hickman, Panagiotis Karkanas, Photeini J. P. McGeorge, James D. Muhly, David S. Reese, Eleni Stravopodi, and Louise Langford-Verstegen
2008 Excavations in the Hagios Charalambos Cave: A Preliminary Report. *Hesperia* 77(4):539–605.

Chondrogianni-Metoki, Areti
2009 *Μη οικιστικές χρήσεις χώρου στους νεολιθικούς οικισμούς: το παράδειγμα της Τούμπας Κρεμαστής Κοιλάδας* [Non Domestic Use of Space in the Neolithic Settlements: The Example of Toumba Kremastis Koiladas]. PhD dissertation, Aristotle University of Thessaloniki.

Chondrou, Danai, Maria Bofill, Haris Procopiou, Roberto Vargiolu, Hassan Zahouani, and Soultana Maria Valamoti
2021 How Do You Like Your Cereal? A Qualitative and Quantitative Use-Wear Analysis on Archaeological Grinding Tools from Prehistoric Greek Sites. *Wear* 2021:203636.

Chondrou, Danai, Soultana Maria Valamoti, Hara Procopiou, and Lamprini Papadopoulou
2018 Grinding Cereals and Pulses in the Neolithic Site of Kleitos: An Experimental Investigation of Microconglomerate Grinding Equipment, Final Products and Use Wear. *Journal of Greek Archaeology* 3:23–45.

Chovanec, Zuzana, Shlomo Bunimovitz, and Zvi Lederman
2015 Is There Opium Here?—Analysis of Cypriot Base Ring Juglets from Tel Beth-Shemesh, Israel. *Mediterranean Archaeology and Archaeometry* 15(2):175–189.

Christides, Vassilios
1963 *Χειμωνιάτικα σιτηρά* [Winter Grain]. [s.n.], Thessaloniki.

Christie, Maria Elisa
2006 Kitchenspace: Gendered Territory in Central Mexico / Espacio de la cocina: Territorio "generizado" en México Central. *Gender, Place, and Culture* 13(6):653–661.
2008 Kitchenspace: Women, Fiestas, and Everyday Life in Central Mexico. University of Texas Press, Austin.

Chrysostomou, Panicos, Tryfon Jagoulis, and Andreas Mäder
2015 The "Culture of Four Lakes": Prehistoric Lakeside Settlements (6th–2nd

millennium BC) in the Amindeon Basin, Western Macedonia, Greece. *Archäologie Schweiz* 38(3):24–32.

Clarke, Michael J.
2001 Akha Feasting: An Ethnoarchaeological Perspective. In *Feasts: Archaeological and Ethnographic Perspectives on Food, Politics, and Power*, edited by Michael Dietler and Brian Hayden, pp. 144–167. Smithsonian Institution Press, Washington, DC.

Clemente, Florian, Martina Unterländer, Olga Dolgova, Carlos Eduardo G. Amorim, Francisco Coroado-Santos, Samuel Neuenschwander, Elissavet Ganiatsou, Diana I. Cruz Dávalos, Lucas Anchieri, Frédéric Michaud, Laura Winkelbach, Jens Blöcher, Yami Ommar Arizmendi Cárdenas, Bárbara Sousa da Mota, Eleni Kalliga, Angelos Souleles, Ioannis Kontopoulos, Georgia Karamitrou-Mentessidi, Olga Philaniotou, Adamantios Sampson, Dimitra Theodorou, Metaxia Tsipopoulou, Ioannis Akamatis, Paul Halstead, Kostas Kotsakis, Dushka Urem-Kotsou, Diamantis Panagiotopoulos, Christina Ziota, Sevasti Triantaphyllou, Olivier Delaneau, Jeffrey D. Jensen, J. Víctor Moreno-Mayar, Joachim Burger, Vitor C. Sousa, Oscar Lao, Anna-Sapfo Malaspinas, and Christina Papageorgopoulou
2021 The Genomic History of the Aegean Palatial Civilizations. *Cell* 184(10):2565–2586.e21.

Cohn, Dan F., and Mordechai E. Kislev
1987 Human Neurolathyrism. *Adler Museum Bulletin* 13(1):5–8.

Colaianni, Gia, Girolamo Fiorentino, Erik Fouache, Angela Pontrandolfo, Alfonso Santoriello, and Francesco Scelza
2012 Palaeovegetational Reconstruction of the Krios Valley, Northern Achaea (Greece): Archaeobotanical Analysis Conducted as Part of the Aegialia Survey Project. *ArcheoSciences, revue d'archéométrie* 36:127–137.

Collard, David
2011 *Altered States of Consciousness and Ritual in Late Bronze Age Cyprus*. PhD dissertation, University of Nottingham.
2012 Drinking with the Dead: Psychoactive Consumption in Cypriote Bronze Age Mortuary Ritual. In *Food and Drink in Archaeology 3*, edited by Dave Collard, Jim Morris, and Elisa Perego, pp. 23–32. University of Nottingham Postgraduate Conference 2009. Prospect Books, Totnes.

Colledge, Sue, and James Conolly
2007 Introduction: Key Themes in Interregional Approaches to the Study of Early Crops and Farming. In *The Origins and Spread of Domestic Plants in Southwest Asia and Europe*, edited by Sue Colledge and James Conolly, pp. xi–xv. Left Coast Press, Walnut Creek.

Colledge, Sue, James Conolly, and Stephen Shennan
2004 Archaeobotanical Evidence for the Spread of Farming in the Eastern Mediterranean. *Current Anthropology* 45, Supplement 4:S35–S58.

Connors, Mary E.
1994 *Native American Arts and Cultures: Exploring Tradition*. Teacher Created Resources, Westminster, California.

Copeland, Tim
2004 Presenting Archaeology to the Public: Constructing Insights On-Site. In *Public Archaeology*, edited by Nick Merriman, pp. 132–144. Routledge, London.

Corbellini, Maria., S. Empilli, Patrizia Vaccino, Andrea Brandolini, Basilio Borghi, Manfred Heun, and Francesco Salamini

1999 Einkorn Characterization for Bread and Cookie Production in Relation to Protein Subunit Composition. *Cereal Chemistry* 76(5):727–733.

Counihan, Carole M.

1999 *The Anthropology of Food and Body: Gender, Meaning, and Power*. Routledge, London.

Cousin, Françoise, and Susana Monzon (editors)

1992 *Cuisines du Monde: Gestes et Recettes*. CNRS, Paris.

Coward, Fiona, Stephen Shennan, Sue Colledge, James Conolly, and Mark Collard

2008 The Spread of Neolithic Plant Economies from the Near East to Northwest Europe: A Phylogenetic Analysis. *Journal of Archaeological Science* 35(1):42–56.

Coxe Stevenson, Matilda

1904 *The Zuñi Indians: Their Mythology, Esoteric Fraternities and Ceremonies*. Twenty-Third Annual Report of the Bureau of American Ethnology to the Secretary of the Smithsonian Institution. Government Printing Office, Washington, DC.

1905 The Zuñi Indians: Their Mythology, Esoteric Fraternities, and Ceremonies. *Twenty-Third Annual Report of the Bureau of American Ethnology, 1901–1902*, 3–608. Bureau of American Ethnology.

Crépon, Katell, Pascal Marget, Corinne Peyronnet, Benoit Carrouée, Paolo Arese, and Gérard Duc

2010 Nutritional Value of Faba Bean (*Vicia faba* L.) Seeds for Feed and Food. *Field Crops Research* 115(3):329–339.

Crewe, Lindy, and Ian Hill

2012 Finding Beer in the Archaeological Record: A Case Study from Kissonerga-Skalia on Bronze Age Cyprus. *Levant* 44(2):205–237.

Cristiani, Emanuela, Anita Radini, Dušan Borić, Giuseppina Mutri, Dragana Filipović, Ethel Allué, Clive Bonsall, Adina Boroneanţ, Giampaolo Dalmeri, Federica Fontana, Domenico Lo Vetro, Fabio Martini, Fabio Negrino, Marco Peresani, Julien Riel-Salvatore, Lucia Sarti, Dario Vujević, and Snežana Vukojicić

2016 The "Hidden Foods" Project: New Research into the Role of Plant Foods in Palaeolithic and Mesolithic Societies of South-East Europe and Italy. *Antiquity* 90(351):Project Gallery.

Crosby, Alfred W., Jr.

2003 *The Columbian Exchange: Biological and Cultural Consequences of 1492*. Reprinted. Praeger, Westport, Connecticut. Originally published 1972, Greenwood Press, Westport, Connecticut.

Crubézy, Éric, Jaroslav Bruzek, Jean Guilaine, Eugenia Cunha, Daniel Rougé, and Jan Jelinek

2001 The Antiquity of Cranial Surgery in Europe and in the Mediterranean Basin. *Comptes Rendus de l'Académie des Sciences—Series IIA—Earth and Planetary Science* 332(6):417–423.

Cullen, Tracey

1995 Mesolithic Mortuary Ritual at Franchthi Cave, Greece. *Antiquity* 69(263):270–289.

Curry, Andrew

2017 Our 9,000-Year Love Affair with Booze. *National Geographic Magazine*, February.

Czajkowska, Beata I., Amy Bogaard, Michael Charles, Glynis Jones, Marianne Kohler-Schneider, Aldona Mueller-Bieniek, and Terence A. Brown
2020 Ancient DNA Typing Indicates That the "New" Glume Wheat of Early Eurasian Agriculture Is a Cultivated Member of the *Triticum timopheevii* Group. *Journal of Archaeological Science* 123:105258.

Dabney, Mary K., Susan E. Allen, Anne Kugler, Anastasia Papathanasiou, and James C. Wright
2020 The Neolithic Settlement on Tsougniza at Ancient Nemea. *Hesperia* 89(1): 1–65.

D'Agata, Anna Lucia
1997 Incense and Perfumes in the Late Bronze Age Aegean. In *Profumi d'Arabia: Atti del Convegno*, edited by Alessandra Avanzini, pp. 85–99. Saggi di Storia Antica 11. L'Erma di Bretschneider, Rome.

Dalby, Andrew
1997 *Siren Feasts: A History of Food and Gastronomy in Greece*. Routledge, London.
2003 *Food in the Ancient World from A to Z*. Routledge, London.

Dalianis, Konstantinos
1983 *Χειμερινά σιτηρά* [Winter Wheat]. Stamoulis, Athens.
1993 *Ψυχανθή για καρπό και σανό* [Leguminosae for Seed and Hay]. Stamoulis, Athens.

Damerow, Peter
2012 Sumerian Beer: The Origins of Brewing Technology in Ancient Mesopotamia. *Cuneiform Digital Library Journal* 2012:2.

D'Andrea, Catherine, and Mitiku Haile
2002 Traditional Emmer Processing in Highland Ethiopia. *Journal of Ethnobiology* 22(2):179–217.

Danforth, Loring M.
1982 *The Death Rituals of Rural Greece*. Princeton University Press, Princeton.

Darlas, Andreas, and Henry De Lumley
1999 Palaeolithic Research in Kalamakia Cave, Areopolis, Peloponnese. In *The Palaeolithic Archaeology of Greece and Adjacent Areas. Proceedings of the ICOPAG Conference, Ioannina, September 1994*, edited by Geoff N. Bailie, Eugenia Adam, Eleni Panagopoulou, Catherine Perlés, and Kostas Zachos, pp. 293–302. British School at Athens Studies 3. The British School at Athens, London.

Davidson, Alan
2002 *The Penguin Companion to Food*. Rev. ed. Penguin, New York.

Davies, Paget
1961 Favism. *Postgraduate Medical Journal* 37:477–480.

De Boeck, Filip
1994 "When Hunger Goes Around the Land": Hunger and Food among the Aluund of Zaire. *Man* 29(2):257–282.

De Boer, Jan G.
2006–2007 Phantom-Mycenaeans in the Black Sea. *Talanta* 38–39:277–302.

D'Egidio, M. G., S. Nardi, and Victor Vallega
1993 Grain, Flour, and Dough Characteristics of Selected Strains of Diploid Wheat, *Triticum monococcum* L. *Cereal Chemistry* 70(3):298–303.

Delatola-Foskolou, Nikoletta
2006 *Κυκλάδων Γεύσεις* [Tastes of the Cyclades]. Mare, Tenos.

Demoule, Jean-Paul, and Catherine Perlès
1993 The Greek Neolithic: A New Review. *Journal of World Prehistory* 7(4):355–416.

Denham, Tim
2007 Early Fig Domestication, or Gathering of Wild Parthenocarpic Figs? *Antiquity* 81(312):457–461.
Dennell, Robin
1976 The Economic Importance of Plant Resources Represented on Archaeological Sites. *Journal of Archaeological Science* 3(3):229–247.
1983 *European Economic Prehistory: A New Approach.* Academic Press, London.
1992 The Origins of Crop Agriculture in Europe. In *The Origins of Agriculture: An International Perspective,* edited by C. Wesley Cowan and Patty Jo Watson, pp. 71–100. Smithsonian Institution Press, Washington, DC.
Dewilde, Bert
1999 *Flax in Flanders throughout the Centuries: History, Technical Evolution, Folklore.* Lanoo, Tielt.
Dietler, Michael
1996 Feasts and Commensal Politics in the Political Economy: Food, Power, and Status in Prehistoric Europe. In *Food and the Status Quest: An Interdisciplinary Perspective,* edited by Polly Wiessner and Wulf Schiefenhovel, pp. 87–125. Berghahn, Oxford.
2001 Theorizing the Feast: Rituals of Consumption, Commensal Politics, and Power in African Contexts. In *Feasts: Archaeological and Ethnographic Perspectives on Food, Politics, and Power,* edited by Michael Dietler and Brian Hayden, pp. 65–114. Smithsonian Institution Press, Washington, DC.
Dietler, Michael, and Brian Hayden (editors)
2001a *Feasts: Archaeological and Ethnographic Perspectives on Food, Politics, and Power.* Smithsonian Institution Press, Washington, DC.
Dietler, Michael, and Brian Hayden
2001b Digesting the Feast: Good to Eat, Good to Drink, Good to Think. In *Feasts: Archaeological and Ethnographic Perspectives on Food, Politics, and Power,* edited by Michael Dietler and Brian Hayden, pp. 1–20. Smithsonian Institution Press, Washington, DC.
Dietler, Michael, and Ingrid Herbich
2001 Feasts and Labor Mobilization: Dissecting a Fundamental Economic Practice. In *Feasts: Archaeological and Ethnographic Perspectives on Food, Politics, and Power,* edited by Michael Dietler and Brian Hayden, pp. 240–264. Smithsonian Institution Press, Washington, DC.
Dietrich, Oliver, Manfred Heun, Jens Notroff, Klaus Schmidt, and Martin Zarnkow
2012 The Role of Cult and Feasting in the Emergence of Neolithic Communities: New Evidence from Göbekli Tepe, South-Eastern Turkey. *Antiquity* 86(333):674–695.
Dilcock, Julie
2001 Returning to the Ancestors? Interpretation of the Mesolithic/Neolithic Transition at Franchthi Cave, Greece. In *Ethnoarchaeology and Hunter-Gatherers: Pictures at an Exhibition,* edited by Kathryn Jane Fewster and Marek Zvelebil, pp. 73–80. British Archaeological Reports International Series 955. Archaeopress, Oxford.
Dillman, A. C.
1928 Daily Growth and Oil Content of Flaxseeds. *Journal of Agricultural Research* 37:357–377.

Dimoula, Anastasia, Zoï Tsirtsoni, Paraskevi Yiouni, Ioannis Stagkidis, Maria Ntinou, Sandra Prevost-Dermarkar, Evanthia Papadopoulou, and Soultana Maria Valamoti
2020 Experimental Investigation of Ceramic Technology and Plant Food Cooking in Neolithic Northern Greece. *STAR: Science and Technology of Archaeological Research* 5(2):269–286.
Dimoula, Anastasia, Sophia Koulidou, Zoï Tsirtsoni, Edward Standall, Oliver Craig, and Soultana Maria Valamoti
2022b Fusion Cuisine in the Shadow of Mount Olympus: An Integrated Study of Mycenaean Cooking Pots. *Greek Archaeology* 7:37–66.
Dimoula, Anastasia, Zoï Tsirtsoni, and Soultana Maria Valamoti
2022a Ceramic Cooking Dishes in the Prehistoric Aegean: An Exploration of Their Variability and Uses. *Hesperia* 91(1):1–61.
Dixit, Anjali A., Kristen M. J. Azar, Christopher D. Gardner, and Latha P. Palaniappan
2011 Incorporation of Whole, Ancient Grains into a Modern Asian Indian Diet to Reduce the Burden of Chronic Disease. *Nutrition Reviews* 69(8):479–488.
Doğer, Ersin, and Yusuf Sezgin
2015 Terebinth and Terebinth Oil Production in the Ancient Period. In *Olive Oil and Wine Production in Eastern Mediterranean during Antiquity. Proceedings of the International Symposium, 17–19 November 2011, Urla—Turkey,* edited by Adnan Diler, Ahmet Kaan Şenol, and Ümit Aydınoğlu, pp. 81–88. Ege Üniversitesi Yayınları, İzmir.
Dong, Guanghui, Yishi Yang, Xinyi Liu, Haiming Li, Yifu Cui, Hui Wang, Guoke Chen, John Dodson, and Fahu Chen
2018 Prehistoric Trans-Continental Cultural Exchange in the Hexi Corridor, Northwest China. *The Holocene* 28(4):621–628.
Douglas, Mary
1966 *Purity and Danger: An Analysis of Concepts of Pollution and Taboo.* Routledge and Keegan Paul, London.
1982 *In the Active Voice.* Routledge and Keegan Paul, London.
Drieu, Léa, Maxime Rageot, Nathan Wales, Ben Stern, Jasmine Lundy, Maximilian Zerrer, Isabella Gaffney, Manon Bondetti, Cynthianne Spiteri, Jane Thomas-Oates, and Oliver E. Craig
2020 Is It Possible to Identify Ancient Wine Production Using Biomolecular Approaches? *STAR: Science and Technology of Archaeological Research* 6:16–29.
Driver, Harold E.
1953 The Acorn in North American Indian Diet. *Proceedings of the Indiana Academy of Science* 62:56–62.
Dubreuil, Laure, and Dani Nadel
2015 The Development of Plant Food Processing in the Levant: Insights from Use-Wear Analysis of Early Epipalaeolithic Ground Stone Tools. *Philosophical Transactions of the Royal Society B: Biological Sciences* 370(1682):20140357.
Ducros, Hélène B.
2015 Review of *Edible Identities: Food as Cultural Heritage,* edited by Ronda L. Brulotte and Michael A. Di Giovine. 2014. Ashgate, Farnham, Surrey. *Gastronomica: The Journal of Critical Food Studies* 15(4):100–102.
Dudley, Robert
2002 Fermenting Fruit and the Historical Ecology of Ethanol Ingestion: Is Alcoholism in Modern Humans an Evolutionary Hangover? *Addiction* 97(4):381–388.

Dueppen, Stephen A., and Daphne Gallagher

2021 Alcohol, Ancestors, and the House: Exploring Ritual Use of Beer at Kirikongo, Burkina Faso. *Journal of Anthropological Archaeology* 64:101353.

Dungel Adalbert, and Josef Szombathy

1890 *Die Tumuli von Gemeinlebarn: Ausgegraben von Adalbert Dungel.* Mitteilungen der Prähistorischen Kommission 1, 2. Tempsky, Vienna.

Efstratiou, Nikos

1996 The Neolithisation of Europe: The Greek Experience. *Documenta Praehistorica* 22:63–82.

2007 The Beginning of the Neolithic in Greece: Probing the Limits of a Grand Narrative. In *Mediterranean Crossroads,* edited by Sophia Antoniadou and Anthony Pace, pp. 123–138. Pierides Foundation, Athens.

Efstratiou, Nikos, Alexandra Karetsou, Eleni S. Banou, and Despina Margomenou

2004 The Neolithic Settlement of Knossos: New Light on an Old Picture. In *Knossos: Palace, City, State. Proceedings of the Conference in Herakleion Organised by the British School at Athens and the 23rd Ephoreia of Prehistoric and Classical Antiquities of Herakleion, in November 2000, for the Centenary of Sir Arthur Evans's Excavations at Knossos,* edited by Gerald Cadogan, Eleni Hatzaki and Adonis Vasilakis, pp. 39–49. British School at Athens Studies 12. The British School at Athens, London.

Eleftherochorinos, Elias

1989 *Ζιζανιολογία. Βιολογία και καταπολέμηση των ζιζανίων* [Weeds: Biology of Weeds and Fighting against Them]. Aristotle University of Thessaloniki.

Eliade, Mircea

1964 *Shamanism: Archaic Techniques of Ecstacy* [Le Chamanisme et les techniques archaïques de l'extase]. Translated by Willard R. Trask. Routledge and Keegan Paul, London.

Enneking, Dirk

1995 *The Toxicity of Vicia Species and Their Utilisation as Grain Legumes.* Centre for Legumes in Mediterranean Agriculture (CLIMA), Occasional Publication 6. University of Western Australia, Nedlands.

Enneking, Dirk, Azzedine Lahlou, Ali Noutfia, and Mustafa Bounejmate

1995 A Note on *Vicia ervilia* Cultivation, Utilisation, and Toxicity in Morocco. *Al Awamia* 89:141–148.

Ertuğ, Füsun

2000 Linseed Oil and Oil Mills in Central Turkey: Flax/*Linum* and *Eruca,* Important Oil Plants of Anatolia. *Anatolian Studies* 50:171–185.

2004 Recipes of Old Tastes with Einkorn and Emmer Wheat. *TÜBA-AR (Turkish Academy of Sciences Journal of Archaeology)* 7:177–188.

2010 The Consumption of Flax: Linseed Oil Production in Turkey. In *Tradiciones y Transformaciones en Etnobotánica. Proceedings of the 5th International Congress of Ethnobotany (ICEB), September 2009, San Carlos de Bariloche, Argentina,* edited by María Lelia Pochettino, Ana H. Ladio, and Patricia M. Arenas, pp. 123–127. Red Iberoamericana de Saberes y Prácticas Locales sobre el Entorno Vegetal (RISAPRET)/Cyted, San Salvador de Jujuy.

Evans, Arthur John

1900–1901 The Palace of Knossos. Provisional Report of the Excavations for the Year 1901. *Annual of the British School at Athens* 7:1–120.

2013 *The Palace of Minos at Knossos: A Comparative Account of the Successive Stages of the Early Cretan Civilization as Illustrated by the Discoveries at Knossos, Vol. 4, Part 2.* Reprinted. Cambridge University Press. Originally published 1935, Macmillan, London.

Evans, John Davies

1964 Excavations in the Neolithic Settlement of Knossos, 1957–60, Part 1. *Annual of the British School at Athens* 59:132–240.

1968 Knossos Neolithic, Part II: Summary and Conclusions. *Annual of the British School at Athens* 63:267–276.

Evershed, Richard P.

2008 Experimental Approaches to the Interpretation of Absorbed Organic Residues in Archeological Ceramics. *World Archaeology* 40(1):26–47.

Facciola, Stephen

1990 *Cornucopia: A Source Book of Edible Plants.* Kampong, Vista, California.

Fappas, Ioannis

2008 The Use of Perfumed Oils during Feasting Activities: A Comparison of Mycenaean and Near Eastern Written Sources. In *DAIS: The Aegean Feast. Proceedings of the 12th International Aegean Conference, University of Melbourne, Centre for Classics and Archaeology, 25–29 March 2008,* edited by Louise Hitchcock, Robert Laffineur, and Janice L. Crowley, pp. 367–377. Aegaeum 29. Université de Liège, Liège/University of Texas at Austin.

2009 *Τα αρωματικά έλαια και οι πρακτικές χρήσης τους στη Μυκηναϊκή Ελλάδα και την Ανατολική Μεσόγειο (14ος–13ος αι. π.Χ.)* [Perfumed Oils and Their Uses in Mycenaean Greece and the Eastern Mediterranean (14th–13th Centuries B.C.)]. PhD dissertation, Aristotle University of Thessaloniki.

2011 Exchange of Ideas in the 14th and 13th Centuries BC East Mediterranean: The Case of Perfumed Oil Use and Ideology. In *Intercultural Contacts in the Ancient Mediterranean. Proceedings of the International Conference at the Netherlands-Flemish Institute in Cairo, 25th to 29th October 2008,* edited by Kim Duistermaat and Ilona Regulski, pp. 483–498. Orientalia Lovaniensia Analecta 202. Peeters, Leuven.

Feldman-Savelsberg, Pamela

1994 Plundered Kitchens and Empty Wombs: Fear of Infertility in the Cameroonian Grassfields. *Social Science and Medicine* 39(4):463–474.

Fernández-Souza, Lilia

2016 Grinding and Cooking: An Approach to Mayan Culinary Technology. In *Cooking Technology: Transformations in Culinary Practice in Mexico and Latin America,* edited by Steffan Igor Ayora-Díaz, pp. 15–28. Bloomsbury, London.

Ferrero Aldo, and Francesco Vidotto

2010 History of Rice in Europe. In *Rice: Origin, Antiquity, and History,* edited by S. D. Sharma, pp. 341–372. Science Publishers/CRC Press, Enfield, New Hampshire.

Fikre, Asnake, Tesgera Negwo, Yu-Haey Kuo, Fernand Lambein, and Seid Ahmedd

2011 Climatic, Edaphic, and Altitudinal Factors Affecting Yield and Toxicity of *Lathyrus sativus* Grown at Five Locations in Ethiopia. *Food and Chemical Toxicology* 49(3):623–630.

Filipović, Dragana, John Meadows, Marta Dal Corso, Wiebke Kirleis, Almuth Alsleben, Örni Akeret, Felix Bittmann, Giovanna Bosi, Beatrice Ciută, Dagmar Dreslerová, Henrike Effenberger, Ferenc Gyulai, Andreas G. Heiss, Monika Hellmund,

Susanne Jahns, Thorsten Jakobitsch, Magda Kapcia, Stefanie Klooß, Marianne Kohler-Schneider, Helmut Kroll, Przemysław Makarowicz, Elena Marinova, Tanja Märkle, Aleksandar Medović, Anna Maria Mercuri, Aldona Mueller-Bieniek, Renato Nisbet, Galina Pashkevich, Renata Perego, Petr Pokorný, Łukasz Pospieszny, Marcin Przybyła, Kelly Reed, Joanna Rennwanz, Hans-Peter Stika, Astrid Stobbe, Tjaša Tolar, Krystyna Wasylikowa, Julian Wiethold, and Tanja Zerl

2020 New AMS ^{14}C Dates Track the Arrival and Spread of Broomcorn Millet Cultivation and Agricultural Change in Prehistoric Europe. *Scientific Reports* 10:13698.

Filipović, Dragana, and Đurđa Obradović

2019 Sideliner or Protagonist? Economic Status of "New Type" Glume Wheat in the Neolithic and Bronze Age Central Balkans. In *Book of Abstracts of 18th Conference of the International Work Group for Palaeoethnobotany, Lecce 3–8 June 2019*, edited by Girolamo Fiorentino and Milena Primavera, pp. 51. Università del Salento, Lecce.

Filipović, Dragana, Đurđa Obradović, and Boban Tripković

2018 Plant Storage in Neolithic Southeast Europe: Synthesis of the Archaeological and Archaeobotanical Evidence from Serbia. *Vegetation History and Archaeobotany* 27(1):31–44.

Fiorentino, Girolamo, and Francesco Solinas

2006 Carboni e carporesti dal palazzo di Monastiraki: Indagini paleoambientali e paleoeconomiche. In *Monastiraki I. Missione Monastiraki, Campagne 2002/2004*, edited by Athanasia Kanta and Massimiliano Marazzi, pp. 123–139. Quaderni della Ricerca Scientifica, Serie Beni Culturali 4. Herder, Rome.

Fitzpatrick, Scott M., and Mark D. Merlin

2018 Introduction: Drugs from a Deep Time Perspective. In *Ancient Psychoactive Substances*, edited by Scott M. Fitzpatrick, pp. 1–19. University Press of Florida, Gainesville.

Flint-Hamilton, Kimberly

1999 Legumes in Ancient Greece and Rome: Food, Medicine, or Poison? *Hesperia* 68(3):371–385.

2016 The Palaeoethnobotany of Vronda. In *Kavousi IIC: The Late Minoan IIIC Settlement at Vronda: Specialist Reports and Analyses*, edited by Geraldine C. Gesell and Leslie Preston Day, pp. 181–194. Prehistory Monographs 52. INSTAP Academic, Philadelphia.

Follieri, Maria

1986 Provviste alimentari vegetali da una casa minoica ad Haghia Triada (Creta). *Annuario della scuola archeologica di Atene e delle missioni italiane in oriente* 57/58, 1979/1980:165–172.

Forbes, Hamish

1996 The Uses of the Uncultivated Landscape in Modern Greece: A Pointer to the Value of the Wilderness in Antiquity? In *Human Landscapes in Classical Antiquity: Environment and Culture*, edited by Graham Shipley and John Salmon, pp. 68–97. Leicester-Nottingham Studies in Ancient Society, Vol. 6. Routledge, London.

Forbes, Mary H. Clark

1976 Gathering in the Argolid: A Subsistence Subsystem in a Greek Agricultural Community. *Annals of the New York Academy of Sciences* 268(1):251–264.

Forste, Kathleen M.
2012 *Agricultural Adaptations during the Late Bronze Age: Archaeobotanical Evidence from Sovjan, Albania, and Tsoungiza, Greece.* Master's thesis, University of Cincinnati.
Foster, C.
1997 Preliminary Report on the Examination of Palaeo-Ethnobotanical Remains from Dimitra, N. Greece. Appendix 13. In *Νεολιθική Μακεδονία* [Neolithic Makedonia], edited by Dimitrios V. Grammenos, pp. 217–219. Publications of the Archaeologiko Deltion 56. Fund of Archaeological Proceeds (TAPA), Athens.
Foster, Ellen D.
1981 The Flax Impost at Pylos and Mycenaean Landholding. *Minos* 17:67–121.
Foxhall, Lin
2007 *Olive Cultivation in Ancient Greece: Seeking the Ancient Economy.* Oxford University Press, Oxford.
Frachetti, Michael D., C. Evan Smith, Cynthia M. Traub, and Tim Williams
2017 Nomadic Ecology Shaped the Highland Geography of Asia's Silk Roads. *Nature* 543(7644):193–198.
French, David H.
1971 An Experiment in Water-Sieving. *Anatolian Studies* 21:59–64.
Frick, Claudia, and Thomas Hebeisen
2005 Mohn als alternative Ölpflanze. *Agrarforschung Schweiz* 12(1):4–9.
Friedrich, Walter L.
1980 Fossil Plants from Weichselian Interstadials, Santorini (Greece) II. In *Thera and the Aegean World II. Papers and Proceedings of the 2nd International Scientific Congress, Santorini, Greece, August 1978,* edited by Christos Doumas, pp. 109–128. London.
Fuller, Dorian Q.
2007 Contrasting Patterns in Crop Domestication and Domestication Rates: Recent Archaeobotanical Insights from the Old World. *Annals of Botany* 100(5):903–924.
2018 Long and Attenuated: Comparative Trends in the Domestication of Tree Fruits. *Vegetation History and Archaeobotany* 27(1): 165–176.
Fuller, Dorian Q., and Robin G. Allaby
2018 Seed Dispersal and Crop Domestication: Shattering, Germination and Seasonality in Evolution under Cultivation. *Annual Plant Reviews online.* Originally published in *Fruit Development and Seed Dispersal,* edited by Lars Østergaard, pp. 238-295. *Annual Plant Reviews* 38(2010). Wiley-Blackwell, Oxford.
Fuller, Dorian Q., Robin G. Allaby, and Chris Stevens
2010 Domestication as Innovation: The Entanglement of Techniques, Technology, and Chance in the Domestication of Cereal Crops. *World Archaeology* 42(1):13–28.
Fuller, Dorian Q., and Emma L. Harvey
2006 The Archaeobotany of Indian Pulses: Identification: Processing and Evidence for Cultivation. *Environmental Archaeology* 11(2):219–246.
Fuller, Dorian Q., and Chris J. Stevens
2019 Between Domestication and Civilization: The Role of Agriculture and

Arboriculture in the Emergence of the First Urban Societies. *Vegetation History and Archaeobotany* 28(3): 263–282.

Fuller, Dorian Q., George Willcox, and Robin G. Allaby
2011 Cultivation and Domestication Had Multiple Origins: Arguments against the Core Area Hypothesis for the Origins of Agriculture in the Near East. *World Archaeology* 43(4):628–652.

Galanidou, Nena
1997 *"Home Is Where the Hearth Is": The Spatial Organisation of the Upper Palaeolithic Rockshelter Occupations at Klithi and Kastritsa in Northwest Greece.* British Archaeological Reports International Series 687. Hedges, Oxford.
2011 Mesolithic Cave Use in Greece and the Mosaic of Human Communities. *Journal of Mediterranean Archaeology* 24(2):219–242.

Galanidou, Nena, and Catherine Perlès (editors)
2003 *The Greek Mesolithic: Problems and Perspectives.* British School at Athens Studies 10. The British School at Athens, London.

Garnier, Nicolas, and Soultana Maria Valamoti
2016 Prehistoric Wine-Making at Dikili Tash (Northern Greece): Integrating Residue Analysis and Archaeobotany. *Journal of Archaeological Science* 74:195–206.

Garrett, J. T.
2003 *The Cherokee Herbal: Native Plant Medicine from the Four Directions.* Bear, Rochester, Vermont.

Gauß, Walter, Michael Lindblom, R. Angus K. Smith, and James C. Wright
2011 *Our Cups Are Full: Pottery and Society in the Aegean Bronze Age. Papers Presented to Jeremy B. Rutter on the Occasion of his 65th Birthday.* Archaeopress, Oxford.

Geller, Markham J.
2010 *Ancient Babylonian Medicine: Theory and Practice.* Wiley-Blackwell, Chichester.

Gennadios, Panagiotis
1997 *Λεξικόν Φυτολογικόν* [Phytological Lexicon]. Reprinted. Trochalia, Athens. Originally published 1914, Leonis, Athens.

Georgiadou, Anastasia
2015 *The Neolithic House at Sosandra.* University Studio Press, Thessaloniki.
2021 *Άψαλος Αλμωπίας. Ο Οικισμός της Ύστερης Εποχής του Χαλκού* [Apsalos Almopias: The Late Bronze Age Settlement]. University Studio Press, Thessaloniki.

Georgiou, Hara S.
1979 Late Minoan Incense Burners. *American Journal of Archaeology* 83(4):427–435.

Germain-Vallée, Cécile, Sandra Prévost-Dermarkar, and Laurent Lespez
2011 Stratégies de prélèvement et de mise en oeuvre de la "terre à bâtir" des structures de combustion néolithiques du site de Dikili Tash (Grèce) à partir d'une étude micromorphologique. *ArchéoSciences/Revue d'Archéométrie* 35:41–63.

Getahun, Haileyesus, Fernand Lambein, Michel Vanhoorne, and Patrick van der Stuyft
2002 Pattern and Associated Factors of the Neurolathyrism Epidemic in Ethiopia. *Tropical Medicine and International Health* 7(2):118–124.

Getahun, Haileyesus, Fernand Lambein, Michel Vanhoorne, and Patrick van der Stuyft
2005 Neurolathyrism Risk Depends on Type of Grass Pea Preparation and on Mixing with Cereals and Antioxidants. *Tropical Medicine and International Health* 10(2):169–178.

Getahun, Haileyesus, Alemayehu Mekonnen, Redda Tekle-Haimanot, and Fernand Lambein
1999 Epidemic of Neurolathyrism in Ethiopia. *The Lancet* 354(9175):306–307.
Ghannadi, Alireza, Ahmad Movahedian, and Zahra Jannesary
2015 Hypocholesterolemic Effects of Balangu (*Lallemantia royleana*) Seeds in the Rabbits Fed on a Cholesterol-Containing Diet. *Avicenna Journal of Phytomedicine* 5(3):167–173.
Ghazanfar, Shaikh M.
2018 Vasco da Gama's Voyages to India: Messianism, Mercantilism, and Sacred Exploits. *Journal of Global Initiatives: Policy, Pedagogy, Perspective 13(1): India: Globalization, Inclusion and Sustainability*:15–40.
Gialousi-Chatzichristou, Efi
2009 *Μυτιληνιές Γεύσεις* [Tastes of Mitilini]. Aeolis, Lesvos.
Gilman, Antonio
1981 The Development of Social Stratification in Bronze Age Europe. *Current Anthropology* 22(1):1–8.
Gimatzidis, Stefanos
2017 Cooking Pots and Ancient Identities: Indicators or Obscurers of Cultural Change? In *ΚΡΑΤΙΣΤΟΣ: Volume in Honour of Professor Peter Delev*, edited by Christo Popov and Julia Tzvetkova, pp. 253–268. Kliment Ohridski University Press, Sofia.
Gimatzidis, Stefanos, Giannis Karliambas, and Stavros Kotsos
2014 Δρακόντιο 2010: Αρχαιολογική έρευνα σε μια τούμπα της λεκάνης του Λαγκαδά [Drakontion 2010: Archaeological Research in a Mound in the Lagadas Basin]. *Proceedings of the Archaeological Work in Macedonia and Thrace (AEMTH)* 24, 2010:189–196.
Girella, Luca, Evi Gorogianni, and Peter Pavúk
2016 Introduction: Methodological Considerations. In *Beyond Thalassocracies: Understanding Processes of Minoanisation and Mycenaeanisation in the Aegean*, edited by Evi Gorogianni, Peter Pavúk, and Luca Girella, pp. 1–14. Oxbow, Oxford.
Girella, Luca, and Peter Pavúk
2016 The Nature of Minoan and Mycenaean Involvement in the Northeastern Aegean. In *Beyond Thalassocracies: Understanding Processes of Minoanisation and Mycenaeanisation in the Aegean*, edited by Evi Gorogianni, Peter Pavúk, and Luca Girella, pp. 15–42. Oxbow, Oxford.
Gkotsinas, Angelos, Angeliki Karathanou, Maria-Fotini Papakonstantinou, George Syrides, and Konstantinos Vouvalidis
2014 Approaching Human Activity and Interaction with the Natural Environment through the Archaeobotanical and Zooarchaeological Remains from Middle Helladic Agia Paraskevi, Central Greece. In *Physis: L'environnement naturel et la relation homme-milieu dans le monde égéen protohistorique. Actes de la 14e rencontre égéenne internationale, Paris, Institut National d'Histoire de l'Art (INHA), 11–14 décembre 2012*, edited by Gilles Touchais, Robert Laffineur, and Françoise Rougemont, pp. 487–491. Aegaeum 37. Peeters, Leuven/Liège.
González-Carretero, Lara, Michèle Wollstonecroft, and Dorian Q. Fuller
2017 A Methodological Approach to the Study of Archaeological Cereal Meals: A

Case Study at Çatalhöyük East (Turkey). *Vegetation History and Archaeobotany* 26(4): 415–432.

González-Turmo, Isabel

2012 The Mediterranean Diet: Consumption, Cuisine, and Food Habits. In *MediTERRA 2012: The Mediterranean Diet for Sustainable Regional Development*, pp. 115–132. CIHEAM/Presses de Sciences Po, Paris.

Goody, Jack

1982 *Cooking, Cuisine, and Class: A Study in Comparative Sociology*. Cambridge University Press, Cambridge.

1998 *Food and Love: A Cultural History of East and West*. Verso, London.

Gopher, Avi, Simcha Lev-Yadun, and Shahal Abbo

2021 *Breaking Ground: Plant Domestication in the Neolithic Levant; The "Core-Area One-Event" Model*. Occasional Publications 5. Tel Aviv University—The Sonia and Marco Nadler Institute of Archaeology/Emery and Claire Yass Publications in Archaeology, Tel Aviv.

Gorny, Ronald L.

1995 Viticulture in Ancient Anatolia. In *The Origins and Ancient History of Wine*, edited by Patrick E. McGovern, Stuart J. Fleming, and Solomon H. Katz, pp. 133–174. Food and Nutrition in History and Anthropology 11. Gordon and Breach, Philadelphia.

Goslar, Tomasz, Justyna Czernik, and Ewa Goslar

2004 Low-Energy ^{14}C AMS in Poznań Radiocarbon Laboratory, Poland. *Nuclear Instruments and Methods in Physics Research Section B: Beam Interactions with Materials and Atoms* 223–224:5–11.

Gosselain, Olivier P.

1999 In Pots We Trust: The Processing of Clay and Symbols in Sub-Saharan Africa. *Journal of Material Culture* 4(2):205–230.

Goudsblom, Johan

1986 The Human Monopoly on the Use of Fire: Its Origins and Conditions. *Human Evolution* 1(6):517–523.

Goulimaris, Dimitrios, Vasilios Serbezis, and Evangelos Albanidis

2005 Dancing Mania as Healing Proceedings in the Context of Religious Worship in Ancient and Contemporary Greece. *Aethlon: The Journal of Sport Literature* 22(2):107–118.

Grammenos, Dimitrios, and Stavros Kotsos (editors)

2002 *Ανασκαφή στον Προϊστορικό Οικισμό «Μεσημεριανή Τούμπα» Τριλόφου, Ν. Θεσσαλονίκης (Ανασκαφικές Περίοδοι 1992, 1994–1996, 2000, 2001)* [Excavations in the Prehistoric Settlement of "Mesimeriani Tumulus" at Trilofos, Thessaloniki (Excavation Seasons of 1992, 1994–1996, 2000, 2001)]. Publications of Archaeological Institute of Northern Greece 1. Archaeological Institute of Northern Greece, Thessaloniki.

Greig, James

1989 *Archaeobotany*. Handbooks for Archaeologists 4. European Science Foundation, Strasbourg.

Greig James, and Judith Turner

1974 Some Pollen Diagrams from Greece and Their Archaeological Significance. *Journal of Archaeological Science* 1(2):177–194.

Groopman, Emily E., Rachel N. Carmody, and Richard Wrangham
2015 Cooking Increases Net Energy Gain from a Lipid-Rich Food. *American Journal of Physical Anthropology* 156(1):11–18.

Grove, Michael D., Gayland F. Spencer, Marjorie V. Wakeman, and Harvey L. Tookey
1976 Morphine and Codeine in Poppy Seed. *Agricultural and Food Chemistry* 24(4):896–897.

Guerra Doce, Elisa
2014 The Origins of Inebriation: Archaeological Evidence of the Consumption of Fermented Beverages and Drugs in Prehistoric Eurasia. *Journal of Archaeological Method and Theory* 22(3):751–782.

Gulfraz, M., R. Parveen, Y. Musseddque, U. Nisar, M. Ithisham, and S. Rehman
2006 Determination of Essential Oil Content of Wild Olive and Its Comparison with Olive Oil. *Ethnobotanical Leaflets* 10:1–12.

Guyot, Jean-Pierre
2012 Cereal-Based Fermented Foods in Developing Countries: Ancient Foods for Modern Research. *International Journal of Food Science and Technology* 47(6):1109–1114.

Hadjichambis, Andreas Ch., Demetra Paraskeva-Hadjichambi, Athena Della, Maria Elena Giusti, Caterina De Pasquale, Cinzia Lenzarini, Elena Censorii, Maria Reyes Gonzalez-Tejero, Cristina Patricia Sanchez-Rojas, Jose M. Ramiro-Gutierrez, Melpomeni Skoula, Chris Johnson, Anaya Sarpaki, Mohamed Hmamouchi, Said Jorhi, Mohamed El-Demerdash, Mustafa El-Zayat, and Andrea Pieroni
2008 Wild and Semi-Domesticated Food Plant Consumption in Seven Circum-Mediterranean Areas. *International Journal of Food Sciences and Nutrition* 59(5):383–414.

Hairfield, Hampton H., Jr., and Elizabeth M. Hairfield
1990 Identification of a Late Bronze Age Resin. *Analytical Chemistry* 62(1):41A–45A.

Hajnalová, Eva
1989 Evidence of the Carbonized Loaf of Bread and Cereals from Bratislava-Devín. *Slovenská Archaeológia* 37(1):89–104.

Hajnalová, Mária, and Dagmar Dreslerová
2010 Ethnobotany of Einkorn and Emmer in Romania and Slovakia: Towards Interpretation of Archaeological Evidence. *Památky archeologické* 101:169–202.

Halstead, Paul
1981 Counting Sheep in Neolithic and Bronze Age Greece. In *Pattern of the Past: Studies in Honour of David Clarke*, edited by Ian Hodder, Glynn Isaac, and Norman Hammond, pp. 307–339. Cambridge University Press, Cambridge.
1990 Waste Not, Want Not: Traditional Responses to Crop Failure in Greece. *Rural History* 1(2):147–164.
1992 Agriculture in the Bronze Age Aegean: Towards a Model of Palatial Economy. In *Agriculture in Ancient Greece. Proceedings of the 7th International Symposium at the Swedish Institute of Athens, 16–17 May 1990*, edited by Berit Wells, pp. 105–117. Paul Åströms, Stockholm.
1994 The North-South Divide: Regional Paths to Complexity in Prehistoric Greece. In *Development and Decline in the Mediterranean Bronze Age*, edited by Clay Mathers and Simon Stoddart, pp. 195–219. Sheffield Archaeological Monographs 8. J. R. Collis, Sheffield.

1996 The Development of Agriculture and Pastoralism in Greece: When, How, Who, and What? In *The Origins and Spread of Agriculture and Pastoralism in Eurasia*, edited by David R. Harris, pp. 296–309. University College London Press, London.

1999 Neighbours from Hell: The Household in Neolithic Greece. In *Neolithic Society in Greece*, edited by Paul Halstead, pp. 77–95. Sheffield Studies in Aegean Archaeology 2. Sheffield Academic Press, Sheffield.

2012 Feast, Food, and Fodder in Neolithic–Bronze Age Greece: Commensality and the Construction of Value. In *Between Feasts and Daily Meals: Toward an Archaeology of Commensal Spaces*, edited by Susan Pollock, pp. 21–51. ETopoi: Journal for Ancient Studies, Special Vol. 2. Exzellenzcluster 264 Topoi, Berlin.

Halstead, Paul, and Glynis Jones

1980 Early Neolithic Economy in Thessaly: Some Evidence from Excavations at Prodromos. *Anthropologika* 1:93–117.

1987 Bioarchaeological Remains from Kalythies Cave, Rhodes. In *Η Νεολιθική Περίοδος στα Δωδεκάνησα* [The Neolithic Period in the Dodecanese], edited by Adamantios Sampson, pp. 135–152. Publications of the Archaeologiko Deltion 35. Fund of Archaeological Proceeds (TAPA), Athens.

1989 Agrarian Ecology in the Greek Islands: Time Stress, Scale, and Risk. *Journal of Hellenic Studies* 109:41–55.

Hamilakis, Yannis

1996 Wine, Oil, and the Dialectics of Power in Bronze Age Crete: A Review of the Evidence. *Oxford Journal of Archaeology* 15(1):1–32.

1998 Eating the Dead: Mortuary Feasting and the Politics of Memory in the Aegean Bronze Age Societies. In *Cemetery and Society in the Aegean Bronze Age*, edited by Keith Branigan, pp. 115–132. Sheffield Studies in Aegean Archaeology 1. Sheffield Academic Press, Sheffield.

1999 Food Technologies/Technologies of the Body: The Social Context of Wine and Oil Production in Bronze Age Crete. *World Archaeology* 31(1):38–54.

2000 The Anthropology of Food and Drink Consumption and Aegean Archaeology. In *Palaeodiet in the Aegean. Papers from a Colloquium Held at the 1993 Meeting of the Archaeological Institute of America in Washington, DC*, edited by Sarah J. Vaughan and William D. E. Coulson, pp. 55–63. Wiener Laboratory Monograph 1. Oxbow, Oxford.

2002 Past as Oral History: Towards an Archaeology of the Senses. In *Thinking through the Body: Archaeologies of Corporeality*, edited by Yannis Hamilakis, Mark Zygmunt Pluciennik, and Sarah Tarlow, pp.121–136. Kluwer Academic/Plenum, New York.

2014 *Archaeology and the Senses: Human Experience, Memory, and Affect.* Cambridge University Press, Cambridge.

Hamilakis, Yannis, and Susan Sherratt

2012 Feasting and the Consuming Body in Bronze Age Crete and Early Iron Age Cyprus. In *Parallel Lives: Ancient Island Societies in Crete and Cyprus: Papers Arising from the Conference in Nicosia Organised by the British School at Athens, the University of Crete, and the University of Cyprus, in November–December 2006*, edited by Gerald Cadogan, Maria Iacovou, Katerina Kopaka, and James Whitley, pp. 187–207. British School at Athens Studies 20. The British School at Athens, London.

Hamon, Caroline, Clarissa Cagnato, Aline Emery-Barbier, and Aurélie Salavert

2021 Food Practices of the First Farmers of Europe: Combined Use-Wear and Microbotanical Studies of Early Neolithic Grinding Tools from the Paris Basin. *Journal of Archaeological Science: Reports* 36:102764.

Hamon, Caroline, and Valerie Le Gall

2013 Millet and Sauce: The Uses and Functions of Querns among the Minyanka (Mali). *Journal of Anthropological Archaeology* 32(1):109–121.

Hanbury, Colin D., C. Leann White, Brendan P. Mullan, and Kadambot H. M. Siddique

2000 A Review of the Potential of *Lathyrus sativus* L. and *Lathyrus cicera* L. Grain for Use as Animal Feed. *Animal Feed Science and Technology* 87(1–2):1–27.

Hancioğlu, Ömre, and Mehmet Karapinar

1997 Microflora of Boza, a Traditional Fermented Turkish Beverage. *International Journal of Food Microbiology* 35(3):271–274.

Hankey, Vronwy

1993 Pottery as Evidence for Trade: The Levant from the Mouth of the River Orontes to the Egyptian Border. In *Wace and Blegen: Pottery as Evidence for Trade in the Aegean Bronze Age, 1939–1989. Proceedings of the International Conference Held at the American School of Classical Studies at Athens, December 2–3, 1989*, edited by Carol Zerner, Peter Zerner, and John Winder, pp. 101–108. J. C. Gieben, Amsterdam.

Hansen, Julie Marie

1988 Agriculture in the Prehistoric Aegean: Data versus Speculation. *American Journal of Archaeology* 92(1):39–52.

1991 *The Palaeoethnobotany of Franchthi Cave. Excavations at Franchthi Cave, Greece, Vol. 7*. Indiana University Press, Bloomington.

1992 Franchthi Cave and the Beginnings of Agriculture in Greece and the Aegean. In *Préhistoire de l'agriculture: Nouvelles approches expérimentales et ethnographiques*, edited by Patricia C. Anderson, pp. 231–247. Monographie du CRA 6. CNRS, Paris.

2000 Palaeoethnobotany and Palaeodiet in the Aegean Region: Notes on Legume Toxicity and Related Pathologies. In *Palaeodiet in the Aegean. Papers from a Colloquium Held at the 1993 Meeting of the Archaeological Institute of America in Washington, DC*, edited by Sarah J. Vaughan and William D. E. Coulson, pp. 13–27. Wiener Laboratory Monograph 1. Oxbow, Oxford.

Hansen, Julie Marie, and Jane M. Renfrew

1978 Palaeolithic-Neolithic Seed Remains at Franchthi Cave, Greece. *Nature* 271(5643):349–352.

Hansson, Ann-Marie

1994 Grain-Paste, Porridge, and Bread: Ancient Cereal Based Food. *Laborativ Arkeologi* 7:5–20.

2002 Pre- and Protohistoric Bread in Sweden: A Definition and a Review. In *Pains, Fours et Foyers Domestiques des Temps Passés—Breads, Ovens, and Hearths of the Past*, edited by Marianne Mesnil and Kaï Fechner, pp. 183–190. *Civilisation* 49(1–2). Institut de Sociologie de l'Université de Bruxelles.

2011 Bread Offerings at Foundation IV, Helgö—Further Influence from the Classical World? In *Excavations at Helgö XVIII: Conclusions and New Aspects*, edited by Birgit Arrhenius and Uaininn O' Meadhra, pp. 53–77. Kungliga Vitterhets Historie och Antikvitets Akademien, Stockholm.

Hardy, Karen, and Lucy Kubiak-Martens

2016a Introduction. In *Wild Harvest: Plants in the Hominin and Pre-Agrarian Human Worlds*, edited by Karen Hardy and Lucy Kubiak-Martens, pp. 2–16. Studying Scientific Archaeology 2. Oxbow, Oxford.

Hardy, Karen, and Lucy Kubiak-Martens (editors)

2016b *Wild Harvest: Plants in the Hominin and Pre-Agrarian Human Worlds*. Studying Scientific Archaeology 2. Oxbow, Oxford.

Harlan, Jack R.

1976 Plant and Animal Distribution in Relation to Domestication. *Philosophical Transactions of the Royal Society B: Biological Sciences* 275(936):13–25.

1992 Wild Grass Seed Harvesting and Implications for Domestication. In *Préhistoire de l'agriculture: Nouvelles approches expérimentales et ethnographiques*, edited by Patricia C. Anderson, pp. 21–27. Monographie du CRA 6. CNRS, Paris.

Harlan, Jack R., and Jan M. J. de Wet

1965 Some Thoughts about Weeds. *Economic Botany* 19(1):16–21.

Harlan, Jack R., and Daniel Zohary

1966 Distribution of Wild Wheats and Barley. *Science* 153(3740):1074–1080.

Harris, David R.

1989 An Evolutionary Continuum of People-Plant Interaction. In *Foraging and Farming: The Evolution of Plant Exploitation*, edited by David R. Harris and Gordon C. Hillman, pp. 11–26. Unwin Hyman, London.

1996 Introduction: Themes and Concepts in the Study of Early Agriculture. In *The Origins and Spread of Agriculture and Pastoralism in Eurasia*, edited by David R. Harris, pp. 1–9. University College London Press, London.

Hartung, Ulrich

2021 Recent Excavations in the Late Predynastic Settlement of Tell el Fara'in/Buto. In *Egypt at Its Origins 6: Proceedings of the Sixth International Conference on Origin of the State, Predynastic and Early Dynastic Egypt, Vienna, 10th–15th September 2017*, edited by E. C. Köhler, N. Kuch, F. Junge, and A. K. Jeske. Orientalia Lovaniensia Analecta. Peeters, Leuven, in press.

Haskell, Halford W.

1985 The Origin of the Aegean Stirrup Jar and Its Earliest Evolution and Distribution (MB III-LBI). *American Journal of Archaeology* 89(2):221–229.

Haskell, Halford W., Richard E. Jones, Peter M. Day, and John T. Killen

2011 *Transport Stirrup Jars of the Bronze Age Aegean and East Mediterranean*. Prehistory Monographs 33. INSTAP Academic, Philadelphia.

Hastorf, Christine A.

1998 The Cultural Life of Early Domestic Plant Use. *Antiquity* 72(278):773–782.

2017 *The Social Archaeology of Food: Thinking about Eating from Prehistory to the Present*. Cambridge University Press, Cambridge.

Hastorf, Christine A., and Sissel Johannessen

1993 Pre-Hispanic Political Change and the Role of Maize in the Central Andes of Peru. *American Anthropologist* 95(1):115–138.

Hayden, Brian

1995 A New Overview of Domestication. In *Last Hunters, First Farmers: New Perspectives on the Prehistoric Transition to Agriculture*, edited Theron Douglas Price and Anne Birgitte Gebauer, pp. 273–299. School of American Research Press, Santa Fe.

2001 Fabulous Feasts: A Prolegomenon to the Importance of Feasting. In *Feasts: Archaeological and Ethnographic Perspectives on Food, Politics, and Power*, edited by Michael Dietler and Brian Hayden, pp. 23–64. Smithsonian Institution Press, Washington, DC.

2003 Were Luxury Foods the First Domesticates? Ethnoarchaeological Perspectives from Southeast Asia. *World Archaeology* 34(3):458–469.

Hayden, Brian, Neil Canuel, and Jennifer Shanse

2013 What Was Brewing in the Natufian? An Archaeological Assessment of Brewing Technology in the Epipaleolithic. *Journal of Archaeological Method and Theory* 20(1):102–150.

Hedrick, Ulysses Prentiss (editor)

1972 *Sturtevant's Edible Plants of the World*. Reprinted. Dover Publications, New York. Originally published 1919, New York Agricultural Experiment Station, Geneva, New York.

Heinrich, Michael

1994 Herbal and Symbolic Medicines of the Lowland Mixe (Oaxaca, Mexico): Disease Concepts, Healer's Roles, and Plant Use. *Anthropos* 89(1–3):73–83.

1998 Indigenous Concepts of Medicinal Plants in Oaxaca, Mexico: Lowland Mixe Plant Classification Based on Organoleptic Characteristics. *Journal of Applied Botany: Angewandte Botanik* 72(3–4):75–81.

Heiss, Andreas G., Marian Berihuete-Azorín, Ferran Antolín, Lucy Kubiak-Martens, Elena Marinova, Elke K. Arendt, Costas G. Biliaderis, Hermann Kretschmer, Athina Lazaridou, Hans-Peter Stika, Martin Zarnkow, Masahiro Baba, Niels Bleicher, Krzysztof M. Ciałowicz, Marek Chłodnicki, Irenäus Matuschik, Helmut Schlichtherle, and Soultana Maria Valamoti

2020 Mashes to Mashes, Crust to Crust: Presenting a Novel Microstructural Marker for Malting in the Archaeological Record. *PLoS ONE* 15(5):e0231696.

Heiss, Andreas G., Ferran Antolín, Marian Berihuete-Azorín, Benedikt Biederer, Rudolf Erlach, Niki Gail, Monika Griebl, Robert Linke, Michaela Lochner, Elena Marinova, Daniel Oberndorfer, Hans-Peter Stika, and Soultana Maria Valamoti

2019 The Hoard of the Rings: "Odd" Annular Bread-Like Objects as a Case Study for Cereal-Product Diversity at the Late Bronze Age Hillfort Site of Stillfried (Lower Austria). *PLoS ONE* 14(6):e0216907.

Heiss, Andreas G., Ferran Antolín, Niels Bleicher, Christian Harb, Stefanie Jacomet, Marlu Kühn, Elena Marinova, Hans-Peter Stika, Soultana Maria Valamoti

2017 State of the (T)art: Analytical Approaches in the Investigation of Components and Production Traits of Archaeological Bread-like Objects, Applied to Two Finds from the Neolithic Lakeshore Settlement Parkhaus Opéra (Zurich, Switzerland). *PLoS ONE* 12(8):e0182401.

Heiss, Andreas G., Nathalie Pouget, Julian Wiethold, Anne Delor-Ahü, and Isabelle Le Goff

2015 Tissue-Based Analysis of a Charred Flat Bread (Galette) from a Roman Cemetery at Saint-Memmie (Dép. Marne, Champagne-Ardenne, North-Eastern France). *Journal of Archaeological Science* 55:71–82.

Heldreich von, Theodor

1909 *Λεξικό των δημωδών ονομάτων των φυτών της Ελλάδος* [Dictionary of Greek Plants]. Miliarakis, Athens.

Henry, Amanda G.
2017 Neanderthal Cooking and the Costs of Fire. *Current Anthropology* 58, Supplement 16:S329–S336.
Herbig, Christoph
2009 Recent Archaeobotanical Investigations into the Range and Abundance of Neolithic Crop Plants in Settlements around Lake Constance and in Upper Swabia (South-West Germany) in Relation to Cultural Influences. *Journal of Archaeological Sciences* 36(6):1277–1285.
Herbig, Christoph, and Ursula Maier
2011 Flax for Oil or Fibre? Morphometric Analysis of Flax Seeds and New Aspects of Flax Cultivation in Late Neolithic Wetland Settlements in Southwest Germany. *Vegetation History and Archaeobotany* 20(6):527–533.
Hidalgo, Alyssa, Andrea Brandolini, Carlo Pompei, and Roberta Piscozzi
2006 Carotenoids and Tocols of Einkorn Wheat (*Triticum monococcum* ssp. *monococcum* L.). *Journal of Cereal Science* 44(2):182–193.
Hill, Albert F.
1952 *Economic Botany: A Textbook of Useful Plants and Plant Products.* Maple Press, York, Pennsylvania.
Hillman, Gordon C.
1981 Reconstructing Crop Husbandry Practices from Charred Remains of Crops. In *Farming Practice in British Prehistory,* edited by Roger Mercer, pp. 123–162. Edinburgh University Press, Edinburgh.
1984 Traditional Husbandry and Processing of Archaic Cereals in Recent Times: The Operations, Products, and Equipment Which Might Feature in Sumerian Texts. Part I: The Glume Wheats. *Bulletin on Sumerian Agriculture* 1:114–152.
1985 Traditional Husbandry and Processing of Archaic Cereals in Recent Times: The Operations, Products, and Equipment That Might Feature in Sumerian texts. Part II: The Free-Threshing Cereals. *Bulletin on Sumerian Agriculture* 2:1–31.
2003 Investigating the Start of Cultivation in Western Eurasia: Studies of Plant Remains from Abu Hureyra on the Euphrates. In *The Widening Harvest: The Neolithic Transition in Europe—Looking Back, Looking Forward,* edited by Albert J. Ammerman and Paolo Biagi, pp. 75–97. AIA Colloquia and Conference Papers 6. Archaeological Institute of America, Boston.
Hillman, Gordon C., and Stuart M. Davies
1992 Domestication Rate in Wild Wheats and Barley under Primitive Cultivation: Preliminary Results and Archaeological Implications of Field Measurements of Selection Coefficient. In *Préhistoire de l'agriculture: Nouvelles approches expérimentales et ethnographiques,* edited by Patricia C. Anderson, pp. 113–158. Monographie du CRA 6. CNRS, Paris.
Hills, Ronald D., Jr., and Emily Erpenbeck
2018 Guide to Popular Diets, Food Choices, and Their Health Outcome. *Health Care Current Reviews* 6(2):1000223.
Hjelmqvist, Hakon
1984 Botanische Analyse einiger Brote. In Birka II: 1. *Systematische Analysen der Gräberfunde,* edited by Greta Arwidsson, pp. 263–272. Almqvist and Wiksell International, Stockholm.
1990 Über die Zusammensetzung einiger prähistorischer Brote. *Fornvännen* 85:9–21.

www.diva-portal.org/smash/get/diva2:1225904/FULLTEXT01.pdf, accessed July 17, 2022.

Hockings, Kimberley J., Nicola Bryson-Morrison, Susana Carvalho, Michiko Fujisawa, Tatyana Humle, William C. McGrew, Miho Nakamura, Gaku Ohashi, Yumi Yamanashi, Gen Yamakoshi, and Tetsuro Matsuzawa

2015 Tools to Tipple: Ethanol Ingestion by Wild Chimpanzees Using Leaf-Sponges. *Royal Society Open Science* 2(6):150150.

Hodder, Ian

1990 *The Domestication of Europe: Structure and Contingency in Neolithic Societies.* Blackwell, Oxford.

Hodkinson, Stephen

1988 Animal Husbandry in the Greek Polis. In *Pastoral Economies in Classical Antiquity,* edited by C. R. Whittaker, pp. 35–74. Cambridge Philological Society, Supplement 14. Cambridge Philological Society.

Hofmanová, Zuzana, Susanne Kreutzer, Garrett Hellenthal, Christian Sell, Yoan Diekmann, David Díez-del-Molino, Lucy van Dorp, Saioa López, Athanasios Kousathanas, Vivian Link, Karola Kirsanow, Lara M. Cassidy, Rui Martiniano, Melanie Strobel, Amelie Scheu, Kostas Kotsakis, Paul Halstead, Sevi Triantaphyllou, Nina Kyparissi-Apostolika, Dushka Urem-Kotsou, Christina Ziota, Fotini Adaktylou, Shyamalika Gopalan, Dean M. Bobo, Laura Winkelbach, Jens Blöcher, Martina Unterländer, Christoph Leuenberger, Çiler Çilingiroğlu, Barbara Horejs, Fokke Gerritsen, Stephen J. Shennan, Daniel G. Bradley, Mathias Currat, Krishna R. Veeramah, Daniel Wegmann, Mark G. Thomas, Christina Papageorgopoulou, and Joachim Burger

2016 Early Farmers from Across Europe Directly Descended from Neolithic Aegeans. *Proceedings of the National Academy of Sciences of the United States of America (PNAS)* 113(25):6886–6891.

Holtorf, Cornelius

2007 *Archaeology Is a Brand! The Meaning of Archaeology in Contemporary Popular Culture.* Left Coast Press, Walnut Creek.

Hopf, Maria

1961 Pflanzenfunde aus Lerna/Argolis. *Der Züchter* 31(5):239–247.

1962a Nutzpflanzen vom Lernäischen Golf. *Jahrbuch des Römisch-Germanischen Zentralmuseums* 9:1–19.

1962b Bericht über die Untersuchung von Samen und Holzkohlenresten von der Argissa-Magula aus den Präkeramischen bis Mittelbronzezeitlichen Schichten. In *Die Deutschen Ausgrabungen auf der Argissa-Magula in Thessalien I: Das präkeramische Neolithikum sowie die Tier- und Pflanzenreste,* edited by Vladimir Milojčič, Joachim Boessneck, and Maria Hopf, pp. 101–110. Beiträge zur Ur- und Frühgeschichtlichen Archäologie des Mittelmeer-Kulturraumes 2. Rudolf Habelt, Bonn.

1971 Appendix: Plant Remains from the Athenian Agora, Neolithic to Byzantine. In *The Athenian Agora XIII: The Neolithic and Bronze Ages,* edited by Sara Anderson Immerwhar, pp. 267–269. The American School of Classical Studies at Athens, Princeton.

Horejs, Barbara

2007 *Das Prähistorische Olynth: Ausgrabungen in der Toumba Agios Mamas 1994–1996. Die spätbronzezeitliche handgemachte Keramik der Schichten 13 bis 1.*

Prähistorische Archäologie in Südosteuropa 21. Marie Leidorf, Rahden/Westfalen.

Horejs, Barbara, Bogdana Milić, Felix Ostmann, Ursula Thanheiser, Bernard P. Weninger, and Alfred Galik

2015 The Aegean in the Early 7th Millennium BC: Maritime Networks and Colonization. *Journal of World Prehistory* 28:289–330.

Hosoya, Leo Aoi

2011 Staple or Famine Food?: Ethnographic and Archaeological Approaches to Nut Processing in East Asian Prehistory. *Archaeological and Anthropological Sciences* 3(1):7–17.

Hourmouziadis, Giorgos H.

2007 Η μικρογραφία του Δισπηλιού ή με αφορμή τα νεολιθικά ειδώλια ενός λιμναίου οικισμού [A Micrography of Dispilio or Considering the Neolithic Figurines of a Lake-Side Settlement]. *EGNATIA (Annals of the Department of History and Archaeology, Aristotle University of Thessaloniki)* 11:51–71.

Housley, Rupert A.

2000 The Carbonized Plant Remains. In *Servia I: Anglo-Hellenic Rescue Excavations 1971–73, directed by Katerina Romiopoulou and Cressida Ridley*, edited by Cressida Ridley, Kenneth A. Wardle, and Catharine A. Mould, pp. 301–329. British School at Athens Supplement 32. The British School at Athens, London.

Howes, David

1987 Olfaction and Transition: An Essay on the Ritual Uses of Smell. *Canadian Review of Sociology/Revue canadienne de sociologie* 24(3):398–416.

Hruby, Julie A.

2006 *Feasting and Ceramics: A View from the Palace of Nestor at Pylos.* PhD dissertation, University of Cincinnati.

2017 Finding Haute Cuisine: Identifying Shifts in Food Styles from Cooking Vessels. In *From Cooking Vessels to Cultural Practices in the Late Bronze Age Aegean*, edited by Julie Hruby and Debra Trusty, pp. 15–26. Oxbow, Oxford.

Hryciuk, Renata Ewa

2019 La Alquimista de los Sabores: Gastronomic Heritage, Gender, and Tourist Imaginary in Mexico. *Revista del CESLA. International Latin American Studies Review* 24:75–100.

Hsu, Elisabeth

2010 Introduction. In *Plants, Health and Healing: On the Interface of Ethnobotany and Medical Anthropology*, edited by Elisabeth Hsu and Stephen Harris, pp. 1–46. Epistemologies of Healing, Vol. 6. Berghahn, New York.

Hu, Frank B., Meir J. Stampfer, JoAnn E. Manson, Eric B. Rimm, Alicja Wolk, Graham A. Golditz, Charles H. Hennekens, and Walter C. Willett

1999 Dietary Intake of Alpha-Linolenic Acid and Risk of Fatal Ischemic Heart Disease among Women. *American Journal of Clinical Nutrition* 69(5):890–897.

Hubbard, Richard N. L.

1979 Appendix 2: Ancient Agriculture and Ecology at Servia. In *Rescue Excavations at Servia 1971–1973: A Preliminary Report*, edited by Mies Wijnen, Cressida Ridley, K. A. Wardle, R. N. L. Hubbard, J. P. N. Watson, and R. E. Jones. *Annual of the British School at Athens* 74:226–228.

Huffman, Michael. A.

1997 Current Evidence for Self-Medication in Primates: A Multidisciplinary Perspective. *American Journal of Physical Anthropology 104, Supplement 25: Yearbook of Physical Anthropology*:S171–S200.

2007 Culture, Religion, and Belief Systems: Animals as Sources of Medicinal Wisdom in Traditional Societies. In *Encyclopedia of Human-Animal Relationships: A Global Exploration of Our Connections with Animals*, edited by Marc Bekoff, pp. 434–441. Greenwood Press, Westport, Connecticut.

2016 An Ape's Perspective on the Origins of Medicinal Plant Use in Humans. In *Wild Harvest: Plants in the Hominin and Pre-Agrarian Human Worlds*, edited by Karen Hardy and Lucy Kubiak-Martens, pp. 55–70. Studying Scientific Archaeology 2. Oxbow, Oxford.

Hugh-Jones, Stephen

1995 Inside-Out and Back-to-Front: The Androgynous House in Northwest Amazonia. In *About the House: Lévi-Strauss and Beyond*, edited by Janet Carsten and Stephen Hugh-Jones, pp. 226–252. Cambridge University Press, Cambridge.

Hughes, Donald J.

2003 Europe as Consumer of Exotic Biodiversity: Greek and Roman Times. *Landscape Research* 28(1):21–31.

Huxley, Anthony, and William Taylor

1989 *Flowers of Greece and the Aegean*. Hogarth Press, London.

Hyman, Mark

2015 Elephant Journal. www.elephantjournal.com.

Iddison, Philip

1996 Arabian Traveller's Observations on Bedouin Food. In *Food on the Move: Proceedings of the Oxford Symposium on Food and Cookery 1996*, edited by Harlan Walker, pp. 177–185. Prospect Books, Totnes.

Ingold, Timothy

1996 Growing Plants and Raising Animals: An Anthropological Perspective on Domestication. In *The Origins and Spread of Agriculture and Pastoralism in Eurasia*, edited by David R. Harris, pp. 12–24. University College London Press, London.

Insoll, Timothy

2011a Substance and Materiality? The Archaeology of Talensi Medicine Shrines and Medicinal Practices. *Anthropology and Medicine* 18(2):181–203.

2011b Introduction. Shrines, Substances and Medicine in Sub-Saharan Africa: Archaeological, Anthropological, and Historical Perspectives. *Anthropology and Medicine* 18(2):145–166.

Insoll, Timothy, Benjamin W. Kankpeyeng, and Samuel N. Nkumbaan

2012 Fragmentary Ancestors? Medicine, Bodies, and Personhood in a Koma Mound, Northern Ghana. In *Archaeology of Spiritualities*, edited by Kathryn Rountree, Christine Morris, and Alan A. D. Peatfield, pp. 25–45. Springer, New York.

Iriarte-Chiapusso, María José, Carlos Alvar Ocete-Pérez, Begoña Hernández-Beloqui, and Rafael Ocete-Rubio

2017 *Vitis vinifera* in the Iberian Peninsula: A Review. *Plant Biosystems* 151(2):245–257.

Ivanova, Maria, Bea De Cupere, Jonathan Ethier, and Elena Marinova

2018 Pioneer Farming in Southeast Europe during the Early Sixth Millennium BC:

Climate-Related Adaptations in the Exploitation of Plants and Animals. *PLoS ONE* 13(8):e0202668.

Jacobsen, Thomas W.
1976 17,000 Years of Greek Prehistory. *Scientific American* 234(6):76–87.

Jacobson-Widding, Anita
1992 Pits, Pots, and Snakes: An Anthropological Approach to Ancient African Symbols. *Nordic Journal of African Studies* 1(1):5–25.

Jacomet, Stefanie, and Christoph Brombacher
2005 Reconstructing Intra-Site Patterns in Neolithic Lakeshore Settlements: The State of Archaeobotanical Research and Future Prospects. In *WES'04—Wetland Economies and Societies. Proceedings of the International Conference in Zurich, 10–13 March 2004*, edited by Philippe Della Casa and Martin Trachsel, pp. 69–94. Collectio Archaeologica 3. Chronos, Zürich.

Jacomet, Stefanie, and Angela Kreuz
1999 *Archäobotanik: Eine Einführung in Aufgaben, Methoden und Ergebnisse vegetations- und agrargeschichtlicher Forschung*. Ulmer, Stuttgart.

Jacquat, Christiane, and Danièle Martinoli
1999 *Vitis vinifera* L.: Wild or Cultivated? Study of the Grape Pips Found at Petra, Jordan, 150 B.C.–A.D. 40. *Vegetation History and Archaeobotany* 8(1–2):25–30.

Jaffrey, Madhur
1985 *A Taste of India*. Pavilion Books, London.

Jahns, Susanne
2005 The Holocene History of Vegetation and Settlement at the Coastal Site of Lake Voulkaria in Acarnania, Western Greece. *Vegetation History and Archaeobotany* 14(1): 55–66.
2009 The Holocene History of Vegetation and Environment of Northern Akarnania, Western Greece. In *Mensch und Umwelt im Spiegel der Zeit. Aspekte geoarchäologischer Forschungen im östlichen Mittelmeergebiet*, edited by Torsten Mattern and Andreas Vött, pp. 5–26. Harrassowitz Verlag, Wiesbaden.

Jannot, Jean-René
2009 The Lotus, Poppy and Other Plants in Etruscan Funerary Contexts. In *Etruscan by Definition: The Cultural, Regional and Personal Identity of the Etruscans. Papers in Honour of Sibylle Haynes*, edited by Judith Swaddling and Philip Perkins, pp. 81–86. British Museum Research Publication 173. British Museum Press, London.

Jarman, H. N.
1972 The Origins of Wheat and Barley Cultivation. In *Papers in Economic Prehistory: Studies by Members and Associates of the British Academy Major Research Project in the Early History of Agriculture*, edited by Eric S. Higgs, pp. 15–26. Cambridge University Press, Cambridge.

Jarman, Michael R., G. N. Bailey, and H. N. Jarman
1982 *Early European Agriculture: Its Foundations and Developments*. Cambridge University Press, Cambridge.

Jennings, Justin, Kathleen L. Antrobus, Sam J. Atencio, Erin Glavich, Rebecca Johnson, German Loffler, and Christine Luu
2005 Drinking Beer in a Blissful Mood: Alcohol Production, Operational Chains, and Feasting in the Ancient World. *Current Anthropology* 46(2):275–303.

Jiang, Hongen, Yongbing Zhang, Enguo Lü, and Changsui Wang
2015 Archaeobotanical Evidence of Plant Utilization in the Ancient Turpan of

Xinjiang, China: A Case Study at the Shengjindian Cemetery. *Vegetation History and Archaeobotany* 24(1):165–177.

Johns, Andreas
1998 Baba Iaga and the Russian Mother. *The Slavic and East European Journal* 42(1):21–36.

Johns, Timothy, and Martin Duquette
1991 Traditional Detoxification of Acorn Bread with Clay. *Ecology of Food and Nutrition* 25(3):221–228.

Johns, Timothy, and Susan L. Keen
1985 Determinants of Taste Perception and Classification among the Aymara of Bolivia. *Ecology of Food and Nutrition* 16(3):253–271.

Jones, Glynis
1981 Crop Processing at Assiros Toumba—A Taphonomic Study. *Zeitschrift für Archäologie* 15(1):105–111.

1983 *The Use of Ethnographic and Ecological Models in the Interpretation of Archaeological Plant Remains: Case Studies from Greece*. PhD dissertation, University of Cambridge.

1984 The LMII Plant Remains. In *The Minoan Unexplored Mansion at Knossos*, edited by Mervyn R. Popham, pp. 303–306. British School at Athens Supplement 17. The British School at Athens, London.

1987 A Statistical Approach to the Archaeological Identification of Crop Processing. *Journal of Archaeological Science* 14(3):311–323.

1992 Ancient and Modern Cultivation of *Lathyrus ochrus* (L.) DC. in the Greek Islands. *Annual of the British School at Athens* 87:211–217.

1995 Charred Grain from Late Bronze Age Gla, Boiotia. *Annual of the British School at Athens* 90:235–238.

1998 Distinguishing Food from Fodder in the Archaeobotanical Record. *Environmental Archaeology* 1(1):95–98.

Jones, Glynis, Amy Bogaard, Paul Halstead, Michael Charles, and Helen Smith
1999 Identifying the Intensity of Crop Husbandry Practices on the Basis of Weed Floras. *Annual of the British School at Athens* 94:167–189.

Jones, Glynis, Mike P. Charles, Martin K. Jones, Sue Colledge, Fiona J. Leigh, Diane A. Lister, Lydia M. J. Smith, Wayne Powell, Terrence A. Brown, and Huw Jones
2013 DNA Evidence for Multiple Introductions of Barley into Europe Following Dispersed Domestications in Western Asia. *Antiquity* 87(337):701–713.

Jones, Glynis, and Paul Halstead
1993a An Early Find of "Fava" from Thebes. *Annual of the British School at Athens* 88:103–104.

1993b Charred Plant Remains from Neolithic-Bronze Age Platia Magoula Zarkou, Thessaly. *Annual of the British School at Athens* 88:1–3.

1995 Maslins, Mixtures, and Monocrops: On the Interpretation of Archaeobotanical Crop Samples of Heterogeneous Composition. *Journal of Archaeological Science* 22(1):103–114.

Jones, Glynis, Paul Halstead, Kenneth A. Wardle, and Diana Wardle
1986 Crop Storage at Assiros. *Scientific American* 254(3):96–103.

Jones, Glynis, and Ann Schofield
2006 Evidence for the Use of Threshing Remains at the Early Minoan Metallurgical Workshop. In *The Chrysokamino Metallurgy Workshop and Its Territory*, edited

by Philip P. Betancourt, pp. 153–154. Hesperia Supplement 36. The American School of Classical Studies at Athens, Princeton.

Jones, Glynis, and Ian Smith
2009 Plant Remains from Block AF. In *Pseira X: The Excavation of Block AF*, edited by Philip P. Betancourt, pp. 125–126. Prehistory Monographs 28. INSTAP Academic, Philadelphia.

Jones, Glynis, and Soultana Maria Valamoti
2005 *Lallemantia*, an Imported or Introduced Oil Plant in Bronze Age Northern Greece. *Vegetation History and Archaeobotany* 14(4):571–577.

Jones, Glynis, Soultana Maria Valamoti, and Michael Charles
2000b Early Crop Diversity: A "New" Glume Wheat from Northern Greece. *Vegetation History and Archaeobotany* 9(3):133–146.

Jones, Martin K.
2007 *Feast: Why Humans Share Food.* Oxford University Press, Oxford.
2009 Moving North: Archaeobotanical Evidence for Plant Diet in Middle and Upper Paleolithic Europe. In *The Evolution of Hominin Diets: Integrating Approaches to the Study of Palaeolithic Subsistence*, edited by Jean-Jacques Hublin and Michael P. Richards, pp. 171–180. Vertebrate Paleobiology and Paleoanthropology. Springer, Dordrecht.

Jones, Martin K., Robin Allaby, and Terry Brown
1998 Wheat Domestication. *Science* 279:302–303.

Jones, Martin K., Harriet Hunt, Catherine Kneale, Emma Lightfoot, Diane Lister, Xinyi Liu, and Giedre Motuzaite-Matuzeviciute
2016 Food Globalisation in Prehistory: The Agrarian Foundations of an Interconnected Continent. *Journal of the British Academy* 4:73–87.

Jørgensen, Grethe
1977 Acorns as a Food-Source in the Later Stone Age. *Acta Archaeologica* 48:233–238.

Joyce, Rosemary A.
2001 Burying the Dead at Tlatilco: Social Memory and Social Identities. In *Social Memory, Identity, and Death: Anthropological Perspectives on Mortuary Rituals*, edited by Meredith S. Chesson, pp. 12–26. Archeological Papers of the American Anthropological Association 10(1):Special Issue.

Jung, Reinhard
2006 Εὔποτον ποτέριον: Mykenische Keramik und mykenische Trinksitten in der Ägäis, in Syrien, Makedonien und Italien. In *Studi di Protostoria in Onore di Renato Peroni*, pp. 407–423. All'Insegna del Giglio, Borgo San Lorenzo.
2012 Can We Say, What's Behind All Those Sherds? Ceramic Innovations in the Eastern Mediterranean at the End of the Second Millennium. In *Materiality and Social Practice: Transformative Capacities of Intercultural Encounters*, edited by Joseph Maran and Philipp W. Stockhammer, pp. 104–120. Oxbow, Oxford.

Kaczanowska, Małgorzata, and Janusz K. Kozłowski
2018 Before the Neolithization: Causes of Mesolithic Diversity in the Southern Balkans. *Acta Archaeologica Academiae Scientiarum Hungaricae* 69(2):253–270.

Kale, R.
1995 Traditional Healers in South Africa: A Parallel Health Care System. *British Medical Journal* 310(6988):1182–1185.

Kalogiropoulou, Evita
2014 Αναζητώντας κοινωνικές ταυτότητες: Η συμβολή των θερμικών κατασκευών

στην οργάνωση του χώρου στη νεολιθική Μακεδονία [In Search of Social Identities: The Contribution of Thermal Structures in the Organisation of Space in Neolithic Macedonia]. In *1912–2012: Εκατό χρόνια έρευνας στην προϊστορική Μακεδονία, Πρακτικά διεθνούς συνεδρίου, Αρχαιολογικό Μουσείο Θεσσαλονίκης, 22–24 Νοεμβρίου 2012* [1912–2012: A Century of Research in Prehistoric Macedonia, International Conference Proceedings, Archaeological Museum of Thessaloniki, 22–24 November 2012], edited by Evaggelia Stefani, Nikos Merousis, and Anastasia Dimoula, pp. 359–372. Archaeological Museum of Thessaloniki.

Kankpeyeng, Benjamin W., Samuel N. Nkumbaan, and Timothy Insoll

2011 Indigenous Cosmology, Art Forms, and Past Medicinal Practices: Towards an Interpretation of Ancient Koma Land Sites in Northern Ghana. *Anthropology and Medicine* 18(2):205–216.

Kapoor, Rakesh, and Yung-Sheng Huang

2006 Gamma Linolenic Acid: An Antiinflammatory Omega-6 Fatty Acid. *Current Pharmaceutical Biotechnology* 7(6):531–534.

Karagiannis, Giorgos

2015 *Φυσικό προζύμι* [Natural Sourdough]. Psihalos, Athens.

Karamitrou-Mentesidi, Georgia, Nikos Efstratiou, Małgorzata Kaczanowska, Janusz K. Kozłowski, Yannis Maniatis, Antonio Curci, Stefania Michalopoulou, Anastasia Papathanasiou, and Soultana Maria Valamoti

2013 New Evidence on the Beginning of Farming in Greece: The Early Neolithic Settlement of Mavropigi in Western Macedonia (Greece). *Antiquity* 87(336):Project Gallery.

Karatasaki, Katerina

2017 *Φαρμακευτικά και ψυχοτρόπα φυτά στο προϊστορικό Αιγαίο: Υποθέσεις και Δεδομένα* [Medicinal and Psychotropic Plants in the Prehistoric Aegean: Hypotheses and Data]. Master's thesis, Aristotle University of Thessaloniki.

Karathanou, Angeliki

2009 *Αρχαιοβοτανικές έρευνες στη νεολιθική θέση Κρεμαστή Κοζάνης* [Archaiobotanical Investigation in the Neolithic Site Kremasti of Kozani]. Master's thesis, Aristotle University of Thessaloniki.

Karathanou, Angeliki, and Soultana Maria Valamoti

2011 Η αρχαιοβοτανική έρευνα στη νεολιθική θέση Κρεμαστή Κοιλάδας [Archaeobotanical Research in the Neolithic Site of Kremastis Koiladas]. *Proceedings of the Archaeological Work in Macedonia and Thrace (AEMTH)* 22, 2008:57–70.

2013a Plant Remains from Ayios Vasileios, a Mycenaean Settlement near Sparta: Preliminary Results. Poster presented at the 16th Conference of the International Work Group for Palaeoethnobotany, 17–22 June 2013, Thessaloniki. Program and Abstracts:34, 222.

2013b The Archaeobotanical Investigation of Human Activity in Late Bronze Age Greece. Poster presented at the 16th Conference of the International Work Group for Palaeoethnobotany, 17–22 June 2013, Thessaloniki. Program and Abstracts:35, 223.

2016 Exploring Intra-Settlement Use of Space in Late Bronze Age Greece: Preliminary Observations on the Archaeobotanical Visibility of Storage and Disposal Strategies during the Late 2nd Millennium BC in the Aegean. Poster presented at the 17th Conference of the International Work Group for Palaeoethnobotany, 4–9 July 2016, Paris. Program and Abstracts:57.

Karg, Sabine

1996 Bronzezeitliche Kulturpflanzen: Eine Kartierung der Dinkelnachweise im nördlichen und südlichen Alpenvorland. In *The Colloquia of the XIII International Congress of Prehistoric and Protohistoric Sciences, Forli (Italia) 8–14 September 1996, 3: Paleoecology*, edited by Lanfredo Castelletti and Mauro Cremasch, pp. 25–33. ABACO. s.r.l., Forli.

Karg, Sabine, and Tanja Märkle

2002 Continuity and Changes in Plant Resources during the Neolithic Period in Western Switzerland. *Vegetation History and Archaeobotany* 11(1–2):169–176.

Karkanas, Panagiotis, Margarita Koumouzelis, Janusz K. Kozłowski, Valéry Sitlivy, Krzysztof Sobczyk, Francesco Berna, and Stephen Weiner

2004 The Earliest Evidence for Clay Hearths: Aurignacian Features in Klissoura Cave 1, Southern Greece. *Antiquity* 78(301):513–525.

Kasapidou, Georgia

2020 *Η επεξεργασία των φυτών στην προϊστορική Ελλάδα μέσα από φυτικά μικροκατάλοιπα: πειραματικά και αρχαιολογικά δεδομένα* [The Processing of Plants in Prehistoric Greece through Plant Microremains: Experimental and Archaeological Data]. Master's thesis, Aristotle University of Thessaloniki.

Katsagelos, Giorgos

1995 *Θρησκευτικά* [Religious]. Published in conjunction with an exhibition of the same title, organized by the Municipality of Thessaloniki for the 30th Dimitria Festival at the French Institute of Thessaloniki, November–December 1995.

Katz, Solomon H.

1987 Fava Bean Consumption: A Case for the Co-evolution of Genes and Culture. In *Food and Evolution: Towards a Theory of Food Habits*, edited by Marvin Harris and Eric B. Ross, pp. 133–159. Temple University Press, Philadelphia.

Katz, Solomon H., and Mary M. Voigt

1986 Bread and Beer: The Early Use of Cereals in the Human Diet. *Expedition* 28(2):23–34.

Kaye, Quetta

2018 Power Over? Ritual Drug-Taking and the Social Context of Power among the Indigenous People of the Caribbean. In *Ancient Psychoactive Substances*, edited by Scott M. Fitzpatrick, pp. 149–175. University Press of Florida, Gainesville.

Kenéz, Árpád, Ákos Pető, and Ferenc Gyulai

2014 Evidence of "New Glume Wheat" from the Late Neolithic (Copper Age) of South-Eastern Hungary (4th Millennium cal. BC). *Vegetation History and Archaeobotany* 23(5):551–566.

Khan, Afshan, Aisha Siddiqui, Anwar Jamal, and Gulnar Fatima

2019 *Lallemantia royleana* Benth. (Balangu): A Compendious Review on Phytochemistry, Pharmacology, and Ethnomedicinal Uses. *Journal of Drug Delivery and Therapeutics* 9(5-S):175–177.

Khokhar, S., J. Frias, K. R. Price, G. R. Fenwick, and C. L. Hedley

1996 Physico-Chemical Characteristics of Khesari Dhal (*Lathyrus sativus*): Changes in α-Galactosides, Monosaccharides, and Disaccharides during Food Processing. *Journal of the Science of Food and Agriculture* 70(4):487–492.

Killen, John T.
1984 The Textile Industries at Pylos and Knossos. In *Pylos Comes Alive: Industry and Administration in a Mycenaean Palace. Papers of a Symposium*, edited by Cynthia W. Shelmerdine and Thomas G. Palaima, pp. 49–63. Fordham University Press, New York.
2004 Wheat, Barley, Flour, Olives, and Figs in Linear B Tablets. In *Food, Cuisine, and Society in Prehistoric Greece*, edited by Paul Halstead and John C. Barrett, pp. 155–173. Sheffield Studies in Aegean Archaeology 5. Oxbow, Oxford.

Kirch, Patrick V.
2001 Polynesian Feasting in Ethnohistoric, Ethnographic, and Archaeological Contexts: A Comparison of Three Societies. In *Feasts: Archaeological and Ethnographic Perspectives on Food, Politics, and Power*, edited by Michael Dietler and Brian Hayden, pp. 168–184. Smithsonian Institution Press, Washington, DC.

Kislev, Mordechai E.
1989a Pre-Domesticated Cereals in the Pre-Pottery Neolithic A Period. In *People and Culture in Change. Proceedings of the Second Symposium on Upper Palaeolithic, Mesolithic, and Neolithic Populations of Europe and the Mediterranean Basin*, edited by Israel Hershkovitz, pp. 147–151. British Archaeological Reports International Series 508. BAR, Oxford.
1989b Origins of the Cultivation of *Lathyrus sativus* and L. *cicera* (Fabaceae). *Economic Botany* 43(2):262–270.

Kislev, Mordechai E., and Ofer Bar-Yosef
1988 The Legumes: The Earliest Domesticated Plants in the Near East? *Current Anthropology* 29(1):175–179.

Kislev, Mordechai E., Anat Hartmann, and Ofer Bar-Yosef
2006 Early Domesticated Fig in the Jordan Valley. *Science* 312(5778):1372–1374.

Kislev, Mordechai E., Orit Simchoni, Yoel Melamed, and Lior Maroz
2011 Flax Seed Production: Evidence from the Early Iron Age Site of Tel Beth-Shean, Israel, and from Written Sources. *Vegetation History and Archaeobotany* 20(6):579–584.

Kiziridou, Thomais
2002 *Ποντίων εδέσματα* [Foods of the Pontic People]. Kyriakidis, Thessaloniki.

Knapp, A. Bernard
1991 Spice, Drugs, Grain, and Grog: Organic Goods in Bronze Age Eastern Mediterranean Trade. In *Bronze Age Trade in the Mediterranean*, edited by Noel H. Gale, pp. 21–68. Studies in Mediterranean Archaeology 90. Paul Åströms, Jonsered.
1993 Thalassocracies in Bronze Age Eastern Mediterranean Trade: Making and Breaking a Myth. *World Archaeology* 24(3):332–347. Ancient Trade: New Perspectives Pages.

Kochilas, Diane
2001 *The Glorious Foods of Greece: Traditional Recipes from the Islands, Cities, and Villages*. HarperCollins, New York.
2003 *Η Ελλάδα της Γεύσης* [The Greece of Taste]. Ellinika Grammata, Athens.

Koh, Andrew J., and Kathleen J. Birney
2017 Organic Compounds and Cultural Continuity: The Penn Museum Late Minoan IIIC Stirrup Jar from Tourloti. *Mediterranean Archaeology and Archaeometry* 17(2):19–33.

Kohler-Schneider, Marianne
2001 *Verkohlte Kultur- und Wildpflanzenreste aus Stillfried an der March als Spiegel spätbronzezeitlicher Landwirtschaft im Weinviertel, Niederösterreich.* Mitteilungen der prähistorischen Kommission 37. Verlag der Österreichischen Akademie der Wissenschaften, Wien.
2003 Contents of a Storage Pit from Late Bronze Age Stillfried, Austria: Another Record of the "New" Glume Wheat. *Vegetation History and Archaeobotany* 12(2):105–111.
Kohler-Schneider, Marianne, and Anita Caneppele
2009 Late Neolithic Agriculture in Eastern Austria: Archaeobotanical Results from Sites of the Baden and Jevišovice Cultures (3600–2800 B.C.). *Vegetation History and Archaeobotany* 18(1):61–74.
Kokkidou, Martha
2018 *Αρχαιοβοτανικά δεδομένα από το Ντικιλί Τας Φιλίππων (Γαλλικός τομέας): προσεγγίζοντας την κοινωνική διάσταση της τροφής σε ένα νεολιθικό οικισμό* [Archaeobotanical Evidence from Dikili Tash, French Sector]. Master's thesis, Aristotle University of Thessaloniki.
Kopaka, Katerina
2011 On Caves and Households in Bronze Age Crete: "Της Ουρανιάς το Φρούδι" Cave at Zakros. In *ΣΤΕΓΑ: The Archaeology of Houses and Households in Ancient Crete*, edited by Kevin T. Glowacki and Natalia Vogeikoff-Brogan, pp. 273–284. Hesperia Supplement 44. The American School of Classical Studies at Athens, Princeton.
Koromila, Georgia, Panagiotis Karkanas, Georgia Kotzamani, Kerry Harris, Yannis Hamilakis, and Nina Kyparissi-Apostolika
2017 Humans, Animals, and the Landscape in Neolithic Koutroulou Magoula, Central Greece: An Approach through Micromorphology and Plant Remains in Dung. In *Communities, Landscapes, and Interaction in Neolithic Greece: Proceedings of the International Conference, Rethymno 29–30 May, 2015*, edited by Apostolos Sarris, Evita Kalogiropoulou, Tuna Kalayci, and Lia Karimali, pp. 269–280. International Monographs in Prehistory: Archaeological Series 20. Berghahn, New York.
Koromila, Georgia, Panagiotis Karkanas, Yannis Hamilakis, Nina Kyparissi-Apostolika, Georgia Kotzamani, and Kerry Harris
2018 The Neolithic Tell as a Multi-Species Monument: Human, Animal, and Plant Relationships through a Micro-Contextual Study of Animal Dung Remains at Koutroulou Magoula, Central Greece. *Journal of Archaeological Science: Reports* 19:753–768.
Kotsachristou, Dimitra
2009 *Οι αρχαιοβοτανικές έρευνες στην προϊστορική Μακεδονία: η συμβολή των δεδομένων της Τούμπας Θεσσαλονίκης* [The Archaeobotanical Studies in Prehistoric Macedonia. The Contribution of Data from Toumba, Thessaloniki]. Master's thesis, Aristotle University of Thessaloniki.
Kotsakis, Kostas
1992 Ο νεολιθικός τρόπος παραγωγής: ιθαγενής ή άποικος [The Neolithic Way of Production: Indigenous or Immigrant?]. In *Πρακτικά του Διεθνούς Συνεδρίου για την Αρχαία Θεσσαλία στη Μνήμη του Δημήτρη Θεοχάρη* [Proceedings of an

International Conference on Ancient Thessaly in Memory of Dimitris Theocharis], edited by Evangelia Kypraiou, pp. 120–135. Publications of the Archaeologiko Deltion 48. Fund of Archaeological Proceeds (TAPA), Athens.

2000 Η αρχή της νεολιθικής στην Ελλάδα [The Beginning of the Neolithic in Greece]. In *Σπήλαιο Θεόπετρας: 12 χρόνια ανασκαφών και έρευνας 1987–1998. Πρακτικά Διεθνούς Συνεδρίου, Τρίκαλα 6–7 Νοεμβρίου 1998* [Theopetra Cave: 12 Years of Excavation and Research 1987–1998. Proceedings of the International Conference, Trikala, 6–7 November 1998], edited by Nina Kyparissi-Apostolika, pp. 173–181. Ministry of Culture, Athens.

2001 Mesolithic to Neolithic in Greece: Continuity, Discontinuity, or Change of Course? *Documenta Praehistorica* 28:63–73.

2003 From the Neolithic Side: The Mesolithic/Neolithic Interface in Greece. In *The Greek Mesolithic: Problems and Perspectives*, edited by Nena Galanidou and Catherine Perlès, pp. 217–221. British School at Athens Studies 10. The British School at Athens, London.

Kotsakis, Kostas

2019 Neolithic Goes West: Concepts and Models on the Neolithisation of the Aegean. In *The Central/Western Anatolian Farming Frontier: Proceedings of the Neolithic Workshop Held at 10th ICAANE in Vienna, April 2016*, edited by Maxime Brami and Barbara Horejs, pp. 223–240. OREA 12. Austrian Academy of Sciences, Vienna.

Kotzamani, Georgia

2010 *Από τη συλλογή στην καλλιέργεια: Αρχαιοβοτανική διερεύνηση των πρώιμων σταδίων εκμετάλλευσης των φυτών και της αρχής της γεωργίας στον ελλαδικό χώρο—σπήλαιο Θεόπετρας, σπήλαιο Σχιστού, Σιδάρι, Ρεβένια* [From Gathering to Cultivation: An Archaeobotanical Investigation of the Early Stages of Plant Exploitation and the Beginnings of Agriculture in Greece—Theopetra cave, Schisto cave, Sidari, Revenia]. PhD dissertation, Aristotle University of Thessaloniki.

Kotzamani, Georgia, and Alexandra Livarda

2014 Plant Resource Availability and Management in Palaeolithic and Mesolithic Greece. In *Physis: L'environnement naturel et la relation homme-milieu dans le monde égéen protohistorique. Actes de la 14e rencontre égéenne internationale, Paris, Institut National d'Histoire de l'Art (INHA), 11–14 décembre 2012*, edited by Gilles Touchais, Robert Laffineur, and Françoise Rougemont, pp. 229–237. Aegaeum 37. Peeters, Leuven/Liège.

2017 Archaeobotanical Remains. In *Ayia Sotira: A Mycenaean Chamber Tomb Cemetery in the Nemea Valley, Greece*, edited by R. Angus K. Smith, Mary K. Dabney, Evangelia Pappi, Sevasti Triantaphyllou, and James C. Wright, pp. 139–145. Prehistory Monographs 56. INSTAP Academic, Philadelphia.

2018 People and Plant Entanglements at the Dawn of Agricultural Practice in Greece: An Analysis of the Mesolithic and Early Neolithic Archaeobotanical Remains. *Quaternary International* 496:80–101.

Koukouli-Chrysanthaki, Chaido

2004 Anthropomorphic Stelae from Greece. *Notizie Archeologiche Bergomensi* 12:85–123.

Koukouli-Chrysanthaki, Chaido, Henrieta Todorova, Ioannis Aslanis, Ivan Vajsov, and Magdalene Valla

2007 Promachon-Topolnitsa: A Greek-Bulgarian Archaeological Project. In *The*

Struma/Strymon River Valley in Prehistory: Proceedings of the International Symposium STRYMON PRAEHISTORIKUS, Kjustendil, Blagoevgrad (Bulgaria) and Serres—Amphipolis (Greece), 27/9–1/10/2004, edited by Henrieta Todorova, Mark Stefanovich, and Georgi Ivanov, pp. 43–78. In the Steps of James Harvey Gaul, Vol. 2. Gerda Henkel Stiftung, Sofia.

Koumouzelis, Margarita, Boleslaw Ginter, Janusz K. Kozłowski, Maciej Pawlikowski, Ofer Bar-Yosef, Rosa Maria Albert, Maria Litynska-Zajac, Ewa Stworzewicz, Piotr Wojtal, Grzegorz Lipecki, Teresa Tomek, Zbigniew M. Bochenski, and Anna Pazdur

2001 The Early Upper Palaeolithic in Greece: The Excavations in Klisoura Cave. *Journal of Archaeological Science* 28(5):515–539.

Kourakou-Dragona, Stavroula

2000 *Τα Κυριακάτικα της Σταυρούλας Κουράκου: Αμπέλια και κρασί στις σελίδες της κυριακάτικης Καθημερινής* [The Sunday Articles of Stavroula Kourakou: Grapevines and Wine in the Pages of Sunday "Kathimerini"]. Stachy, Athens.

Kourou, Nota

2001 The Sacred Tree in Greek Art: Mycenaean versus Near Eastern Traditions. In *La questione delle influenze vicino-orientali sulla religione greca: Stato degli studi e prospettive della ricerca. Atti del Colloquio Internazionale, Roma, 20–22 maggio 1999*, edited by Sergio Ribichini, Maria Rocchi and Paolo Xella, pp. 31–53. Consiglio Nationale delle Ricerche, Roma.

Kraft, John C., George Rapp Jr., and Stanley E. Aschenbrenner

1975 Late Holocene Paleogeography of the Coastal Plain of the Gulf of Messenia, Greece, and Its Relationships to Archaeological Settings and Coastal Change. *Geological Society of America Bulletin* 86(9):1191–1208.

1980 Late Holocene Palaeogeomorphic Reconstructions in the Area of the Bay of Navarino: Sandy Pylos. *Journal of Archaeological Science* 7(3):187–210.

Krăsteva-Blagoeva, Evgenija

2008 Tasting the Balkans: Food and Identity. *Ethnologia Balkanica* 12:25–36.

Kreuz, Angela, Tanja Märkle, Elena Marinova, Manfred Rösch, Eva Schäfer, Silke Schamuhn, and Tanja Zerl

2014 The Late Neolithic Michelsberg Culture—Just Ramparts and Ditches? A Supraregional Comparison of Agricultural and Environmental Data. *Praehistorische Zeitschrift* 89(1):72–115.

Kreuz, Angela, and Elena Marinova

2017 Archaeobotanical Evidence of Crop Growing and Diet within the Areas of the Karanovo and the Linear Pottery Cultures: A Quantitative and Qualitative Approach. *Vegetation History and Archaeobotany* 26(6):639–657.

Krikou, Ioanna

2019 Γκορτσιά: Ένα παμπάλαιο ελληνικό δένδρο—Τα φύλλα της χρησιμοποιούνταν στη ιατρική από τα αρχαία χρόνια [Gortzia: A Very Old Greek Tree—Its Leaves Were Being Used in Medicine since Ancient Times]. *TheSecretRealTruth* (blog), July 8.

Kritikos, Panagiotis, and Stella Papadaki

1963 Μήκωνος και οπίου ιστορία και εξάπλωσις εν τη περιοχή της Ανατολικής Μεσογείου κατά την αρχαιότητα [The History of the Poppy and of Opium and Their Expansion in Antiquity in the Eastern Mediterranean Area]. *Αρχαιολογική Εφημερίς* [Archaeologiki Ephimeris] 102: 80–150 (in Greek, English translation in *Bulletin of Narcotics* 19[3], 1967:17–38).

Kroll, Helmut
1979 Kulturpflanzen aus Dimini. In *Festschrift Maria Hopf zum 65. Geburtstag am 14. September 1979*, edited by Udelgard Körber-Grohne, pp. 173–189. Archaeo-Physika 8. Rheinland-Verlag, Köln.
1981 Thessalische Kulturpflanzen. *Zeitschrift für Archaeologie* 15:97–103.
1982 Kulturpflanzen von Tiryns. *Archäologischer Anzeiger* 25:467–485.
1983 *Kastanas. Ausgrabungen in einem Siedlungshügel der Bronze- und Eisenzeit Makedoniens 1975–1979: Die Pflanzenfunde.* Prähistorissche Archäologie in Südosteuropa 2. Volker Spiess, Berlin.
1991 Südosteuropa. In *Progress in Old World Palaeoethnobotany: A Retrospective View on the Occasion of 20 Years of the International Work Group for Palaeoethnobotany*, edited by Willem Van Zeist, Krystyna Wasylikowa, and Karl-Ernst Behre, pp. 161–177. Balkema, Rotterdam.
1993 Kulturpflanzen von Kalapodi. *Archäologischer Anzeiger* 2:161–182.
1998 Die Kultur- und Naturlandschaften des Titeler Plateaus im Spiegel der metallzeitlichen Pflanzenreste von Feudvar. In *Feudvar: Ausgrabungen und Forschungen in einer Mikroregion am Zusammenfluss von Donau und Theiss*, edited by Bernhard Hänsel and Predrag Medović, pp. 305–317. Prähistorische Archäologie in Südosteuropa 13. Oetker/Voges, Kiel.
2003 Rural Plenty: The Result of Hard Work—Rich Middle Bronze Age Plant Remains from Agios Mamas, Chalkidike. In *Troia and the Troad: Scientific Approaches*, edited by Günther A. Wagner, Ernst Pernicka, and Hans-Peter Uerpmann, pp. 293–301. Springer, Berlin-Heidelberg.

Kronen Zeitung
2017 Brotkultur in Europa. 15 August:50. Vienna.

Krug, Antje
1997 *Αρχαία ιατρική: Επιστημονική και θρησκευτική ιατρική στην αρχαιότητα* [Heilkunst und Heilkult: Medizin in der Antike]. Translated by Eleni Manakidou. Papadimas, Athens.

Kubiak-Martens, Lucy, and Jerzy J. Langer
2008 Predynastic Brewing Based on Botanical and Physicochemical Evidence from Tell el-Farkha, Eastern Delta. In *Egypt at Its Origins 2: Proceedings of the International Conference "Origin of the State, Predynastic and Early Dynastic Egypt," Toulouse (France), 5th–8th September 2005*, edited by Béatrix Midant-Reynes, Yann Tristant, Joanne Rowland, and Stan Hendrickx, pp. 425–439. Orientalia Lovaniensia Analecta 172. Peeters, Leuven.

Kunkel, Günther
1984 *Plants for Human Consumption: An Annotated Checklist of the Edible Phanerogams and Ferns.* Koeltz Scientific Books, Koenigstein.

Kyparissi-Apostolika, Nina
1999 The Palaeolithic Deposits of Theopetra Cave in Thessaly (Greece). In *The Palaeolithic Archaeology of Greece and Adjacent Areas. Proceedings of the ICOPAG Conference, Ioannina, September 1994*, edited by Geoff N. Bailie, Eugenia Adam, Eleni Panagopoulou, Catherine Perlés, and Kostas Zachos, pp. 232–239. British School at Athens Studies 3. The British School at Athens, London.
2003 The Mesolithic in Theopetra Cave: New Data on a Debated Period of Greek Prehistory. In *The Greek Mesolithic: Problems and Perspectives*, edited by Nena

Galanidou and Catherine Perlès, pp. 189–198. British School at Athens Studies 10. The British School at Athens, London.

Kyparissi-Apostolika, Nina (editor)

2000 *Σπήλαιο Θεόπετρας: 12 χρόνια ανασκαφών και έρευνας 1987–1998. Πρακτικά Διεθνούς Συνεδρίου, Τρίκαλα 6–7 Νοεμβρίου 1998* [Theopetra Cave: 12 Years of Excavation and Research 1987–1998. Proceedings of the International Conference, Trikala, 6–7 November 1998]. Ministry of Culture, Athens.

Kyrieleis, Helmut

1983 *Το Ηραίο της Σάμου* [The Hereon at Samos]. Translated by E. Lazaridou and Alexandros Kesisoglou. Krene, Athens.

Ladizinsky, Gideon

1989 Origin and Domestication of the Southwest Asian Grain Legumes. In *Foraging and Farming: The Evolution of Plant Exploitation,* edited by David R. Harris and Gordon C. Hillman, pp. 374–389. Unwin Hyman, London.

Ladizinsky, Gideon, D. Braun, and F. J. Muehlbauer

1983 Evidence for Domestication of *Lens nigricans* (M. Bieb.) Godron in S Europe. *Botanical Journal of the Linnean Society* 87(2):169–176.

Lambein, Fernand

2000 Homeopathy, Longevity, and *Lathyrus sativus* Toxicity. *Lathyrus Lathyrism Newsletter* 1:4–5.

Lambraki, Mirsini

1997 *Τα χόρτα* [The Greens]. Trochalia, Athens.

Langgut, Dafna

2017 The Citrus Route Revealed: From Southeast Asia into the Mediterranean. *HortScience* 52(6):814–822.

Langgut, Dafna, Rachid Cheddadi, Josè Sebastián Carrión, Mark Cavanagh, Daniele Colombaroli, Warren John Eastwood, Raphael Greenberg, Thomas Litt, Anna Maria Mercuri, Andrea Miebach, C. Neil Roberts, Henk Woldring, and Jessie Woodbridge

2019 The Origin and Spread of Olive Cultivation in the Mediterranean Basin: The Fossil Pollen Evidence. *The Holocene* 29(5):902–922.

Lannoy, Sylvie, Philippe Marinval, Alain Buleon, Hubert Chiron, Philippe Mejanelle, Serge Pin, Joselyne Rech, and Alain Tchapla

2002 Etudes de "pains/galettes" archéologiques français. In *Pains, Fours et Foyers Domestiques des Temps Passés—Breads, Ovens, and Hearths of the Past,* edited by Marianne Mesnil and Kaï Fechner, pp. 119–160. *Civilisation* 49(1–2). Institut de Sociologie de l' Université de Bruxelles.

Lardos, Andreas, José Prieto-Garcia, and Michael Heinrich

2011 Resins and Gums in Historical Iatrosophia Texts from Cyprus—A Botanical and Medico-Pharmacological Approach. *Frontiers in Pharmacology* 2:Article 32.

Larsson, Åsa Maria

2003 Secondary Burial Practices in the Middle Neolithic: Causes and Consequences. *Current Swedish Archaeology* 11:153–170.

Lathiras, Pavlos

2020 *Archaeobotanical Remains from the Neolithic Site of Kyparissi.* Master's thesis, Aristotle University of Thessaloniki.

Lawson, Ian, S. Al-Omari, Polychronis Tzedakis, Charlotte Bryant, and K. Christanis
2005 Lateglacial and Holocene Vegetation History at Nisi Fen and the Boras Mountains, Northern Greece. *The Holocene* 15(6):873–887.
Lawson, Ian, Mick Frogley, Charlotte Bryant, Richard Preece, and Polychronis Tzedakis
2004 The Lateglacial and Holocene Environmental History of the Ioannina Basin, North-West Greece. *Quaternary Science Reviews* 23(14–15):1599–1625.
Le Clerc, J. A., L. H. Bailey, and H. L. Wessling
1918 Milling and Baking Test of Einkorn, Emmer, Spelt, and Polish Wheat. *Journal of American Society of Agronomy* 10(5):215–217.
Lehner, Mark
2009 Enigma of the Pedestal Building: Hypotheses Desert Refrigerator? Malting Machine? Or Both? In *Beyond the Horizon: Studies in Egyptian Art, Archaeology, and History in Honour of Barry J. Kemp*, edited by Salima Ikram and Aidan Dodson, pp. 189–222. Supreme Council of Antiquities, Cairo.
Leitzmann, Claus
2014 Vegetarian Nutrition: Past, Present, Future. *American Journal of Clinical Nutrition* 100, Supplement 1:496S–502S.
Lentjes, Daphne, and Giulia Saltini-Semerari
2016 Big Debates over Small Fruits: Wine and Oil Production in Protohistoric Southern Italy (ca. 1350–750 BC). *BABesch* 91:1–16.
Leonard, Albert, Jr.
1994 *An Index to the Late Bronze Age Aegean Pottery from Syria-Palestine*. Studies in Mediterranean Archaeology 114. Paul Åströms, Jonsered.
Leonard, Albert, Jr., and Eric H. Cline
1998 The Aegean Pottery at Megiddo: An Appraisal and Reanalysis. *Bulletin of the American Schools of Oriental Research* 309:3–39.
Leonti, Marco, Sabine Nebel, Diego Rivera, and Michael Heinrich
2006 Wild Gathered Food Plants in the European Mediterranean: A Comparative Analysis. *Economic Botany* 60(2):130–142.
Leslie, Charles, and Allan Young (editors)
1992 *Paths to Asian Medical Knowledge*. University of California Press, Berkeley.
Lévi-Strauss, Claude
1962 *Le totémisme aujourd'hui*. Presses Universitaires de France, Paris.
1964 *Le cru et le cuit*. Plon, Paris.
1965 Le Triangle Culinaire. *L'Arc* 26:19–29.
1977 *Άγρια σκέψη* [La pensée sauvage]. Translated by Eva Kalpourtzi. Papazisis, Athens.
2001 *Το Ωμό και το Μαγειρεμένο* [Le cru et le cuit]. Translated by Marina Lomi. Arsenidis, Athens.
Levin, Gregory M.
2006 *Pomegranate*. Third Millennium, Tempe, Arizona.
Lev-Tov Chattah, Netta, and P. Smith
2006 Variation in Occlusal Dental Wear of Two Chalcolithic Populations in the Southern Levant. *American Journal of Physical Anthropology* 130(4):471–479.
Lewicka, Katarzyna, Michele Mallie, and Jean-Marie Bastide
1995 Genetic Variability in the *Saccharomyces* sensu stricto Complex Revealed by Multilocus Enzyme Electrophoresis. *International Journal of Systematic and Evolutionary Biology* 45(3):538–543.

Li, Yuqi
2021 Agriculture and Palaeoeconomy in Prehistoric Xinjiang, China (3000–200 BC). *Vegetation History and Archaeobotany* 30:287–303.

Lidström-Holmberg, Cecilia
1998 Prehistoric Grinding Tools as Metaphorical Traces of the Past. *Current Swedish Archaeology* 6:123–142.

Liu, Li, Jiajing Wang, Danny Rosenberg, Hao Zhao, György Lengyel, and Dani Nadele
2018 Fermented Beverage and Food Storage in 13,000 Y-Old Stone Mortars at Raqefet Cave, Israel: Investigating Natufian Ritual Feasting. *Journal of Archaeological Science: Reports* 21:783–793.

Liu, Xinru
1988 *Ancient India and Ancient China: Trade and Religious Exchanges AD 1–600*. Oxford University Press, Delhi.

Liu, Xinyi, and Martin K. Jones
2014 Food Globalisation in Prehistory: Top Down or Bottom Up? *Antiquity* 88(341):956–963.

Liu, Xinyi, Penelope J. Jones, Giedre Motuzaite Matuzeviciute, Harriet V. Hunt, Diane L. Lister, Ting An, Natalia Przelomska, Catherine J. Kneale, Zhijun Zhao, and Martin K. Jones
2019 From Ecological Opportunism to Multi-Cropping: Mapping Food Globalisation in Prehistory. *Quaternary Science Reviews* 206:21–28.

Liu, Xinyi, Emma Lightfoot, Tamsin C. O'Connell, Hui Wang, Shuicheng Li, Liping Zhou, Yaowu Hu, Giedre Motuzaite-Matuzeviciute, and Martin K. Jones
2014 From Necessity to Choice: Dietary Revolutions in West China in the Second Millennium BC. *World Archaeology* 46(5):661–680.

Livarda, Alexandra
2012 The Archaeobotanical Evidence of the Late Bronze Age and Protogeometric Occupation under the Roman Villa Dionysus, Knossos, Crete, and an Overview of the Protogeometric Data of Greece. *Annual of the British School at Athens* 107:189–209.
2014 Archaeobotany in Greece. *Archaeological Reports* 60:106–116.

Livarda, Alexandra, and Georgia Kotzamani
2013 The Archaeobotany of Neolithic and Bronze Age Crete: Synthesis and Prospects. *Annual of the British School at Athens* 108:1–29.
2020 Plant Resources and Subsistence in the Late Minoan Mountain 'Villa' at Zominthos, Crete. *Archaeological and Anthropological Sciences* 12(10): Article 237.

Livarda, Alexandra, Hector A. Orengo, Nuria Cañellas-Boltà, Santiago Riera-Mora, Llorenç Picornell-Gelabert, Vasiliki Tzevelekidi, Rena Veropoulidou, Ricard Marlasca Martín, and Athanasia Krahtopoulou
2021 Mediterranean Polyculture Revisited: Olive, Grape, and Subsistence Strategies at Palaikastro, East Crete, between the Late Neolithic and Late Bronze Age. *Journal of Anthropological Archaeology* 61:101271.

Logothetis, Vasilios
1962 Αι αυτοφυείς άμπελοι [*Vitis vinifera* L. ssp. *silvestris* Gmel.] ως πρωτογενές αμπελουργικόν υλικόν εν Ελλάδι [Wild Grapevines (*Vitis vinifera* L. ssp. *silvestris* Gmel.) as Primary Winemaking Material in Greece]. In *Επιστημονική Επετηρίς της Γεωπονικής και Δασολογικής Σχολής του Αριστοτελείου Πανεπιστημίου*

Θεσσαλονίκης 5 [Scientific Yearbook of the Department of Agriculture and Forestry of Aristotle University of Thessaloniki 5]:1–43.

1970 Η εξέλιξις της αμπέλου και της αμπελουργίας εις την Ελλάδα κατά τα αρχαιολογικά ευρήματα της περιοχής [The Evolution of Vine and Wine in Greece according to the Archaeological Evidence]. In *Επιστημονική Επετηρίς της Γεωπονικής και Δασολογικής Σχολής του Αριστοτελείου Πανεπιστημίου Θεσσαλονίκης 13* [Scientific Yearbook of the Department of Agriculture and Forestry of Aristotle University of Thessaloniki 13]: 2–83.

1974 Η συμβολή της αμπέλου και του οίνου στον πολιτισμό της Ελλάδος και της Ανατολικής Μεσογείου [The Contribution of Vine and Wine to the Culture of Greece and Eastern Mediterranean]. In *Επιστημονική Επετηρίς της Γεωπονικής Σχολής και Δασολογικής Σχολής του Αριστοτελείου Πανεπιστημίου Θεσσαλονίκης 17* [Scientific Yearbook of the Department of Agriculture and Forestry of Aristotle University of Thessaloniki 17]:1–286.

López-Bellido, Luis

1994 Grain Legumes for Animal Feed. In *Neglected Crops: 1492 from a Different Perspective,* edited by J. Esteban Hernández Bermejo and Jens León, pp. 273–288. Plant Production and Protection Series 26. Food and Agriculture Organization of the United Nations (FAO), Rome.

Lu, Houyuan, Xiaoyan Yang, Maolin Ye, Kam-Biu Liu, Zhengkai Xia, Xiaoyan Ren, Linhai Cai, Naiqin Wu, and Tung-Sheng Liu

2005 Millet Noodles in Late Neolithic China. *Nature* 437(7061):967–968.

Lu, Minxia, Liang Chen, Jinxiu Wang, Ruiliang Liu, Yang Yang, Meng Wei, and Guanghui Dong

2019 A Brief History of Wheat Utilization in China. *Frontiers of Agricultural Science and Engineering* 6(3):288–295.

Luce, Jean-Marc, and Philippe Marinval

2008 L'alimentation végétale. In *L'aire du pilierdes Rhodiens (fouille 1990–1992): À la frontier du profane et du sacré,* edited by Jean-Marc Luce, pp. 230–256. Fouilles de Delphes 2, Topographie et Architecture 13. Ecole française d'Athènes, Athens.

Łuczaj, Łukasz

2016 Ethnobotanical Review of Wild Edible Plants of Slovakia. *Acta Societatis Botanicorum Poloniae* 81(4):245–255.

Lust, John

1983 *The Herb Book.* Bantam, New York.

Lyons, Diane, and Catherine D'Andrea

2003 Griddles, Ovens, and Agricultural Origins: An Ethnoarchaeological Study of Bread Baking in Highland Ethiopia. *American Anthropologist* 105(3):515–530.

Maltas, Tom, Vasıf Şahoğlu, Hayat Erkanal, and Rıza Tuncel

2022 Prehistoric Farming Settlements in Western Anatolia: Archaeobotanical Insights into the Late Chalcolithic of the Izmir Region, Turkey. *Journal of Mediterranean Archaeology* 34(2): 252–277.

McCombie-Young, Thomas Charles

1927 A Field Study of Lathyrism. *Indian Journal of Medical Research* 15(2):453–479.

McGee, Harold

1984 *On Food and Cooking: The Science and Lore of the Kitchen.* Scribner, New York.

McGovern, Patrick E.
2003 *Ancient Wine: The Search for the Origins of Viniculture.* Princeton University Press, Princeton.

McGovern, Patrick E., Stuart J. Fleming, and Solomon H. Katz
1995 *The Origins and Ancient History of Wine.* Food and Nutrition in History and Anthropology 11. Gordon and Breach, Philadelphia.

McGovern, Patrick E., Donald L. Glusker, Lawrence J. Exner, and Mary M. Voigt
1996 Neolithic Resinated Wine. *Nature* 381(6582):480–481.

McGovern, Patrick E., Mindia Jalabadze, Stephen Batiuk, Michael P. Callahan, Karen E. Smith, Gretchen R. Hall, Eliso Kvavadze, David Maghradze, Nana Rusishvili, Laurent Bouby, Osvaldo Failla, Gabriele Cola, Luigi Mariani, Elisabetta Boaretto, Roberto Bacilieri, Patrice This, Nathan Wales, and David Lordkipanidze
2017 Early Neolithic Wine of Georgia in the South Caucasus. *Proceedings of the National Academy of Sciences of the United States of America (PNAS)* 114(48):E10309–E10318.

McGovern, Patrick E., and Rudolph H. Michel
1995 The Analytical and Archaeological Challenge of Detecting Ancient Wine: Two Case Studies from the Ancient Near East. In *The Origins and Ancient History of Wine,* edited by Patrick E. McGovern, Stuart J. Fleming, and Solomon H. Katz, pp. 57–65. Food and Nutrition in History and Anthropology 11. Gordon and Breach, Philadelphia.

McGrew, William C.
2011 Natural Ingestion of Ethanol by Animals: Why? In *Liquid Bread: Beer and Brewing in Cross-Cultural Perspective,* edited by Wulf Schiefenhövel and Helen Macbeth, pp. 13–20. Berghahn, New York.

Maier, Ursula
2001 Archäobotanische Untersuchungen in der neolithischen Ufersiedlung Hornstaad-Hörnle IA am Bodensee. In *Siedlungsarchäologie im Alpenvorland VI: Archäobotanische Untersuchungen in Hornstaad-Hörnle IA,* edited by Ursula Maier and Richard Vogt, pp. 9–384. Forschungen und Berichte zur Vor- und Frühgeschichte in Baden-Württemberg 74. Konrad Theiss, Stuttgart.
2004 Archäobotanische Untersuchungen in jung- und endneolithischen Moorsiedlungen am Federsee (mit einem Beitrag von Richard Vogt). In *Ökonomischer und ökologischer Wandel am vorgeschichtlichen Federsee: Archäologische und naturwissenschaftliche Untersuchungen,* edited by Joachim Köninger and Helmut Schlichtherle, pp. 71–159. Hemmenhofener Skripte 5. Janus-Verlag, Freiburg im Breisgau.

Majno, Guido
1975 *The Healing Hand: Man and Wound in the Ancient World.* Harvard University Press, Cambridge, Massachusetts.

Malapani, Evangelia
2015 *Η νεκρόπολη των Ελληνικών Ανθείας στο πλαίσιο της θεώρησης των θαλαμωτών τάφων της Μεσσηνίας κατά την υστεροελλαδική περίοδο* [The Ellinika Necropolis at Antheia in the Framework of Studying the Messenian Chamber Tombs during the Late Helladic Period]. PhD dissertation, University of Ioannina.

Maltezou, Ioanna
2004 *Σαμοθράκη: Γαστρονομία, Ταυτότητα, Κληρονομημένη Γνώση* [Samothrace: Gastronomy, Identity, Inherited Knowledge]. Epikoinonia A. E., Samothrace.

Mangafa, Maria

1993 Αρχαιοβοτανική μελέτη του σπηλαίου Σκοτεινής στα Θαρρούνια Ευβοίας [The Archaeobotanical Remains of Skoteini Cave, Tharrounia, Euboea]. In *Σκοτεινή Θαρρουνίων: Το σπήλαιο, ο οικισμός και το νεκροταφείο* [Skoteini, Tharrounia: The Cave, the Settlement and the Cemetery], edited by Adamantios Sampson, pp. 360–369. Konstantopoulos, Athens.

2000 Η εκμετάλλευση των φυτών από τη Μέση Παλαιολιθική έως τη Νεολιθική περίοδο: από την καρποσυλλογή στην καρποκαλλιέργεια: Αρχαιοβοτανική μελέτη του σπηλαίου Θεόπετρα [Plant Exploitation from the Middle Palaeolithic to the Neolithic: From Food Gathering to Farming. Archaeobotanical Study of Theopetra Cave]. In *Σπήλαιο Θεόπετρας: 12 χρόνια ανασκαφών και έρευνας 1987–1998. Πρακτικά Διεθνούς Συνεδρίου, Τρίκαλα 6–7 Νοεμβρίου 1998* [Theopetra Cave: 12 Years of Excavation and Research 1987–1998. Proceedings of the International Conference, Trikala, 6–7 November 1998], edited by Nina Kyparissi-Apostolika, pp. 135–137. Ministry of Culture, Athens.

2002 Η αρχαιοβοτανική μελέτη του οικισμού [The Archaeobotanical Study of the Settlement]. In *Δισπηλιό: 7500 Χρόνια Μετά* [Dispilio: 7500 Years After], edited by Giorgos Hourmouziadis, pp. 115–134. University Studio Press, Thessaloniki.

Mangafa, Maria, and Kostas Kotsakis

1996 A New Method for the Identification of Wild and Cultivated Charred Grape Seeds. *Journal of Archaeological Science* 23(3):409–418.

Mangafa, Maria, Kostas Kotsakis, and Stelios Andreou

1998 Αμπελοκαλλιέργεια στην προϊστορική Μακεδονία: Τα δεδομένα της προϊστορικής Τούμπας Θεσσαλονίκης [Viticulture in Prehistoric Macedonia: Evidence from the Prehistoric Mound of Thessaloniki]. In *Αμπελοοινική ιστορία στο χώρο της Μακεδονίας και της Θράκης* [The Viticultural History of Macedonia and Thrace]. *Proceedings of the Three-Day Working Meeting on the Viticultural History of Macedonia and Thrace, Naoussa, 17–19 September 1993*, pp. 158–169. ETVA Cultural and Technological Foundation, Athens.

Mangafa, Maria, Kostas Kotsakis, and John Anastasios Stratis

2001 Πειραματική απανθράκωση προϊόντων της αμπέλου και διερεύνηση της χρήσης της κατά την αρχαιότητα [Experimental Charring of Vine Products and Investigation on Its Uses in Antiquity]. In *Αρχαιομετρικές Μελέτες για την Ελληνική Προϊστορία και Αρχαιότητα* [Archaeometry Issues in Greek Prehistory and Antiquity], edited by Yannis Basiakos, Eleni Aloupi, and Yorgos Fakorellis, pp. 495–510. Hellenic Society of Archaeometry/Society of Messenian Archaeological Studies, Athens.

Maniatis, Yiannis

2014 Χρονολόγηση με άνθρακα-14 των μεγάλων πολιτισμικών αλλαγών στην προϊστορική Μακεδονία: πρόσφατες εξελίξεις [Radiocarbon Dating of the Major Cultural Changes in Prehistoric Macedonia: Recent Developments]. In *1912–2012: Εκατό χρόνια έρευνας στην προϊστορική Μακεδονία, Πρακτικά διεθνούς συνεδρίου, Αρχαιολογικό Μουσείο Θεσσαλονίκης, 22–24 Νοεμβρίου 2012* [1912–2012: A Century of Research in Prehistoric Macedonia, International Conference Proceedings, Archaeological Museum of Thessaloniki, 22–24 November 2012], edited by Evaggelia Stefani, Nikos Merousis, and Anastasia Dimoula, pp. 205–222. Archaeological Museum of Thessaloniki.

Manios, Yannis, Vivian Detopoulou, Francesco Visioli, and Claudio Galli

2006 Mediterranean Diet as a Nutrition Education and Dietary Guide: Misconceptions and the Neglected Role of Locally Consumed Foods and Wild Green Plants. In *Local Mediterranean Food Plants and Nutraceuticals*, edited by Michael Heinrich, Walter E. Mueller, and Claudio Galli, pp. 154–170. Forum of Nutrition 59. Karger, Basel.

Maran, Joseph

2012 Ceremonial Festing Equipment, Social Space, and Interculturality in Post-Palatial Tiryns. In *Materiality and Social Practice: Transformative Capacities of Intercultural Encounters*, edited by Joseph Maran and Philipp Stockhammer, pp. 121–136. Oxbow, Oxford.

Maran, Joseph, and Alkestis Papadimitriou

2006 Forschungen im Stadtgebiet von Tiryns 1999–2002. *Archaeologischer Anzeiger*:97–169.

Marcopoulos, C. A.

1965 Seed and Seed Oil of *Pistacia terebinthus* and *P. lentiscus*. *Journal of the American Oil Chemists' Society* 42(1):1–2.

Margaritis, Evi

2002 Αρχαιοβοτανικά δεδομένα από το νεολιθικό οικισμό της Σταυρούπολης: Μια πρώτη προσέγγιση [Archaeobotanical Remains from the Neolithic Site of Stavroupoli Thessaloniki: A Preliminary Report]. In *Σωστικές ανασκαφές στο Νεολιθικό οικισμό Σταυρούπολης Θεσσαλονίκης* [Rescue Excavations at the Neolithic Settlement of Stavroupoli, Thessaloniki], edited by Dimitrios Grammenos and Stavros Kotsos, pp. 805–827. Publications of the Archaeological Institute of Northern Greece 2. Archaeological Institute of Northern Greece, Thessaloniki.

2007 Αρχαιοβοτανικές μελέτες στο Νεολιθικό οικισμό Αυγής Καστοριάς, 2005–2007 [Archaeobotanical Research in the Neolithic Settlement at Avgi, Kastoria, 2005–2007]. Website accessed March 19, 2020.

2011 Storage, Gathering, and Lathyrism? at Dispilio. *Anaskamma* 5:113–123.

2013a Foodstuffs, Fruit Tree Cultivation, and Occupation Patterns at Dhaskalio. In *The Sanctuary on Keros and the Origins of Aegean Ritual Practice: The Excavations of 2006–2008, Vol. 1: The Settlement at Dhaskalio*, edited by Colin Renfrew, Olga Philaniotou, Neil Brodie, Giorgos Gavalas, and Michael J. Boyd, pp. 389–404. McDonald Institute for Archaeological Research, Cambridge.

2013b Distinguishing Exploitation, Domestication, Cultivation, and Production: The Olive in the Third Millennium Aegean. *Antiquity* 87(337):746–757.

2013c Arboriculture at Aphrodite's Kephali. In *Aphrodite's Kephali: An Early Minoan I Defensive Site in Eastern Crete*, edited by Philip P. Betancourt, pp. 111–114. Prehistory Monographs 41. INSTAP Academic, Philadelphia.

2014 Acts of Destruction and Acts of Preservation: Plants in the Ritual Landscape of Prehistoric Greece. In *Physis: L'environnement naturel et la relation homme-milieu dans le monde égéen protohistorique. Actes de la 14e rencontre égéenne internationale, Paris, Institut National d'Histoire de l'Art (INHA), 11–14 décembre 2012*, edited by Gilles Touchais, Robert Laffineur, and Françoise Rougemont, pp. 279–285. Aegaeum 37. Peeters, Leuven/Liège.

2017 The Plant Remains of the House Tombs at Petras: Acts of Destruction, Transformation, and Preservation. In *Petras, Siteia: The Pre- and Proto-Palatial Cemetery in Context. Acts of a Two-Day Conference Held at the Danish Institute at Athens, 14–15 February 2015 Athens*, edited by Metaxia Tsipopoulou, pp. 225–235. Monographs of the Danish Institute at Athens 21. Aarhus University Press, Aarhus, Denmark.

2018 The Plant Remains from Alepotrypa Cave: Use, Discard, and Structured Deposition. In *Neolithic Alepotrypa Cave in the Mani, Greece. In Honor of George Papathanassopoulos*, edited by Anastasia Papathanasiou, William A. Parkinson, Daniel J. Pullen, Michael L. Galaty and Panagiotis Karkanas, pp. 316–326. Oxbow, Oxford.

Margaritis, Evi, Katie Demakopoulou, and Ann-Louise Schallin

2014 The Archaeobotanical Samples from Midea: Agricultural Choices in the Mycenaean Argolid. In *Physis: L'environnement naturel et la relation homme-milieu dans le monde égéen protohistorique. Actes de la 14e rencontre égéenne internationale, Paris, Institut National d'Histoire de l'Art (INHA), 11–14 décembre 2012*, edited by Gilles Touchais, Robert Laffineur, and Françoise Rougemont, pp. 271–278. Aegaeum 37. Peeters, Leuven/Liège.

Margaritis, Evi, and Martin K. Jones

2006 Beyond Cereals: Crop Processing and *Vitis vinifera* L. Ethnography, Experiment, and Charred Grape Remains from Hellenistic Greece. *Journal of Archaeological Science* 33(6):784–805.

2008 Olive Oil Production in Hellenistic Greece: The Interpretation of Charred Olive Remains from the Site of Platania, Macedonia, Greece. *Vegetation History and Archaeobotany* 17(4):393–401.

Marinatos, Spyridon N.

1937 Αι Μινωϊκαί θεαί του Γάζι [The Minoan Goddesses of Gazi]. *Αρχαιολογική Εφημερίς* [Archaeologiki Ephimeris] 76:278–291.

Marini Govigli, Valentino, John R. Healey, Jennifer L. G. Wong, Kalliopi Stara, Rigas Tsiakiris, and John M. Halley

2020 When Nature Meets the Divine: Effect of Prohibition Regimes on the Structure and Tree Species Composition of Sacred Forests in Northern Greece. *Web Ecology* 20(2):53–86.

Marinova, Elena

2006 *Vergleichende paläoethnobotanische Untersuchung zur Vegetationsgeschichte und zur Entwicklung der prähistorischen Landnutzung in Bulgarien.* Dissertationes Botanicae 401. J. Cramer in der Gebrüder Borntraeger Verlagsbuchhandlung, Berlin-Stuttgart.

2007 Archaeobotanical Data from the Early Neolithic of Bulgaria. In *The Origins and Spread of Domestic Plants in Southwest Asia and Europe*, edited by Sue Colledge and James Conolly, pp. 93–109. Left Coast Press, Walnut Creek.

Marinova, Elena, Bea De Cupere, and Vassil Nikolov

2016 Preliminary Results of the Bioarchaeological Research at the Neolithic site of Mursalevo (Southwest Bulgaria): Evidence on Food Storage, Processing, and Consumption from Domestic Contexts. In *Southeast Europe and Anatolia in Prehistory: Essays in Honor of Vassil Nikolov on his 65th Anniversary*, edited by Krum Băčvarov and Ralf Gleser, pp. 509–517. Universitätsforschungen zur prähistorischen Archäologie 293. Rudolf Habelt, Bonn.

Marinova, Elena, and Maria Ntinou

2018 Neolithic Woodland Management and Land-Use in South-Eastern Europe: The

Anthracological Evidence from Northern Greece and Bulgaria. *Quaternary International* 496:51–67.

Marinova, Elena, and Tzvetana Popova
2008 *Cicer arietinum* (Chick Pea) in the Neolithic and Chalcolithic of Bulgaria: Implications for Cultural Contacts with the Neighbouring Regions? *Vegetation History and Archaeobotany* 17, Supplement 1:S73-S80.

Marinova, Elena, and Soultana Maria Valamoti
2014 Crop Diversity and Choice in Prehistoric Southeastern Europe: Cultural and Environmental Factors Shaping the Archaeobotanical Record of Northern Greece and Bulgaria. In *Plants and People: Choices and Diversity through Time*, edited by Alexandre Chevalier, Elena Marinova, and Leonor Peña-Chocarro, pp. 64–74. Early Agricultural Remnants and Technical Heritage (EARTH): 8,000 Years of Resiliance and Innovation, Vol. 1. Oxbow, Oxford.

Marinova, Elena, Jan M. A. van der Valk, Soultana Maria Valamoti, and Joachim Bretschneider
2011 An Experimental Approach for Tracing Olive Processing Residues in the Archaeobotanical Record, with Preliminary Examples from Tell Tweini, Syria. *Vegetation History and Archaeobotany* 20(5):471–478.

Märkle, Tanja, and Manfred Rösch
2008 Experiments on the Effects of Carbonization on Some Cultivated Plant Seeds. *Vegetation History and Archaeobotany* 17, Supplement 1:S257–S263.

Martinez, Gustavo J.
2010 Los criterios terapeuticos en la farmacopea natural de los tobas bermejenos del Chaco Central (Argentina). In *Tradiciones y Transformaciones en Etnobotanica. Proceedings of the 5th International Congress of Ethnobotany (ICEB), September 2009, San Carlos de Bariloche, Argentina*, edited by María Lelia Pochettino, Ana H. Ladio, and Patricia M. Arenas, pp. 213–218. Red Iberoamericana de Saberes y Prácticas Locales sobre el Entorno Vegetal (RISAPRET)/Cyted, San Salvador de Jujuy.

Martinoli, Danièle, and Stefanie Jacomet
2004 Identifying Endocarp Remains and Exploring Their Use at Epipalaeolithic Öküzini in Southwest Anatolia, Turkey. *Vegetation History and Archaeobotany* 13(1):45–54.

Marvelli, Silvia, S. De' Siena, Elisabetta Rizzoli, and Marco Marchesini
2013 The Origin of Grapevine Cultivation in Italy: The Archaeobotanical Evidence. *Annali di Botanica* 3:155–163.

Mason, Sarah
1995 Acornutopia? Determining the Role of Acorns in Past Human Subsistence. In *Food in Antiquity: Studies in Ancient Society and Culture*, edited by John Wilkins, David Harvey, and Michael J. Dobson, pp. 12–24. University of Exeter Press, Exeter.

Mason, Sarah, and Mark Nesbitt
2009 Acorns as Food in Southeast Turkey: Implications for Prehistoric Subsistence in Southwest Asia. In *From Foragers to Farmers. Papers in Honour of Gordon C. Hillman*, edited by Andrew S. Fairnbairn and Ehud Weiss, pp. 71–85. Oxbow, Oxford.

Matalas, Antonia-Leda
2006a Dietary Patterns in Pre-World War II Greece. Disparities within Peasant and Urban Foodways. *Food and History* 4(1):239–253.

2006b Disparities within Traditional Mediterranean Food Patterns: An Historical Approach of the Greek Diet. *International Journal of Food Sciences and Nutrition* 57(7–8):529–536.

Mathieson, Iain, Songül Alpaslan-Roodenberg, Cosimo Posth, Anna Szécsényi-Nagy, Nadin Rohland, Swapan Mallick, Iñigo Olalde, Nasreen Broomandkhoshbacht, Francesca Candilio, Olivia Cheronet, Daniel Fernandes, Matthew Ferry, Beatriz Gamarra, Gloria González Fortes, Wolfgang Haak, Eadaoin Harney, Eppie Jones, Denise Keating, Ben Krause-Kyora, Isil Kucukkalipci, Megan Michel, Alissa Mittnik, Kathrin Nägele, Mario Novak, Jonas Oppenheimer, Nick Patterson, Saskia Pfrengle, Kendra Sirak, Kristin Stewardson, Stefania Vai, Stefan Alexandrov, Kurt W. Alt, Radian Andreescu, Dragana Antonović, Abigail Ash, Nadezhda Atanassova, Krum Bacvarov, Mende Balázs Gusztáv, Hervé Bocherens, Michael Bolus, Adina Boroneanţ, Yavor Boyadzhiev, Alicja Budnik, Josip Burmaz, Stefan Chohadzhiev, Nicholas J. Conard, Richard Cottiaux, Maja Čuka, Christophe Cupillard, Dorothée G. Drucker, Nedko Elenski, Michael Francken, Borislava Galabova, Georgi Ganetsovski, Bernard Gély, Tamás Hajdu, Veneta Handzhyiska, Katerina Harvati, Thomas Higham, Stanislav Iliev, Ivor Janković, Ivor Karavanić, Douglas J. Kennett, Darko Komšo, Alexandra Kozak, Damian Labuda, Martina Lari, Catalin Lazar, Maleen Leppek, Krassimir Leshtakov, Domenico Lo Vetro, Dženi Los, Ivaylo Lozanov, Maria Malina, Fabio Martini, Kath McSweeney, Harald Meller, Marko Menđušić, Pavel Mirea, Vyacheslav Moiseyev, Vanya Petrova, T. Douglas Price, Angela Simalcsik, Luca Sineo, Mario Šlaus, Vladimir Slavchev, Petar Stanev, Andrej Starović, Tamás Szeniczey, Sahra Talamo, Maria Teschler-Nicola, Corinne Thevenet, Ivan Valchev, Frédérique Valentin, Sergey Vasilyev, Fanica Veljanovska, Svetlana Venelinova, Elizaveta Veselovskaya, Bence Viola, Cristian Virag, Joško Zaninović, Steve Zäuner, Philipp W. Stockhammer, Giulio Catalano, Raiko Krauß, David Caramelli, Gunita Zariņa, Bisserka Gaydarska, Malcolm Lillie, Alexey G. Nikitin, Inna Potekhina, Anastasia Papathanasiou, Dušan Borić, Clive Bonsall, Johannes Krause, Ron Pinhasi, and David Reich

2018 The Genomic History of Southeastern Europe. *Nature* 555(7695):197–203.

Matsas, Dimitris

1991 Samothrace and the Northeastern Aegean: The Minoan Connection. *Studia Troica* 1:159–179.

1995 Minoan Long-Distance Trade: A View from the Northern Aegean. In *POLITEIA: Society and State in the Aegean Bronze Age. Proceedings of the 5th International Aegean Conference, University of Heidelberg, Archäologisches Institut, 10–13 April 1994*, edited by Robert Laffineur and Wolf-Dietrich Niemeier, pp. 235–247. Aegaeum 12. Université de Liège/University of Texas at Austin.

Matsas, Dimitris, Luca Girella, and Peter Pavúk

2011 Minoanization or Hybridization? Aspects of the Minoan Presence at Mikro Vouni (Samothrace) in the Middle Bronze Age. Unpublished paper presented at the 11th International Cretological Congress, 21–27 October 2011, Rethymno.

Matta, Raúl

2019 Celebrity Chefs and the Limits of Playing Politics from the Kitchen. In *Globalized Eating Cultures: Mediatization and Mediation*, edited by Jörg Dürrschmidt and York Kautt, pp. 183–201. Palgrave Macmillan, Cham, Switzerland.

Matterne, Véronique
1993 *Etude des macrorestes et du paysage d'un tell de Macedoine orientale au Neolithique Recent: Dikili Tash.* D.E.A. Universite de Paris, Vol. 1. Sorbonne.

Mavroidi, Ioanna
2012 *Ο οικισμός του Αγίου Αθανασίου, νομού Θεσσαλονίκης: Οι ενδοκοινοτικές σχέσεις σε μια κοινότητα της Πρώιμης Εποχής του Χαλκού μέσα από τη μελέτη της κεραμικής και της αρχιτεκτονικής οργάνωσης* [The Settlement of Agios Athanasios, Thessaloniki: Intra- Community Relations in an Early Bronze Age Site through the Study of the Pottery and the Architectural Organization]. PhD dissertation, Aristotle University of Thessaloniki.

Mavroidi, Ioanna
2014 Η Πρώιμη Εποχή του Χαλκού στη Μακεδονία: Μια συνθετική επαναπροσέγγιση των αρχαιολογικών δεδομένων [The Early Bronze Age in Macedonia: A Synthetic Re-Approach of the Archaeological Evidence]. In *1912–2012: Εκατό χρόνια έρευνας στην προϊστορική Μακεδονία, Πρακτικά διεθνούς συνεδρίου, Αρχαιολογικό Μουσείο Θεσσαλονίκης, 22–24 Νοεμβρίου 2012* [1912–2012: A Century of Research in Prehistoric Macedonia, International Conference Proceedings, Archaeological Museum of Thessaloniki, 22–24 November 2012], edited by Evaggelia Stefani, Nikos Merousis, and Anastasia Dimoula, pp. 261–270. Archaeological Museum of Thessaloniki.

Mavroidi Ioanna, Stelios Andreou, and Maria Pappa
2008 Οικισμός της Πρώιμης Εποχής του Χαλκού στον Άγιο Αθανάσιο Θεσσαλονίκης: Προκαταρκτική έκθεση της μελέτης του υλικού [An Early Bronze Age Settlement at Agios Athanasios Thessalonikis: A Preliminary Report on the Study of the Archaeological Material]. *Proceedings of the Archaeological Work in Macedonia and Thrace (AEMTH)* 20, 2006:479–490.

Mavromati, Antigoni
2017 Landscape and Wood-Fuel in Akrotiri (Thera, Greece) during the Bronze Age. *Quaternary International* 458:44–55.

Mazow, Laura, Susanne Grieve, and Anthony Kennedy
2014 Contamination in Organic Residue Analysis: A Cautionary Tale. *Journal of Eastern Mediterranean Archaeology and Heritage Studies* 2(2):90–109.

Medović, Aleksandar
2003 Archäobotanische Untersuchungen in der metallzeitlichen Siedlung Židovar, Vojvodina/Jugoslawien. Ein Vorbericht. *Starinar* 52:181–190.

Mee, Christopher
2008 Mycenaean Greece, the Aegean and Beyond. In *The Cambridge Companion to the Aegean Bronze Age*, edited by Cynthia W. Shelmerdine, pp. 362–386. Cambridge University Press, Cambridge.

Megaloudi, Fragkiska
2004 Funeral Plant Offerings from Greek Historical Sites: A Preliminary Study. In *Spatial Analysis of Funerary Areas*, edited by Ladislav Šmejda and Jan Turek, pp. 109–114. University of West Bohemia, Plzeň.
2005 Burnt Sacrificial Plant Offerings in Hellenistic Times: An Archaeobotanical Case Study from Messene, Peloponnese, Greece. *Vegetation History and Archaeobotany* 14(4):329–340.
2006 *Plants and Diet in Greece from Neolithic to Classic Periods: The Archaeobotanical*

Remains. British Archaeological Reports International Series 1516. Archaeopress, Oxford.

Megaloudi, Fragkiska, and Philippe Marinval

2002 Donnés préliminaires sur l'économie végétale du site de Ftelia, Cyclades (Grèce) au Néolithique recent. In *The Neolithic Settlement at Ftelia, Mykonos*, edited by Adamantios Sampson, pp. 191–200. University of the Aegean, Rhodes.

Megaloudi, Fragkiska, Stratis Papadopoulos, and Marina Sgourou

2007 Plant Offerings from the Classical Necropolis of Limenas, Thasos, Northern Greece. *Antiquity* 81(314):933–943.

Megas, Georgios A.

1988 *Ελληνικές Γιορτές και Έθιμα της Λαϊκής Λατρείας* [Greek Festivals and Customs of Folk Worship]. Odusseus, Athens.

Melena, José Luis

1983 Olive Oil and Other Sorts of Oil in Mycenaean Tablets. *Minos* 18:98–123.

Menevisoglou, Pavlos

1972 *Το Άγιον Μύρον εν τη Ορθοδόξω, Ανατολική Εκκλησία* [The Holy Myrrh in the Orthodox, East Church]. Patriarxikon Idryma Paterikon Meleton, Thessaloniki.

Mercuri, Anna Maria

2009 PaCE: A Project for Europe. In *Plants and Culture: Seeds of the Cultural Heritage of Europe*, edited by Jean-Paul Morel and Anna Maria Mercuri, pp. 11–16. Edipuglia, Bari.

Merlin, Mark D.

2003 Archaeological Evidence for the Tradition of Psychoactive Plant Use in the Old World. *Economic Botany* 57(3):295–323.

2018 Prehistoric Intoxicants of North America. In *Ancient Psychoactive Substances*, edited by Scott M. Fitzpatrick, pp. 112–127. University Press of Florida, Gainesville.

Merousis, Nikos

2016 Did the Minoans Consume Only Olive Oil? ki-ta-no in the Knossos Tablets Reconsidered. *Pasiphae* X:177–186.

Merrillees, Robert S.

1962 Opium Trade in the Bronze Age Levant. *Antiquity* 36(144):287–292.

Merrillees, Robert S., and John Davies Evans

1989 Highs and Lows in the Holy Land: Opium in Biblical Times (with a Report by John Evans). In *Yigael Yadin Memorial Volume*, edited by Amnon Ben-Tor, Jonas Greenfield, and Abraham Malamat, pp. 148–154. Eretz Israel 20. Israel Exploration Society, Jerusalem.

Meskell, Lynn

1995 Goddesses, Gimbutas, and "New Age" Archaeology. *Antiquity* 69(262):74–86.

Mesnil, Marianne, and Assia Popova

2002 L'offrande céréalière dans les rituels funéraires du sud-est européen. In *Pains, Fours et Foyers Domestiques des Temps Passés—Breads, Ovens, and Hearths of the Past*, edited by Marianne Mesnil and Kaï Fechner, pp. 101–117. Civilisation 49(1–2). Institut de Sociologie de l' Université de Bruxelles.

Meurers-Balke, Jutta, and Jens Lüning

1992 Some Aspects and Experiments Concerning the Processing of Glume Wheats.

In *Préhistoire de l'agriculture: Nouvelles approches expérimentales et ethnographiques*, edited by Patricia C. Anderson, pp. 341–362. Monographie du CRA 6. CNRS, Paris.

Michaelides, Demetrios (editor)
2014 *Medicine and Healing in the Ancient Mediterranean World*. Oxbow, Oxford.

Micha-Lampaki, Aspasia
1984 *Η διατροφή των Αρχαίων Ελλήνων κατά τους αρχαίους Κωμωδιογράφους* [The Nutrition of Ancient Greeks according to the Poets of Ancient Greek Comedy]. PhD dissertation, National and Kapodistrian University of Athens.

Mielke, Horst, and Bernd Rodemann
2007 Der Dinkel, eine besondere Weizenart: Anbau, Pflanzenschutz, Ernte und Verarbeitung. *Nachrichtenblatt des Deutschen Pflanzenschutzdienstes* 59(2):40–45.

Miller, Jay
2000 Indien Personhood II: Baby in the Oven Sparks Being in the World. *American Indian Culture and Research Journal* 24(3):155–160.

Miller, Naomi F.
2008 Sweeter Than Wine? The Use of the Grape in Early Western Asia. *Antiquity* 82(318):937–946.

Mills, Barbara J.
2007 Performing the Feast: Visual Display and Suprahousehold Commensalism in the Puebloan Southwest. *American Antiquity* 72(2):210–239.

Mills, John S., and Raymond White
1989 The Identity of the Resins from the Late Bronze Age Shipwreck at Ulu Burun (Kaş). *Archaeometry* 31(1):37–44.

Mishra, Vijay Nath, Chandra Bhushan Tripathi, A. Kumar, V. Nandmer, A. Z. Ansari, B. Kumar, Nath Rameshwar Chaurasia, and Deepika Joshi
2014 Lathyrism: Has the Scenario Changed in 2013? *Neurological Research* 36(1):38–40.

Mobley-Tanaka, Jeannette L.
1997 Gender and Ritual Space during the Pithouse to Pueblo Transition: Subterranean Mealing Rooms in the North American Southwest. *American Antiquity* 62(3):437–448.

Mohagheghzadeh, Abdolali, Pouya Faridi, Mohammadreza Shams-Ardakani, and Younes Ghasemi
2006 Medicinal Smokes. *Journal of Ethnopharmacology* 108(2):161–184.

Mohan, V. S., Vidhya Nagarajan, and C. A. Gopalan
1966 Simple Practical Procedures for the Removal of Toxic Factors in *Lathyrus sativus* (Khesari Dhal). *Indian Journal of Medical Research* 54(4):410–414.

Mol, Jean-Jacques van
2002 Aspects de la culture d'un blé vêtu, l'épeautre. In *Pains, Fours et Foyers Domestiques des Temps Passés—Breads, Ovens, and Hearths of the Past*, edited by Marianne Mesnil and Kaï Fechner, pp. 161–167. Civilisation 49(1–2). Institut de Sociologie de l' Université de Bruxelles.

Molares, Soledad, and Ana Ladio
2010 Criterios ambientales y organolepticos en los patrones de seleccion y uso de plantas medicinales en una comunidad Mapuche de la Patagonia semiarida Argentina. In *Tradiciones y Transformaciones en Etnobotanica. Proceedings of the*

5th International Congress of Ethnobotany (ICEB), September 2009, San Carlos de Bariloche, Argentina, edited by María Lelia Pochettino, Ana H. Ladio, and Patricia M. Arenas, pp. 286–290. Red Iberoamericana de Saberes y Prácticas Locales sobre el Entorno Vegetal (RISAPRET)/Cyted, San Salvador de Jujuy.

Molina-Cano, José Luis, Joanne R. Russell, Marian A. Moralejo, José Luis Escacena, G. Arias, and Wayne Powell

2005 Chloroplast DNA Microsatellite Analysis Supports a Polyphyletic Origin for Barley. *Theoretical and Applied Genetics* 110:613–619.

Molloy, Barry, Jo Day, Sue Bridgford, Valasia Isaakidou, Eleni Nodarou, Georgia Kotzamani, Marina Milić, Tristan Carter, Polly Westlake, Vera Klontza-Jaklova, Ellinor Larsson, and Barbara J. Hayden

2014 Life and Death of a Bronze Age House: Excavation of Early Minoan I Levels at Priniatikos Pyrgos. *American Journal of Archaeology* 118(2):307–358.

Mommersteeg, H. J. P. M., Marie-France Loutre, Regis Young, T. A. Wijmstra, and Henry Hooghiemstra

1995 Orbital Forced Frequencies in the 975000 Year Pollen Record from Tenagi Philippon (Greece). *Climate Dynamics* 11(1):4–24.

Monah, Dan

2002 Découvertes de pains et de restes d'aliments céréaliers en Europe de l'Est et Centrale. In *Pains, Fours et Foyers Domestiques des Temps Passés—Breads, Ovens, and Hearths of the Past*, edited by Marianne Mesnil and Kaï Fechner, pp. 67–76. Civilisation 49(1–2). Institut de Sociologie de l' Université de Bruxelles.

Moniaki, Aikaterini

2009 *Η αρχαιοβοτανική έρευνα στον Άγιο Αθανάσιο Θεσσαλονίκης* [The Archaeobotanical Investigation of Aghios Athanassios Thessaloniki]. Master's thesis, Aristotle University of Thessaloniki.

Moore, Jerry D.

1989 Pre-Hispanic Beer in Coastal Peru: Technology and Social Context of Prehistoric Production. *American Anthropologist New Series* 91(3):682–695.

Moreno-Larrazabal, Aitor, Andrés Teira-Brión, Itsaso Sopelana-Salcedo, Amaia Arranz-Otaegui, and Lydia Zapata

2015 Ethnobotany of Millet Cultivation in the North of the Iberian Peninsula. *Vegetation History and Archaeobotany* 24(4):541–554.

Morris, Christine, and Allan Peatfield

2014 Health and Healing on Cretan Bronze Age Peak Sanctuaries. In *Medicine and Healing in the Ancient Mediterranean World*, edited by Demetrios Michaelides, pp. 54–63. Oxbow, Oxford.

Morrison, Frank B.

1957 *Feeds and Feeding: A Hand-Book for the Student and Stockman*. Morrison, New York.

Motuzaite-Matuzeviciute, Giedre, Richard A. Staff, Harriet V. Hunt, Xinyi Liu, and Martin K. Jones

2013 The Early Chronology of Broomcorn Millet (*Panicum miliaceum*) in Europe. *Antiquity* 87(338):1073–1085.

Moundrea-Agrafioti, Antikleia

2007 Λίθινα εργαλεία [Stone Tools]. In *Ακρωτήρι Θήρας, Δυτική Οικία: Τράπεζες, λίθινα, μετάλλινα, ποικίλα* [Akrotiri, Thera, The West House: Tables—Stone, Metal, and Various Finds], edited by Christos Doumas, pp. 71–112. Monographs 246. The Library of the Archaeological Society at Athens.

Mountjoy, Penelope A.
1993 *Mycenaean Pottery: An Introduction*. Oxford University Committee for Archaeology, Oxford.
Mountrakis, Constantine, Stamatina Georgaki, and Sotiris K. Manolis
2011 A Trephined Late Bronze Age Skull from Peloponnesus, Greece. *Mediterranean Archaeology and Anthropology* 11(1):1–8.
Muhly-Metaxa, Polymnia
1984 Minoan Hearths. *American Journal of Archaeology* 88(2):107–122.
Mukherjee, Anna J., Matthew A. James, Peter Pfälzner, and Richard P. Evershed
2007 Biomolecular Analysis of Ceramic Containers, Skeletal Remains, Anthropogenic Sediments, and Organic Artefacts from the Royal Tomb at Qatna. In *Urban and Natural Landscapes of an Ancient Syrian Capital: Settlement and Environment at Tell Mishrifeh/Qatna and in Central-Western Syria. Proceedings of the International Conference held in Udine, 9–11 December 2004*, edited by Daniele Morandi Bonacossi, pp. 189–197. Studi archeologici su Qatna 1, Documents d'archéologie syrienne 12. Forum, Udine.
Mukherjee, Pulok K.
2001 Evaluation of Indian Traditional Medicine. *Therapeutic Innovation and Regulatory Science* 35(2):623–632.
Mulder-Heymans, Noor
2002 Archaeology, Experimental Archaeology, and Ethnoarchaeology on Bread Ovens in Syria. In *Pains, Fours et Foyers Domestiques des Temps Passés—Breads, Ovens, and Hearths of the Past*, edited by Marianne Mesnil and Kaï Fechner, pp. 197–221. Civilisation 49(1–2). Institut de Sociologie de l' Université de Bruxelles.
Müller, Noémi S., Anno Hein, Vassilis Kilikoglou, and Peter M. Day
2013 Bronze Age Cooking Pots: Thermal Properties and Cooking Methods. *Préhistoires Méditerranéennes*. Open Edition Journals 4.
Murti, V. V. S, T. R. Seshadri, and T. A. Venkitasubramanian
1964 Neurotoxic Compounds of the Seeds of *Lathyrus sativus*. *Phytochemistry* 3(1):73–78.
Mylonas, George E.
1929 *Excavations at Olynthus I: The Neolithic Settlement*. Johns Hopkins University Studies in Archaeology 6. Johns Hopkins Press, Baltimore.
1959 *Aghios Kosmas: An Early Bronze Age Settlement and Cemetery in Attica*. Princeton University Press, Princeton.
Nadel, Dani, Dolores R. Piperno, Irene Holst, Ainit Snir, and Ehud Weiss
2012 New Evidence for the Processing of Wild Cereal Grains at Ohalo II, a 23000-Year-Old Campsite on the Shore of the Sea of Galilee, Israel. *Antiquity* 86(334):990–1003.
Naghibi, Farzaneh, Mahmoud Mosaddegh, Saeed Mohammadi Motamed, and Abdolbaset Ghorbani
2005 Labiatae Family in Folk Medicine in Iran: From Ethnobotany to Pharmacology. *Iranian Journal of Pharmaceutical Research* 2005(2):63–79.
Nedelcheva, Anely
2013 An Ethnobotanical Study of Wild Edible Plants in Bulgaria. *EurAsian Journal of BioSciences* 7:77–94.
Neef, Reinder
1990 Introduction, Development, and Environmental Implications of Olive Culture:

The Evidence from Jordan. In *Man's Role in the Shaping of the Eastern Mediterranean Landscape. Proceedings of the INQUA/BAI Symposium on the Impact of Ancient Man on the Landscape of the Eastern Mediterranean Region and the Near East, Groningen, Netherlands, 6–9 March 1989*, edited by Sytze Bottema, Gertie Entjes-Nieborg, and Willem van Zeist, pp. 295–306. Balkema, Rotterdam.

2001 Getreide im Silokomplex an der Poternenmauer (Boğazköy)—Erste Aussagen zur Landwirtschaft. In *Die Ausgrabungen in Boğazköy-Hattusa 2000*, edited by Jürgen Seeher. *Archäologischer Anzeiger* 3:335–341.

Nelson, Max

2005 *The Barbarian's Beverage: A History of Beer in Ancient Europe*. Routledge, London.

Nesbitt, Mark

2001 Wheat Evolution: Integrating Archaeological and Biological Evidence. In *Wheat Taxonomy: The Legacy of John Percival*, edited by Peter D. S. Caligari and Peter E. Brandham, pp. 37–59. The Linnean: Special Issue 3. Academic Press, London.

2002 When and Where Did Domesticated Cereals First Occur in Southwest Asia? In *The Dawn of Farming in the Near East*, edited by René T. J. Cappers and Sytze Bottema, pp. 113–132. Studies in Early Near Eastern Production, Subsistence, and Environment 6. Ex oriente, Berlin.

Nesbitt, Mark, Gordon C. Hillman, Leonor Peña-Chocarro, Delwen Samuel, and A. T. Szabó

1996 Checklist for Recording the Cultivation and Uses of Hulled Wheats. In *Hulled Wheat: Proceedings of the First International Workshop on Hulled Wheats, 21–22 July 1995, Castelvecchio Pascoli, Tuscany, Italy*, edited by Stefano Padulosi, Karl Hammer, and Joachim Heller, pp. 234–246. Promoting the Conservation and Use of Underutilised and Neglected Crops 4. International Plant Genetic Resources Institute, Rome.

Nesbitt, Mark, and Delwen Samuel

1996 From Staple Crop to Extinction? The Archaeology and History of the Hulled Wheats. In *Hulled Wheat: Proceedings of the First International Workshop on Hulled Wheats, 21–22 July 1995, Castelvecchio Pascoli, Tuscany, Italy*, edited by Stefano Padulosi, Karl Hammer, and Joachim Heller, pp. 40–99. Promoting the Conservation and Use of Underutilised and Neglected Crops 4. International Plant Genetic Resources Institute, Rome.

Nestle, Marion

1995 Mediterranean Diets: Historical and Research Overview. *American Journal of Clinical Nutrition* 61(6):1313S–1320S.

Newman, C. Walt, and Rosemary K. Newman

2006 A Brief History of Barley Foods. *Cereal Foods World* 51(1):4–7.

Newton, Claire, Christine Lorre, Caroline Sauvage, Sarah Ivorra, and Jean-Frédéric Terral

2014 On the Origins and Spread of *Olea europaea* L. (Olive) Domestication: Evidence for Shape Variation of Olive Stones at Ugarit, Late Bronze Age, Syria—a Window on the Mediterranean Basin and on the Westward Diffusion of Olive Varieties. *Vegetation History and Archaeobotany* 23(5):567–575.

Nielsen, Kjeld

1986 *Incense in Ancient Israel*. Supplement Vetus Testamentum 38. E. J. Brill. Leiden.

Nielsen, Thomas Heine

1999 The Concept of Arkadia—The People, Their Land and Their Organization. In *Defining Ancient Arkadia, Symposium, April 1–4, 1998, Acts of the Copenhagen*

Polis Center, Vol. 6, edited by Thomas Heine Nielsen and James Roy, pp. 16–79. The Royal Danish Academy of Sciences and Letters, Copenhagen.

Ninou, Ismini, Nikos Efstratiou, and Soultana Maria Valamoti
2022 Grinding and Pounding in Early Neolithic SE Europe: Culinary Preferences and Social Dimensions of Plantfood Processing. In *Cooking with Plants in Ancient Europe and Beyond,* edited by Soultana Maria Valamoti, Anastasia Dimoula, and Maria Ntinou. Sidestone Press, in press.

Nitsch, Erika, Stelios Andreou, Aurélien Creuzieux, Armelle Gardeisen, Paul Halstead, Valasia Isaakidou, Angeliki Karathanou, Dimitra Kotsachristou, Daphne Nikolaidou, Aikaterini Papanthimou, Chryssa Petridou, Sevi Triantaphyllou, Soultana Maria Valamoti, Anastasia Vasileiadou, and Amy Bogaard
2017 A Bottom-Up View of Food Surplus: Using Stable Carbon and Nitrogen Isotope Analysis to Investigate Agricultural Strategies and Diet at Bronze Age Archontiko and Thessaloniki Toumba, Northern Greece. *World Archaeology* 49(1):105–137.

Nitsch, Erika, Glynis Jones, Anaya Sarpaki, Mette Marie Hald, and Amy Bogaard
2019 Farming Practice and Land Management at Knossos, Crete: New Insights from $\delta^{13}C$ and $\delta^{15}N$ Analysis of Neolithic and Bronze Age Crop Remains. In *Country in the City: Agricultural Functions in Protohistoric Urban Settlements (Aegean and Western Mediterranean),* edited by Dominique Garcia, Raphaël Orgeolet, Maia Pomadère, and Julien Zurbach, pp. 159–174. Archaeopress, Oxford.

Nixon-Darcus, Laurie A., and Catherine D'Andrea
2017 Necessary for Life: Studies of Ancient and Modern Grinding Stones in Highland Ethiopia. *African Archaeological Review* 34(2):193–223.

Ntinou, Maria, Angeliki Karathanou, Clémence Pagnoux, and Soultana Maria Valamoti
2022 Land Management and Food Resources in Bronze Age Central Greece: Insights from Archaeobotanical Assemblages from the Sites of Agia Paraskevi, Kynos and Mitrou (Phthiotida). In *Cooking with Plants in Ancient Europe and Beyond,* edited by Soultana Maria Valamoti, Anastasia Dimoula, and Maria Ntinou. Sidestone Press.

Ntinou, Maria, and Nina Kyparissi-Apostolika
2016 Local Vegetation Dynamics and Human Habitation from the Last Interglacial to the Early Holocene at Theopetra Cave, Central Greece: The Evidence from Wood Charcoal Analysis. *Vegetation History and Archaeobotany* 25(2):191–206.

Nunn, John F.
2002 *Ancient Egyptian Medicine.* British Museum Press, London.

Oflaz, Aslı, Walter Dörfler, and Mara Weinelt
2019 An Overview of Olive Trees in the Eastern Mediterranean during the Mid- to Late Holocene: Selective Exploitation or Established Arboriculture? In *How's Life? Living Conditions in the 2nd and 1st Millennia BCE,* edited by Marta Dal Corso, Wiebke Kirleis, Jutta Kneisel, Nicole Taylor, Magdalena Wieckowska-Lüth, and Marco Zanon, pp.131–165. Scales of Transformation in Prehistoric and Archaic Societies 4. Sidestone Press, Leiden.

Ólafsdóttir, Sigrún
2013 Social Construction and Health. In *Medical Sociology on the Move: New Directions in Theory,* edited by William C. Cockerham, pp. 41–59. Springer, Dordrecht.

Olmo, Harold P.
1995 The Origin and Domestication of the Vinifera Grape. In *The Origins and Ancient History of Wine*, edited by Patrick E. McGovern, Stuart J. Fleming, and Solomon H. Katz, pp. 31–43. Food and Nutrition in History and Anthropology 11. Gordon and Breach, Philadelphia.

Omidi, Heshmat, Hoda Shams, Mehdi Seif Sahandi, and Tayebeh Rajabian
2018 Balangu (*Lallemantia* sp.) Growth and Physiology under Field Drought Conditions Affecting Plant Medicinal Content. *Plant Physiology and Biochemistry* 130:641–646.

Opler, Morris E.
1963 The Cultural Definition of Illness in Village India. *Human Organization* 22(1):32–35.

Oybak Dönmez, Emel
2005 Environmental Archaeology Early Bronze Age Crop Plants from Yenibademli Höyük (Gökçeada), Western Turkey. *Journal of Human Palaeoecology* 10(1):39–49.

Özcan, Mehmet Musa
2004 Characteristics of Fruit and Oil of Terebinth (*Pistacia terebinthus* L.) Growing Wild in Turkey. *Journal of the Science of Food and Agriculture* 84(6):517–520.

Özcan, Mehmet Musa, and Çiğdem Atalay
2006 Determination of Seed and Oil Properties of Some Poppy (*Papaver somniferum* L.) Varieties. *Grasas y Aceites* 57(2):169–174.

Pagnoux, Clémence
2016 *Émergence, développement et diversification de l'arboriculture en Grèce du Néolithique à l'époque romaine: Confrontation des données archéobotaniques, morphométriques, épigraphiques et littéraires*. PhD dissertation, Université Paris 1.

Pagnoux, Clémence, Vincent Bonhomme, Laurent Bouby, Sarah Ivorra, Susan E. Allen, and Soultana Maria Valamoti
2022 Early Viticulture in Neolithic and Bronze Age Greece: Looking for the Best Traditional Morphometric Method to Distinguish Wild and Domestic Grape Pips. In *Cooking with Plants in Ancient Europe and Beyond*, edited by Soultana Maria Valamoti, Anastasia Dimoula, and Maria Ntinou. Sidestone Press, in press.

Pagnoux, Clémence, Laurent Bouby, Sarah Ivorra, Christophe Petit, Soultana Maria Valamoti, Thierry Pastor, Sandrine Picq, and Jean-Frédéric Terral
2015 Inferring the Agrobiodiversity of *Vitis vinifera* L. (grapevine) in Ancient Greece by Comparative Shape Analysis of Archaeological and Modern Seeds. *Vegetation History and Archaeobotany* 24(1):75–84.

Pagnoux, Clémence, Laurent Bouby, Soultana Maria Valamoti, Vincent Bonhomme, Sarah Ivorra, Eugenia Gkatzogia, Angeliki Karathanou, Dimitra Kotsachristou, Helmut Kroll, and Jean-Frédéric Terral
2021 Local Domestication or Diffusion? Insights into Viticulture in Greece from Neolithic to Archaic Times, Using Geometric Morphometric Analyses of Archaeological Grape Seeds. *Journal of Archaeological Science* 125:105263.

Pagnoux, Clémence, Alessandra Celant, Sylvie Coubray, Girolamo Fiorentino, and Véronique Matterne
2013 The Introduction of Citrus to Italy, with Reference to the Identification Problems of Seed Remains. *Vegetation History and Archaeobotany* 22(5):421–438.

Pagnoux Clémence, Sylvie Coubray, and Véronique Matterne
2016 L'introduction des agrumes en Méditerranée occidentale: Données archéobotaniques, iconographiques et littéraires. *Food and History* 14(2–3):107–129.

Palmer, Carol
2002 Milk and Cereals: Identifying Food and Food Identity among Fallāhīn and Bedouin in Jordan. *Levant* 34(1):173–195.

Palmer, Ruth
1995 Wine and Viticulture in the Linear A and B Texts of the Bronze Age Aegean. In *The Origins and Ancient History of Wine*, edited by Patrick E. McGovern, Stuart J. Fleming, and Solomon H. Katz, pp. 269–285. Food and Nutrition in History and Anthropology 11. Gordon and Breach, Philadelphia.

Panagiotakopulu, Eva, Paul C. Buckland, Peter M. Day, and Christos Doumas
1995 Natural Insecticides and Insect Repellents in Antiquity: A Review of the Evidence. *Journal of Archaeological Science* 22(5):705–710.

Panagiotidis, Sampson, and Maria L. Papadopoulou
2016 Human-Landscape Interactions in Halkidiki (NC Greece) over the Last 3.5 Millennia, Revealed through Palynological, and Archaeological-Historical Archives. *Journal of Archaeological Science: Reports* 7:138–145.

Panagiotopoulos, Konstantinos, Anne Aufgebauer, Frank Schäbitz, and Bernd Wagner
2013 Vegetation and Climate History of the Lake Prespa Region since the Lateglacial. *Quaternary International* 293:157–169.

Pankin, Artem, Janine Altmüller, Christian Becker, and Maria von Korff
2018 Targeted Resequencing Reveals Genomic Signatures of Barley Domestication. *New Phytologist* 218(3):1247–1259.

Papadimitriou, Konstantinos, Bruno Pot, and Effie Tsakalidou
2015 How Microbes Adapt to a Diversity of Food Niches. *Current Opinion in Food Science* 2:29–35.

Papadopoulou, Evanthia
2010 *Οι πηλοκατασκευές του προϊστορικού οικισμού στο Αρχοντικό Γιαννιτσών: Συμβολή στη μελέτη της τεχνολογίας της τροφής* [The Built Clay Facilities of the Prehistoric Settlement of Archontiko Giannitson: Contribution to the Study of Food Technology]. PhD dissertation, Aristotle University of Thessaloniki.

Papadopoulou, Evanthia, and Yannis Maniatis
2013 Reconstructing Thermal Food Processing Techniques: The Application of FTIR Spectroscopy in the Analysis of Clay Thermal Structures from EBA Archontiko. In *Diet, Economy, and Society in the Ancient Greek World: Towards a Better Integration of Archaeology and Science. Proceedings of the International Conference Held at the Netherlands Institute at Athens on 22–24 March 2010*, edited by Sofia Voutsaki and Soultana Maria Valamoti, pp. 113–122. Pharos Supplement 1. Peeters, Leuven.

Papadopoulou, Evanthia, and Sandra Prévost-Dermarkar
2007 Il n'y a pas de cuisine sans feu: Une approche des techniques culinaires au Néolithique et à l'Âge du Bronze Ancien à travers les structures de combustion en Grèce du Nord. In *Cooking Up the Past: Food and Culinary Practices in the Neolithic and Bronze Age Aegean*, edited by Christopher Mee and Josette Renard, pp. 123–135. Oxbow, Oxford.

Papaefthymiou-Papanthimou, Aikaterini, and Evi Papadopoulou
2014 Αρχοντικό Γιαννιτσών: Ένας οικισμός της Πρώιμης Εποχής του Χαλκού στη

Μακεδονία [Archondiko Giannitson: An Early Bronze Age Settlement in Macedonia]. In *1912–2012: Εκατό χρόνια έρευνας στην προϊστορική Μακεδονία, Πρακτικά διεθνούς συνεδρίου, Αρχαιολογικό Μουσείο Θεσσαλονίκης, 22–24 Νοεμβρίου 2012* [1912–2012: A Century of Research in Prehistoric Macedonia, International Conference Proceedings, Archaeological Museum of Thessaloniki, 22–24 November 2012], edited by Evaggelia Stefani, Nikos Merousis, and Anastasia Dimoula, pp. 271–280. Archaeological Museum of Thessaloniki.

Papaefthymiou-Papanthimou, Aikaterini, Angeliki Pilali-Papastergiou, Despoina Basogianni, Evanthia Papadopoulou, Evangelia Tsagaraki, and Ioannis Fappas

2002 Αρχοντικό 2000: Τυπολογική παρουσίαση και ερμηνευτικά προβλήματα των πηλόκτιστων κατασκευών [Archondiko 2000: A Typological Presentation and Problems Relating to the Function of the Clay Structures]. *Proceedings of the Archaeological Work in Macedonia and Thrace (AEMTH)* 14, 2000:421–433.

Papaefthymiou-Papanthimou, Aikaterini, Soultana Maria Valamoti, Evi Papadopoulou, Evangelia Tsagkaraki, and Evangelia Voulgari

2013 Food Storage in the Context of an Early Bronze Age Household Economy: New Evidence from Archontiko Giannitson. In *Diet, Economy, and Society in the Ancient Greek World: Towards a Better Integration of Archaeology and Science. Proceedings of the International Conference Held at the Netherlands Institute at Athens on 22–24 March 2010*, edited by Sofia Voutsaki and Soultana Maria Valamoti, pp. 103–111. Pharos Supplement 1. Peeters, Leuven.

Papagrigorakis, Manolis J., Panagiotis Toulas, Manolis G. Tsilivakos, Antonis A. Kousoulis, Despoina Skorda, George Orfanidis, and Philippos N. Synodinos

2014 Neurosurgery during the Bronze Age: A Skull Trepanation in 1900 BC Greece. *World Neurosurgery* 81(2):431–435.

Papathanasiou, Anastasia

2001 *Bioarchaeological Analysis of Neolithic Alepotrypa Cave, Greece*. British Archaeological Reports International Series 961. Hedges, Oxford.

2011 Health, Diet, and Social Implications in Neolithic Greece from the Study of Human Osteological Material. In *Human Bioarchaeology of the Transition to Agriculture*, edited by Ron Pinhasi and Jay T. Stock, pp. 87–106. Wiley-Blackwell, Oxford.

2012 Evidence of Trauma in Neolithic Greece. In *Sticks, Stones, and Broken Bones: Neolithic Violence in a European Perspective*, edited by Rick J. Schulting and Linda Fibiger, pp. 249–263. Oxford University Press, Oxford.

Pappa, Maria

2008 *Οργάνωση του χώρου και οικιστικά στοιχεία στους νεολιθικούς οικισμούς της Κεντρικής Μακεδονίας* [Organization of Communal and Domestic Space in the Neolithic Settlements of Central Macedonia]. PhD dissertation, Aristotle University of Thessaloniki.

Pappa, Maria, and Manthos Besios

1999 The Neolithic Settlement at Makriyalos, Northern Greece: Preliminary Report on the 1993–1995 Excavations. *Journal of Field Archaeology* 26(2):177–195.

Pappa, Maria, Paul Halstead, Kostas Kotsakis, and Dushka Urem-Kotsou

2004 Evidence for Large-Scale Feasting at Late Neolithic Makriyalos, N Greece. In *Food, Cuisine, and Society in Prehistoric Greece*, edited by Paul Halstead and John C. Barrett, pp. 16–44. Sheffield Studies in Aegean Archaeology 5. Oxbow, Oxford.

Pappa, Maria, and Stratos Nanoglou
2019 The Rescue Excavation Project at the Neolithic Settlement of Vasilika Kyparissi in Central Macedonia, Greece. *Antiquity* 93(369):e16.

Pappa, Maria, and Rena Veropoulidou
2011 The Neolithic Settlement at Makriyalos, Northern Greece: Evidence from the *Spondylus gaederopus* Artifacts. In *Spondylus in Prehistory: New Data and Approaches; Contributions to the Archaeology of Shell Technologies*, edited by Fotis Ifantidis and Marianna Nikolaidou, pp. 105–121. British Archaeological Reports International Series 2216. Archaeopress, Oxford.

Paraskevopoulou, Pelagia
2019 *Αρχαιοβοτανικά δεδομένα από το Νεολιθικό οικισμό στη λίμνη Κορώνεια (Archaeobotanical Remains from the Neolithic Settlement at Lake Koroneia)*. Master's thesis, Aristotle University of Thessaloniki.

Parker-Pearson, Mike
2003 Food, Identity, and Culture: An Introduction and Overview. In *Food, Culture, and Identity in the Neolithic and Early Bronze Age*, edited by Mike Parker-Pearson, pp. 1–30. British Archaeological Reports International Series 1117. Archaeopress, Oxford.

Parkhurst-Ferguson, Priscilla
2010 Culinary Nationalism. *Gastronomica* 10(1):102–109.

Parsons, James J.
1962 The Acorn-Hog Economy of the Oak Woodlands of Southwestern Spain. *Geographical Review* 52(2):211–235.

Patil, Rhambo T., and Shahab Sokhansanj
2003 Dehulling and Splitting Pulses. In *Handbook of Postharvest Technology: Cereals, Fruits, Vegetables, Tea, and Spices*, edited by Amalendu Chakraverty, Arun S. Mujumdar, G. S. Vijaya Raghavan, and Hosahalli S. Ramaswamy, pp. 397–426. Books in Soils, Plants, and the Environment, Vol. 93. Marcel Dekker, New York.

Pavlov, Ivan
2001 *Prisastvia na hraneneto po balgarskite zemi prez XV–XIX vek* [Presence of Food in the Bulgarian Lands in XV–XIX Century]. Prof. Marin Drinov Publishing House, Sofia.

Payne, Sebastian
1982 Faunal Evidence for Environmental/Climatic Change at Franchthi Cave (Southern Argolid, Greece), 25,000 BP–5000 BP: Preliminary Results. In *Palaeoclimates, Palaeoenvironments, and Human Communities in the Eastern Mediterranean Region in Later Prehistory*, edited by John L. Bintliff and Willem van Zeist, pp. 133–136. British Archaeological Reports International Series 133(1–2). BAR, Oxford.

Peachey, Claire Patricia
1995 *Terebinth Resin in Antiquity: Possible Uses in the Late Bronze Age Aegean Region.* Master's thesis, Texas A&M University.

Pearsall, Deborah M.
1989 *Paleoethnobotany: A Handbook of Procedures.* Academic Press, San Diego.

Pearson, James L.
2002 *Shamanism and the Ancient Mind: A Cognitive Approach to Archaeology.* AltaMira Press, Walnut Creek.

Pecci, Alessandra, Gianluca Giorgi, Laura Salvini, Miguel Ángel, and Cau Ontiveros

2013 Identifying Wine Markers in Ceramics and Plasters Using Gas Chromatography–Mass Spectrometry: Experimental and Archaeological Materials. *Journal of Archaeological Science* 40(1):109–115.

Peña-Chocarro, Leonor

1996 In Situ Conservation of Hulled Wheat Species: The Case of Spain. In *Hulled Wheat: Proceedings of the First International Workshop on Hulled Wheats, 21–22 July 1995, Castelvecchio Pascoli, Tuscany, Italy*, edited by Stefano Padulosi, Karl Hammer, and Joachim Heller, pp. 128–146. Promoting the Conservation and Use of Underutilised and Neglected Crops 4. International Plant Genetic Resources Institute, Rome.

1999 *Prehistoric Agriculture in Southern Spain during the Neolithic and the Bronze Age: The Application of Ethnographic Models.* British Archaeological Reports International Series 818. BAR, Oxford.

2007 Early Agriculture in Central and Southern Spain. In *The Origins and Spread of Domestic Plants in Southwest Asia and Europe*, edited by Sue Colledge and James Conolly, pp. 173–187. Left Coast Press, Walnut Creek.

Peña-Chocarro, Leonor, and Lydia Zapata-Peña

1999 History and Traditional Cultivation of *Lathyrus sativus* L. and *Lathyrus cicera* L. in the Iberian Peninsula. *Vegetation History and Archaeobotany* 8(1–2):49–52.

Peña-Chocarro, Leonor, Lydia Zapata-Peña, Jesús Emilio González Urquijo, and Juan José Ibáñez Estévez

2009 Einkorn (*Triticum monococcum* L.) Cultivation in Mountain Communities of the Western Rif (Morocco): An Ethnoarchaeological Project. In *From Foragers to Farmers: Papers in Honour of Gordon C. Hillman*, edited by Andrew S. Fairnbairn and Ehud Weiss, pp. 103–111. Oxbow, Oxford.

Pennacchio, Marcello, Lara Jefferson, and Kayri Havens

2010 *Uses and Abuses of Plant-Derived Smoke: Its Ethnobotany as Hallucinogen, Perfume, Incense, and Medicine.* Oxford University Press, Oxford.

Pérez-Jordà, Guillem, Leonor Peña-Chocarro, Marcos García Fernández, and Juan Carlos Vera Rodríguez

2017 The Beginnings of Fruit Tree Cultivation in the Iberian Peninsula: Plant Remains from the City of Huelva (Southern Spain). *Vegetation History and Archaeobotany* 26(5):527–538.

Perissoratis, Constantine, and Nikolaos Conispoliatis

2003 The Impacts of Sea-Level Changes during Latest Pleistocene and Holocene Times on the Morphology of the Ionian and Aegean Seas (SE Alpine Europe). *Marine Geology* 196(3–4):145–156.

Perlès, Catherine

1977 *Préhistoire du Feu.* Masson, Paris.

1999 Long-Term Perspectives on the Occupation of the Franchthi Cave: Continuities and Discontinuities. In *The Palaeolithic Archaeology of Greece and Adjacent Areas. Proceedings of the ICOPAG Conference, Ioannina, September 1994*, edited by Geoff N. Bailie, Eugenia Adam, Eleni Panagopoulou, Catherine Perlés, and Kostas Zachos, pp. 311–318. British School at Athens Studies 3. The British School at Athens, London.

2001 *The Early Neolithic in Greece: The First Farming Communities in Europe.* Cambridge University Press, Cambridge.

2003 The Mesolithic at Franchthi: An Overview of the Data and Problems. In *The Greek Mesolithic: Problems and Perspectives*, edited by Nena Galanidou and Catherine Perlès, pp. 79–87. British School at Athens Studies 10. The British School at Athens, London.

2016 Food and Ornaments: Diachronic Changes in the Exploitation of Littoral Resources at Franchthi Cave (Argolid, Greece) during the Upper Palaeolithic and the Mesolithic (39,000–7000 cal BC). *Quaternary International* 407(B):45–58.

Perlès, Catherine, Anita Quiles, and Hélène Valladas

2013 Early Seventh-Millennium AMS Dates from Domestic Seeds in the Initial Neolithic at Franchthi Cave (Argolid, Greece). *Antiquity* 8(338):1001–1015.

Perlès, Catherine, and Karen D. Vitelli

1999 Craft Specialization in the Neolithic of Greece. In *Neolithic Society in Greece*, edited by Paul Halstead, pp. 96–107. Sheffield Studies in Aegean Archaeology 2. Sheffield Academic Press, Sheffield.

Pesexidis, S.

1981 *Τα Κυριότερα Ζιζάνια της Ελλάδος* [The Main Weeds of Greece]. Greek Sugar Industry, Thessaloniki.

Petridou, Chryssa

2016 *Χρήσεις των φυτών στο Αρχοντικό Γιαννιτσών κατά την Εποχή του Χαλκού: Μια αρχαιοβοτανική μελέτη* [Plant Use at Archondiko Giannitson in Bronze Age: An Archaeobotanical Study]. Master's thesis, Aristotle University of Thessaloniki.

2022 *Plant Food Remains from Prehistoric Greece: An Interdisciplinary Approach*. PhD dissertation, Aristotle University of Thessaloniki, in preparation.

Petrov, L., N. Dzhelepov, E. Yordanov, and S. Uzunova

1978 *Balgarska nacionalna kuhnja* [Bulgarian National Cuisine]. Zemizdat, Sofia.

Philippa-Touchais, Anna, Gilles Touchais, Oreste Decavallas, Armelle Gardeisen, Matthieu Ghilardi, Evi Margaritis, Odysseas Metaxas, Sevi Triantaphyllou, and Efi Tsiolaki

2014 Environnement, alimentation, hygiène et mode de vie dans le Grèce mésohelladique: Le cas de l'Aspis d'Argos. In *Physis: L'environnement naturel et la relation homme-milieu dans le monde égéen protohistorique. Actes de la 14e rencontre égéenne internationale, Paris, Institut National d'Histoire de l'Art (INHA), 11–14 décembre 2012*, edited by Gilles Touchais, Robert Laffineur, and Françoise Rougemont, pp. 531–538. Aegaeum 37. Peeters, Leuven/Liège.

Photiades, J., and G. Alexandrou

1979 Food Legume Research and Production in Cyprus. In *Food Legume Improvement and Development: Proceedings of a Workshop Held at the University of Aleppo, Syria, 2–7 May 1978*, edited by Geoffrey C. Hawtin and George J. Chancellor, pp. 75–79. International Development Research Centre, Ottawa.

Pieroni, Andrea

2021 Wild Foods: A Topic for Food Pre-History and History or a Crucial Component of Future Sustainable and Just Food Systems? *Foods* 10(4):827.

Pieroni, Andrea, Maria Elena Giusti, Caterina de Pasquale, Cinzia Lenzarini, Eleonora Censorii, María Reyes Gonzáles-Tejero, Cristina Patricia Sánchez-Rojas, Jose M. Ramiro-Gutiérrez, Melpomeni Skoula, Chris Johnson, Anaya Sarpaki, Athena Della, Demetra Paraskeva-Hadijchambi, Andreas Hadjichambis, Mohammed

Hmamouchi, Said El-Jorhi, Mohamed El-Demerdash, Mustafa El-Zayat, Omar Al-Shahaby, Zahia Houmani, and Mekious Scherazed
2006 Circum-Mediterranean Cultural Heritage and Medicinal Plant Uses in Traditional Animal Healthcare: A Field Survey in Eight Selected Areas within the RUBIA Project. *Journal of Ethnobiology and Ethnomedicine* 2:Article 16.
Pieroni, Andrea, and Bren Torry
2007 Does Taste Matter? Taste and Medicinal Perceptions Associated with Five Selected Herbal Drugs among Three Ethnic Groups in West Yorkshire, Northern England. *Journal of Ethnobiology and Ethnomedicine* 3:Article 21.
Pilcher, Jeffrey M.
1996 Tamales or Timbales: Cuisine and the Formation of Mexican National Identity, 1821–1911. *The Americas* 53(2):193–216.
Pilgeram, Alice L., David C. Sands, Darrin Boss, Nick Dale, David Wichman, Peggy Lamb, Chaofu Lu, Rick Barrows, Mathew Kirkpatrick, Brian Thompson, and Duane L. Johnson
2007 *Camelina sativa*, A Montana Omega-3 and Fuel Crop. In *Issues in New Crops and New Uses*, edited by Jules Janick and Anna Whipkey, pp. 129–131. ASHS Press, Alexandria, Virginia.
Piperno, Dolores R., Ehud Weiss, Irene Holst, and Dani Nadel
2004 Processing of Wild Cereal Grains in the Upper Palaeolithic Revealed by Starch Grain Analysis. *Nature* 430(7000):670–673.
Piskov, Ivaylo
2010 *The Bulgarian Cookbook* (CreateSpace e-book).
Platon, Lefteris, and Katerina Kopaka
1993 Linoi Minoikoi: Installations minoennes de traitement des produits liquides. *Bulletin de Correspondance Hellénique* 177:35–101.
Polunin, Oleg
1969 *Flowers of Europe: A Field Guide*. Oxford University Press, Oxford.
1987 *Flowers of Greece and the Balkans: A Field Guide*. Oxford University Press, Oxford.
Polunin, Oleg, and Antony Huxley
1987 *Flowers of the Mediterranean*. Hogarth Press, London.
Popper-Gideon, Ariela, and Jonathan Ventura
2008 Claiming Power through Hardship: Initiation Narratives of Palestinian Traditional Women Healers in Israel. *Social Science and Medicine* 67(11):1807–1816.
Poulakou-Rebelakou, Effie, Marianna Karamanou, and George Androutsos
2015 The Impact of Ancient Greek Medicine in India: The Birth of Unani Medicine. *Acta Medico-Historica Adriatica* 13(2):323–328.
Powell, Marvin A.
1995 Wine and the Vine in Ancient Mesopotamia: The Cuneiform Evidence. In *The Origins and Ancient History of Wine*, edited by Patrick E. McGovern, Stuart J. Fleming, and Solomon H. Katz, pp. 97–122. Food and Nutrition in History and Anthropology 11. Gordon and Breach, Philadelphia.
Power, Robert C., Domingo C. Salazar-García, Lawrence G. Straus, Manuel R. González Morales, and Amanda G. Henry
2015 Microremains from El Mirón Cave Human Dental Calculus Suggest a Mixed Plant–Animal Subsistence Economy during the Magdalenian in Northern Iberia. *Journal of Archaeological Science* 60:39–46.

Pratt, Catherine E.
2021 *Oil, Wine, and the Cultural Economy of Ancient Greece: From the Bronze Age to the Archaic*. Cambridge University Press, Cambridge.

Preece, John E., and Jeff Moersfelder
2016 Pomegranate: The Grainy Apple. *Journal of the American Pomological Society* 70(4):187–193.

Preucel, Robert W., and Stephen A. Mrozowski
2010 The New Pragmatism. In *Contemporary Archaeology in Theory: The New Pragmatism*, edited by Robert W. Preucel and Stephen A. Mrozowski, pp. 1–50. 2nd ed. Wiley-Blackwell, Chichester.

Prévost-Dermarkar, Sandra
2002 Les foyers et les fours domestiques en Egée au Néolithique et à l'Age du Bronze. In *Pains, Fours et Foyers Domestiques des Temps Passés—Breads, Ovens, and Hearths of the Past*, edited by Marianne Mesnil and Kaï Fechner, pp. 223–237. *Civilisation* 49(1–2). Institut de Sociologie de l' Université de Bruxelles.
2003 Les fours néolithiques de Dikili Tash (Macédoine, Grèce): Une approche expérimentale des techniques de construction des voûtes en terre à bâtir. In *Le feu domestique et ses structures au Néolithique et aux Ages des métaux (Actes du colloque de Bourg-en-Bresse et Beaune, 7–8 Octobre 2000)*, edited by Marie-Chantal Frère-Sautot, pp. 215–223. Monique Mergoil. Montagnac.

Price, T. Douglas, Joachim Wahl, and R. Alexander Bentley
2006 Isotopic Evidence for Mobility and Group Organization among Neolithic Farmers at Talheim, Germany, 5000 BC. *European Journal of Archaeology* 9(2–3):259–284.

Primavera, Milena, and Girolamo Fiorentino
2013 Acorn Gatherers: Fruit Storage and Processing in South-East Italy during the Bronze Age. *Origini* 35:211–227.

Procopiou, Hara
2003 Les techniques de décorticagedans le monde égéen: Étude ethnoarchéologiquedans les Cyclades. In *Le traitement des récoltes: Un regard sur la diversité, du Néolithique au Présent; Actes des XXIIIe Rencontres internationales d'archéologie et d'histoire d'Antibes, 17–19 octobre 2002*, edited by Patricia C. Anderson, Linda S. Cummings, Thomas K. Schippers, and Bernard Simonel, pp. 115–136. APDCA, Antibes.

Protonotariou-Deilaki, Evangelia
1973 Αρχαιότητες καὶ μνημεία Αργολιδος-Κορινθίας 1971–1972—Ακροναυπλία [Antiquities and Momuments of Argolis-Korinthia 1971–1972—Akronafplia]. *Αρχαιολογικόν Δελτίον (B1 Χρονικά)* [*Archaeologiko Deltion (B1 Chronika)*] 28:90–94.

Pryor, Alexander J. E., Madeline Steele, Martin K. Jones, Jiří Svoboda, and David G. Beresford-Jones
2013 Plant Foods in the Upper Palaeolithic at Dolní Věstonice? Parenchyma Redux. *Antiquity* 87(338):971–984.

Psilakis, Nikos, and Maria Psilaki
2001 *Το ψωμί των Ελλήνων και τα γλυκίσματα της λαϊκής μας παράδοσης* [The Bread of the Greeks and the Sweets of Our Popular Tradition]. Karmanor, Herakleion.

Puumalainen, Tuija, Heli Nykopp, and Hely Tuorila
2002 Old Product in a New Context: Importance of the Type of Dish for the

Acceptance of Grünkern, a Spelt-Based Traditional Cereal. *Lebensmittel, Wissenschaft und Technologie* 35(6):549–553.

Pyrros, Dionysios
1838 *Βοτανική Πρακτική, προσηρμοσμένη εἰς τήν ἰατρικήν καί οἰκονομίαν, μεθ' ὅσης ἐπιμελείας συνερανισθεῖσα, ἐκ διαφόρων συγγραφέων, καί συντεθεῖσα κατ' ἀλφάβητον καί κατά τό σύστημα τοῦ σοφοῦ Λινναίου.* [Botanical Practice Applied to Medicine and Economy]. Aggelidis, Athens.

Rackham, Oliver
1996 Ecology and Pseudo-Ecology: The Example of Ancient Greece. In *Human Landscapes in Classical Antiquity: Environment and Culture*, edited by Graham Shipley and John Salmon, pp. 16–43. Leicester-Nottingham Studies in Ancient Society 6. Routledge, London.

Raftopoulou, Eleni
1996 Η πώληση και η χρήση του ελαιόλαδου και του αρωματικού λαδιού στην Αρχαϊκή Ελλάδα [The Sale and Use of Olive Oil and Perfumed Oil in Archaic Greece]. In *Ελιά και Λάδι* [Olive and Oil]. *Proceedings of the 4th Three-Day Working Meeting on Olive and Oil, Kalamata, 7–13 May 1993*, pp. 110–117. ETVA Cultural and Technological Foundation, Athens.

Ragir, Sonia
2000 Diet and Food Preparation: Rethinking Early Hominid Behavior. *Evolutionary Anthropology* 9(4):153–155.

Rahimi, Amir, Mustafa Kıralan, Neşet Arslan, Ali Bayrak, and Sezen Doğramacı
2011 Variation in Fatty Acid Composition of Registered Poppy (*Papaver sumniferum* L.) Seed in Turkey. *Akademik Gıda* 9(3):22–25.

Rambach, Jörg
2018 Romanos-Navarino Dunes in the Pylia: The Early Helladic II Settlement and the Case of the Early Helladic II Well. In *Pottery Technologies and Sociocultural Connections between the Aegean and Anatolia during the 3rd Millennium BC*, edited by Eva Alram-Stern and Barbara Horejs, pp. 215–248. OREA 10. Austrian Academy of Sciences, Vienna.

Rao, S. L. N., P. R. Adiga, and P. S. Sarma
1964 The Isolation and Characterization of β-N-Oxalyl-L-α,β-Diaminopropionic Acid: A Neurotoxin from the Seeds of *Lathyrus sativus**. *Biochemistry* 3(3):432–436.

Recchia, Giulia, and Alberto Cazzella
2017 Permeable Boundaries in the Late 3rd Millennium BC Central Mediterranean: Contacts and Mobility between the Balkans, Greece, Southern Italy, and Malta. In *Έσπερος/Hesperos: The Aegean Seen from the West. Proceedings of the 16th International Aegean Conference, University of Ioannina, Department of History and Archaeology, Unit of Archaeology and Art History, 18–21 May 2016*, edited by Michael Fotiadis, Robert Laffineur, Yannos Lolos, and Andreas Vlachopoulos, pp. 93–104. Aegaeum 41. Peeters, Leuven/Liège.

Reddy, B. Sanjeeva
2006 Indigenous Technical Knowledge on Pulses Storage and Processing Practices in Andhra Pradesh. *Indian Journal of Traditional Knowledge* 5(1):87–94.

Reimer, Paula J., William E. N. Austin, Edouard Bard, Alex Bayliss, Paul G. Blackwell, Christopher Bronk Ramsey, Martin Butzin, Hai Cheng, R. Lawrence Edwards,

Michael Friedrich, Pieter M. Grootes, Thomas P. Guilderson, Irka Hajdas, Timothy J. Heaton, Alan G. Hogg, Konrad A. Hughen, Bernd Kromer, Sturt W. Manning, Raimund Muscheler, Jonathan G. Palmer, Charlotte Pearson, Johannes van der Plicht, Ron W. Reimer, David A. Richards, E. Marian Scott, John R. Southon, Christian S. M. Turney, Lukas Wacker, Florian Adolphi, Ulf Büntgen, Manuela Capano, Simon M. Fahrni, Alexandra Fogtmann-Schulz, Ronny Friedrich, Peter Köhler, Sabrina Kudsk, Fusa Miyake, Jesper Olsen, Frederick Reinig, Minoru Sakamoto, Adam Sookdeo, and Sahra Talamo

2020 The IntCal20 Northern Hemisphere Radiocarbon Age Calibration Curve (0–55 cal kBP). *Radiocarbon* 62(4):725–757.

Reingruber, Agathe

2018 Geographical Mobility and Social Motility in the Aegean before and after 6600 BC. *Praehistorische Zeitschrift* 93(1):1–24.

Renfrew, Colin

1972 *The Emergence of Civilisation: The Cyclades and the Aegean in the Third Millennium BC*. Methuen, London.

Renfrew, Jane M.

1966 A Report on Recent Finds of Carbonized Cereals and Seeds from Prehistoric Thessaly. *Thessalika* 5:31–36.

1968 Appendix X: The Cereal Remains. In *Excavations at Saliagos Near Antiparos*, edited by John Davies Evans and Colin Renfrew, pp. 139–141. British School at Athens Supplement 5. Thames and Hudson, London.

1972 The Plant Remains. In *Myrtos: An Early Bronze Settlement in Crete*, edited by Peter M. Warren, pp. 315–317. British School at Athens Supplement 7. Thames and Hudson, London.

1973a *Palaeoethnobotany: The Prehistoric Food Plants of the Near East and Europe*. Columbia University Press, New York.

1977 Appendix 3: Seeds from Area K. In *Keos 1: Kephala, A Late Neolithic Settlement and Cemetery*, edited by John E. Coleman, pp. 127–128. The American School of Classical Studies at Athens, Princeton.

1982 Early Agriculture in Melos. In *An Island Polity: The Archaeology of Exploitation in Melos*, edited by Colin Renfrew and Malcom Wagstaff, pp. 156–160. Cambridge University Press, Cambridge.

1989 Carbonized Grains and Seeds. In *Achilleion: A Neolithic Settlement in Thessaly, Greece, 6400–5600 B.C.*, edited by Marija Gimbutas, Shan M. M. Winn, and Daniel M. Shimabuku, pp. 307–310. Monumenta Archaeologica 14. Institute of Archaeology, University of California, Los Angeles.

1995 Palaeoethnobotanical Finds of *Vitis* from Greece. In *The Origins and Ancient History of Wine*, edited by Patrick E. McGovern, Stuart J. Fleming, and Solomon H. Katz, pp. 255–267. Food and Nutrition in History and Anthropology 11. Gordon and Breach, Philadelphia.

1997 Plant Husbandry at Prehistoric Dimitra, Appendix 14. In *Νεολιθική Μακεδονία* [Neolithic Makedonia], edited by Dimitrios V. Grammenos, pp. 220–226. Publications of the Archaeologiko Deltion 56. Fund of Archaeological Proceeds (TAPA), Athens.

2003 Grains, Seeds, and Fruits from Prehistoric Sitagroi. In *Prehistoric Sitagroi: Excavations in Northeast Greece, 1968–1970, Vol. 2: The Final Report*, edited by

Ernestine S. Elster and Colin Renfrew, pp. 1–29. Monumenta Archaeologica 20. Institute of Archaeology, University of California, Los Angeles.

2006 The Carbonized Seeds. In *Markiani, Amorgos: An Early Bronze Age Fortified Settlement; Overview of the 1985–1991 Investigations,* edited by Lila Marangou, Colin Renfrew, Christos Doumas, and Giorgos Gavalas, pp. 245–246. British School at Athens Supplement 40. The British School at Athens, London.

Renna, Massimiliano, Vito Antonio Rinaldi, and Maria Gonnella

2015 The Mediterranean Diet between Traditional Foods and Human Health: The Culinary Example of Puglia (Southern Italy). *International Journal of Gastronomy and Food Science* 2(2):63–71.

Revedin, Anna, Biancamaria Aranguren, Roberto Becattini, Laura Longo, Emanuele Marconi, Marta Mariotti-Lippi, Natalia Skakun, Andrey Sinitsyn, Elena Spiridonova, and Jiří Svoboda

2010 Thirty-Thousand-Year-Old Evidence of Plant Food Processing. *Proceedings of the National Academy of Sciences of the United States of America (PNAS)* 107(44):18815–18819.

Revedin, Anna, Laura Longo, Marta Mariotti-Lippi, Emanuele Marconi, Annamaria Ronchitelli, Jiří Svoboda, Eva Anichini, Matilde Gennai, and Biancamaria Aranguren

2015 New Technologies for Plant Food Processing in the Gravettian. *Quarternay International* 359/360:77–88.

Riehl, Simone, and Elena Marinova

2008 Mid-Holocene Vegetation Change in the Troad (W Anatolia): Man-Made or Natural? *Vegetation History and Archaeobotany* 17:297–312.

Riley, Frank-Richard

2002 Olive Oil Production on Bronze Age Crete: Nutritional Properties, Processing Methods, and Storage Life of Minoan Olive Oil. *Oxford Journal of Archaeology* 21(1):63–75.

Rivera-Nuñez, Diego, and Concepción Obón de Castro

1989 La dieta cereal prehistorica y su supevivencia en el area Mediterranea. *Trabajos de Prehistoria* 46:247–254.

Rivera-Nuñez, Diego, and Michael J. Walker

1989 A Review of Palaeobotanical Findings of Early *Vitis* in the Mediterranean and of the Origins of Cultivated Grape-Vines, with Special Reference to New Pointers to Prehistoric Exploitation in the Western Mediterranean. *Review of Palaeobotany and Palynology* 61(3–4):205–237.

Rizzotti, Giovanni

1952 Il latirismo in Etiopia [Lathyrism in Ethiopia]. *Archivio Italiano di Scienze Mediche Tropicali e di Parassitologia* 33(9):493–500.

Robitaille, Jérôme

2016 The Ground Stone Industry of the Mursi of Maki, Ethiopia: Ethnoarchaeological Research on Milling and Crushing Equipment (Technique and Function). *Journal of Lithic Studies* 3(3):429–456.

Robkin, A. L. H.

1979 The Agricultural Year, the Commodity SA, and the Linen Industry of Mycenaean Pylos. *American Journal of Archaeology* 83(4):469–474.

Rode, Janko
2002 Study of Autochthon *Camelina sativa* (L.) Crantz in Slovenia. *Journal of Herbs, Spices, and Medical Plants* 9(4):313–318.
Romano, Patrizia, Angela Capece, and Lene Jespersen
2006 Taxonomic and Ecological Diversity of Food and Beverage Yeasts. In *Yeasts in Food and Beverages*, edited by Amparo Querol and Graham H. Fleet, pp. 13–53. Springer, Berlin-Heidelberg.
Rösch, Manfred
2016 Weinbau am Bodensee im Spiegel der Rebpollen. In *Seewein: Weinkultur am Bodensee*, edited by Thomas Knubben and Andreas Schmauder, pp. 51–59. Jan Thorbecke, Ostfildern.
Rosengarten, Frederic, Jr.
1984 *The Book of Edible Nuts*. Walker, New York.
Rossignol-Strick, Martine
1995 Sea-Land Correlation of Pollen Records in the Eastern Mediterranean for the Glacial-Interglacial Transition: Biostratigraphy versus Radiometric Time-Scale. *Quaternary Science Reviews* 14(9):893–915.
Rosso, Ana María
2010 Poppy and Opium in Ancient Times: Remedy or Narcotic? *Biomedicine International* 1:81–87.
Rottoli, Mauro, and Andrea Pessina
2007 Neolithic Agriculture in Italy: An Update of Archaeobotanical Data with Particular Emphasis on Northern Settlements. In *The Origins and Spread of Domestic Plants in Southwest Asia and Europe*, edited by Sue Colledge and James Conolly, pp. 141–153. Left Coast Press, Walnut Creek.
Rowan, Erica
2015 Olive Oil Pressing Waste as a Fuel Source in Antiquity. *American Journal of Archaeology* 119(4):465–482.
Rudgley, Richard
1993 *The Alchemy of Culture: Intoxicants in Society*. British Museum Press, London.
Runnels, Curtis N.
1981 *A Diachronic Study and Economic Analysis of Millstones from the Argolid*. PhD dissertation, Indiana University, Bloomington.
2001 The Stone Age of Greece from the Palaeolithic to the Advent of the Neolithic with an Addendum. In *Aegean Prehistory: A Review*, edited by Tracey Cullen, pp. 225–258. American Journal of Archaeology Supplement 1. Archaeological Institute of America, Boston.
2003 The Origins of the Greek Neolithic: A Personal View. In *The Widening Harvest: The Neolithic Transition in Europe—Looking Back, Looking Forward*, edited by Albert J. Ammerman and Paolo Biagi, pp. 121–132. AIA Colloquia and Conference Papers 6. Archaeological Institute of America, Boston.
Runnels, Curtis N., and Julie Hansen
1986 The Olive in the Prehistoric Aegean: The Evidence for Domestication in the Early Bronze Age. *Oxford Journal of Archaeology* 5(3):299–308.
Runnels, Curtis N., and Tjeerd H. van Andel
1988 Trade and the Origins of Agriculture in the Eastern Mediterranean. *Journal of Mediterranean Archaeology* 1(1):83–109.

2003 The Early Stone Age of the Nomos of Preveza: Landscape and Settlement. In *Landscape Archaeology in Southern Epirus, Greece I*, edited by James Wiseman and Konstantinos Zachos, pp. 47–134. Hesperia Supplement 32. The American School of Classical Studies at Athens, Princeton.

Rutter, Jeremy B.

2008 The Anatolian Roots of Early Helladic III Ceremonial Drinking Behaviour. In *The Aegean in the Neolithic, Chalcolithic, and the Early Bronze Age: Proceedings of the International Symposium, October 13th–19th 1997, Urla-Izmir (Turkey)*, edited by Hayat Erkanal, Harald Hauptmann, Vasıf Şahoğlu, and Rıza Tuncel, pp. 461–481. Ankara University Press, Ankara.

2012 Migrant Drinking Assemblages in Aegean Bronze Age Settings. In *Materiality and Social Practice: Transformative Capacities of Intercultural Encounters*, edited by Joseph Maran and Philipp Stockhammer, pp. 73–88. Oxbow, Oxford.

2017 Partying in Prehistory: Social Drinking Behaviors in the Bronze Age Aegean, ca. 2600–1400 BCE. Uppsala University online document.

Sadvari, Joshua W., Christina Tsoraki, Lilian Dogiama, and Christopher J. Knüsel

2015 Reading the Bones, Reading the Stones: An Integrated Approach to Reconstructing Activity Patterns at Neolithic **Çatalhöyük**. In *Assembling Çatalhöyük*, edited by Ian Hodder and Arkadiusz Marciniak, pp. 59–74. Themes in Contemporary Archaeology 1. Maney Publishing, Leeds.

Saeidnia, Soodabeh, and Ahmad Reza Gohari

2012 Importance of *Brassica napus* as a Medicinal Food Plant. *Journal of Medicinal Plants Research* 6(14):2700–2703.

Saijirahu, Buyanchuglagin

2008 Folk Medicine among the Mongols in Inner Mongolia. *Asian Medicine* 4(2):338–356.

Sakelarakis, Yannis

2005 *Γεύση μιᾶς προϊστορικῆς ἐλιᾶς* [A Taste of a Prehistoric Olive]. Ikaros, Athens.

Salavert, Aurélie

2010 Le pavot (*Papaver somniferum*) à la fin du 6e millénaire av. J.-C. en Europe occidentale. *Anthropobotanica* 1(3):3–16.

Salavert, Aurélie, Lucie Martin, Ferran Antolín, and Antoine Zazzo

2018 The Opium Poppy in Europe: Exploring Its Origin and Dispersal during the Neolithic. *Antiquity* 92(364):e1, Project Gallery.

Salavert, Aurélie, Antoine Zazzo, Lucie Martin, Ferran Antolín, Caroline Gauthier, François Thil, Olivier Tombret, Laurent Bouby, Claire Manen, Mario Mineo, Aldona Mueller-Bieniek, Raquel Piqué, Mauro Rottoli, Núria Rovira, Françoise Toulemonde, and Ivana Vostrovská

2020 Direct Dating Reveals the Early History of Opium Poppy in Western Europe. *Scientific Reports* 10:20263.

Samadi, Soheila, Masoud Khaiyami, and Abdollah Hasanzadeh Goorut Tappe

2007 A Comparison of Important Physical and Chemical Characteristics of Six *Lallemantia iberica* (Bieb.) Fisch. and Mey. Varieties. *Pakistan Journal Nutrition* 6(4):387–390.

Sampson, Adamantios

2006 *Προϊστορία του Αιγαίου: Παλαιολιθική, Μεσολιθική, Νεολιθική* [The Prehistory of the Aegean Basin: Palaeolithic, Mesolithic, Neolithic]. Perivolaki-Atrapos, Athens.

Sampson, Adamantios, Janusz K. Kozłowski, and Małgorzata Kaczanowska

2003 Mesolithic Chipped Stone Industries from the Cave of Cyclops on the Island of Youra (Northern Sporades). In *The Greek Mesolithic: Problems and Perspectives*, edited by Nena Galanidou and Catherine Perlès, pp. 123–130. British School at Athens Studies 10. The British School at Athens, London.

Sampson, Adamantios, and Fragkiska Megaloudi

2006 Το σπήλαιο του Σαρακηνού στη Βοιωτία. Συμβολή στην πολιτιστική ακολουθία και στην παλαιοοικονομία της Νεολιθικής στην Κεντρική Ελλάδα [The Sarakenos Cave in Boeotia. Contribution to the Cultural Sequence and the Paleoeconomy of the Neolithic in Central Greece]. In *Αρχαιολογικό Έργο Θεσσαλίας και Στερεάς Ελλάδας. Πρακτικά επιστημονικής συνάντησης, Βόλος, 27/2–2/3/2003* [Proceedings of the 1st Archaeological Work in Thessaly and Central Greece, Volos, 27/2–2/3/2003], edited by Alexandros Mazarakis-Ainian, pp. 585–598. University of Thessaly, Volos.

Samuel, Delwen

1993 Ancient Egyptian Cereal Processing: Beyond the Artistic Record. *Cambridge Archaeological Journal* 3(2):276–283.

1996 Archaeology of Ancient Egyptian Beer. *Journal of the American Society of Brewing Chemists* 54(1):3–12.

2000 Brewing and Baking. In *Ancient Egyptian Materials and Technology*, edited by Paul T. Nicholson and Ian Shaw, pp. 537–576. Cambridge University Press, Cambridge.

2006 Modified Starch. In *Ancient Starch Research*, edited by Robin Torrence and Huw Barton, pp. 205–216. Left Coast Press, Walnut Creek.

Sarkar, Preetam, Lohith Kumar D. H., Chanda Dhumal, Shubham Subrot Panigrahi, and Ruplal Choudhary

2015 Traditional and Ayurvedic Foods of Indian Origin. *Journal of Ethnic Foods* 2(3):97–109.

Sarpaki, Anaya

1987a *The Palaeoethnobotany of the West House Akrotiri, Thera: A Case Study*. PhD dissertation, University of Sheffield.

1987b Παλαιοεθνοβοτανικές παρατηρήσεις από το Σπήλαιο Άγιος Γεώργιος στις Καλυθιές Ρόδου [Paleoethnobotanical Observations from the Cave Agios Georgios in Kalythies, Rhodes]. In *Η Νεολιθική Περίοδος στα Δωδεκάνησα* [The Neolithic Period in the Dodecanese], edited by Adamantios Sampson, pp. 131–155. Publications of the Archaeologiko Deltion 35. Fund of Archaeological Proceeds (TAPA), Athens.

1989 Excavations at Palaikastro, 1988:5. Preliminary Environmental Report. *Annual of the British School at Athens* 84:435–436.

1992a The Palaeoethnobotanical Approach: The Mediterranean Triad or Is It a Quartet? In *Agriculture in Ancient Greece: Proceedings of the 7th International Symposium at the Swedish Institute of Athens, 16–17 May 1990*, edited by Berit Wells, pp. 61–76. Paul Åströms, Stockholm.

1992b A Palaeoethnobotanical Study of the West House, Akrotiri, Thera. *Annual of the British School at Athens* 87:219–230.

1995 Toumba Balomenou, Chaeronia: Plant Remains from the Early and Middle Neolithic Levels. In *Res Archaeobotanicae: International Work Group for*

Palaeoethnobotany. Proceedings of the 9th Symposium, Kiel 1992, edited by Helmut Kroll and Rainer Pasternak, pp. 281–300. Oetker-Voges, Kiel.

1999a The Archaeobotanical Study of Tzambakas House, Rethymno, Crete. In *Minoans and Mycenaeans: Flavours of Their Time. Exhibition Catalogue, National Archaeological Museum of Athens, 12 July–27 November 1999*, edited by Yannis Tzedakis and Holley Martlew, pp. 40–41. Kapon, Athens.

1999b Σύκα [Figs]. In *Minoans and Mycenaeans: Flavours of Their Time. Exhibition Catalogue, National Archaeological Museum of Athens, 12 July–27 November 1999*, edited by Yannis Tzedakis and Holley Martlew, pp. 102. Kapon, Athens.

2000 The Study of Palaeodiet in the Aegean: Food for Thought. In *Palaeodiet in the Aegean. Papers from a Colloquium Held at the 1993 Meeting of the Archaeological Institute of America in Washington, DC*, edited by Sarah J. Vaughan and William D. E. Coulson, pp. 115–121. Wiener Laboratory Monograph 1. Oxbow, Oxford.

2001a Processed Cereals and Pulses from the Late Bronze Age Site of Akrotiri, Thera—Preparations Prior to Consumption: A Preliminary Approach to Their Study. *Annual of the British School at Athens* 96:27–40.

2001b Condiments, Perfume, and Dye Plants in Linear B: A Look at the Textual and Archaeobotanical Evidence. In *Manufacture and Measurement: Counting, Measuring, and Recording Craft Items in Early Aegean Societies*, edited by Anna Michailidou, pp. 195–265. Μελετήματα 33. National Hellenic Research Foundation, Research Center for Greek and Roman Antiquity, Athens.

2007a Archaeobotanical Observations. In *Palaikastro: Two Late Minoan Wells*, edited by Joseph Alexander MacGillivray, Hugh Sackett, and Jan M. Driessen, pp. 211–216. British School at Athens Supplement 43. The British School at Athens, London.

2007b Résultats archéobotaniques preliminaries dans divers secteurs de Malia. *Bulletin de Correspondance Hellénique* 131:882–884.

2009a Review of *Η αρχαιοβοτανική έρευνα της διατροφής στην προϊστορική Ελλάδα* [An Archaeobotanical Investigation of Diet in Prehistoric Greece], by Soultana Maria Valamoti. University Studio Press, Thessaloniki. *Mediterranean Archaeology and Archaeometry* 9(2):135–147.

2009b Τα αρχαιοβοτανικά υπολείμματα [The Archaeobotanical Remains]. In *Ζάκρος: "Της Ουρανιάς το Φρούδι." Το σπήλαιο και οι φυσικές κόγχες* [Zakros: "Tis Ouranias to Froudi." The Cave and the Natural Cavities], edited by Katerina Kopaka, pp. 61–68. [s.n.], Athens.

2009c The Archaeobotanical (Seed) Remains: A Preliminary Report. www.drakainacave.gr/index.php?option=com_content&view=article&id=55&Itemid=56&lang=en.

2011 Archaeobotanical Seed Remains. In *The Cave of the Cyclops: Mesolithic and Neolithic Networks in the Northern Aegean, Greece, Volume II: Bone Tool Industries, Dietary Resources and the Paleoenvironment, and Archeometrical Studies*, edited by Adamantios Sampson, pp. 315–324. Prehistory Monographs 31. INSTAP Academic, Philadelphia.

2013 The Economy of Neolithic Knossos: The Archaeobotanical Data. In *The Neolithic Settlement of Knossos in Crete: New Evidence for the Early Occupation of Crete and the Aegean Islands*, edited by Nikos Efstratiou, Alexandra Karetsou, and Maria Ntinou, pp. 63–94. Prehistory Monographs 42. INSTAP Academic, Philadelphia.

2016 The Archaeobotanical Remains from EM to LM IIIC at the Greek-Swedish

Excavations. In *The Greek-Swedish Excavations at the Agia Aikaterini Square Kastelli, Khania, 1970–1987, 2001, 2005 and 2008: Results of the Excavations under the Direction of Yannis Tzedakis and Carl-Gustaf Styrenius. Vol. V: 1 (text). The Late Minoan IIIA, 1 and II Settlements,* edited by Erik Hallager and Birgitta P. Hallager, pp. 418–446. Eddy.se ab, Stockholm.

2019 The Archaeobotanical Remains. In *Palaikatsro: Building 1,* edited by Joseph A. MacGillivray and Hugh Sackett, pp. 349–370. British School at Athens Supplement 48. The British School at Athens, London.

2021 The Archaeology of Garlic (*Allium sativum*): The Find at Akrotiri, Thera, Greece. Documenta Praehistorica 48:432–445.

Sarpaki, Anaya, and Joanna Bending

2004 Archaeobotanical Assemblages. In *Mochlos IC: Period III. Neopalatial Settlement on the Coast: The Artisans' Quarter and the Farmhouse at Chalinomouri. The Small Finds,* edited by Jeffrey S. Soles and Costis Davaras, pp. 126–137. Prehistory Monographs 9. INSTAP Academic, Philadelphia.

Sarpaki, Anaya, and Glynis Jones

1990 Ancient and Modern Cultivation of *Lathyrus clymenum* L. in the Greek Islands. *Annual of the British School at Athens* 85:363–368.

Sarpaki, Anaya, and Athanasia Kanta

2011 Monastiraki in the Amari Region of Crete: Some Interim Archaeobotanical Insights into Middle Bronze Age Subsistence. In *Πεπραγμένα 1ου Διεθνούς Κρητολογικού Συνεδρίου (Χανιά, 1–8 Οκτωβρίου 2006), Τόμ. A1* [Proceedings of the 10th Cretological Congress (Chania, 1–8 October 2006), Vol. A1], edited by Maria Andreadaki-Vlazaki and Eleni Papadopoulou, pp. 247–270. Literary Society "Chryssostomos," Chania.

Savvopoulou, Thomi

2012 Πύραυνοι αρχαϊκών χρόνων από την Αξιούπολη Κιλκίς [Archaic Pyraunoi from Axioupolis, Kilkis]. In *Archaic Pottery of the Northern Aegean and Its Periphery (700–480 BC): Proceedings of the Archaeological Meeting, Thessaloniki, 19–22 May 2011,* edited by Michalis Tiverios, Vasiliki Misailidou-Despotidou, Eleni Manakidou, and Anna Arvanitaki, pp. 151–162. Publications of the Archaeological Institute of Macedonian and Thracian Studies 11. Ziti, Thessaloniki.

Scarborough, John

1982 Beans, Pythagoras, Taboos, and Ancient Dietetics. *Classical World* 75(6):355–358.

1991 The Pharmacology of Sacred Plants, Herbs, and Roots. In *Magika Hiera: Ancient Greek Magic and Religion,* edited by Christopher A. Faraone and Dirk Obbink, pp. 138–174. Oxford University Press, New York.

Scherer, Andrew K.

2015 *Mortuary Landscapes of the Classic Maya: Rituals of Body and Soul.* University of Texas Press, Austin.

Schlesier, Renate (editor)

2011 A Different God? Dionysos and Ancient Polytheism. De Gruyter, Berlin.

Schönfelder, Ingrid, and Peter Schönfelder

1990 *Collins Photoguide to the Wild Flowers of the Mediterranean.* Collins, London.

Scott, David

1983 *Traditional Arab Cookery.* Rider, London.

Sebald, Sidney V., Anastasia Papathanasiou, and Gisela Grupe
2022 Changing Subsistence Economies in the Course of the Neolithic Transition: Isotopic Sourcing of Collagen Isotopic Ratios in Human Skeletons from Early Neolithic Anatolia and Greece. *Journal of Archaeological Science: Reports* 43:103450.
Séfériadès, Michel Louis
1993 The European Neolithisation Process. *Documenta Praehistorica* 21:137–162.
Seremetakis, Constantina-Nadia (editor)
1994 *The Senses Still: Perception and Memory as Material Culture in Modernity*. Westview Press, Boulder, Colorado.
Serpico, Margaret, and Raymond White
2000 Oil, Fat, and Wax. In *Ancient Egyptian Materials and Technology*, edited by Paul T. Nicholson and Ian Shaw, pp. 390–429. Cambridge University Press, Cambridge.
Sfikas, George
1983 Δέντρα και θάμνοι της Ελλάδας [Trees and Shrubs of Greece]. Efstathiadis, Athens.
1995 Φαρμακευτικά φυτά της Ελλάδας [Medicinal Plants of Greece]. Efstathiadis, Athens.
Shackleton, Judith C.
1988 *Marine Molluscan Remains from Franchthi Cave: Excavations at Franchthi Cave, Greece, Vol. 4*. Indiana University Press, Bloomington.
Shay, Jennifer M., and C. Thomas Shay
1978 Modern Vegetation and Fossil Plant Remains. In *Excavations at Nichoria in Southwest Greece, Vol. I: Site, Environs, and Techniques*, edited by George Rapp Jr. and Stanley E. Aschenbrenner, pp. 41–59. University of Minnesota Press, Minneapolis.
Shay, C. Thomas, Jennifer M. Shay, Katherine A. Frego, and Janusz Zwiazek
1995 The Modern Flora and Plant Remains from Bronze Age Deposits at Kommos. In *Kommos: An Excavation on the South Coast of Crete, Vol. I: The Kommos Region and Houses of the Minoan Town. Part I: The Kommos Region, Ecology, and Minoan Industries*, edited by Joseph W. Shaw and Maria C. Shaw, pp. 91–162. Princeton University Press, Princeton.
Shay, C. Thomas, Jennifer M. Shay, and Margaret R. M. Kapinga
1998 The Bronze Age Plant and Insect Remains and Modern Vegetation. In *Excavations on the Acropolis of Midea: Results of the Greek-Swedish Excavations under the Direction of Katie Demakopoulou and Paul Åström, Vol. I:1 (text). The Excavations on the Lower Terraces, 1985–1991*, edited by Gisela Walberg, pp. 295–357. Paul Åströms, Stockholm.
Sheen, Barbara
2012 *Foods of Poland*. KidHaven, Detroit.
Shelmerdine, Cynthia W.
1984 The Perfumed Oil Industry at Pylos. In *Pylos Comes Alive: Industry and Administration in a Mycenaean Palace. Papers of a symposium*, edited by Cynthia W. Shelmerdine and Thomas G. Palaima, pp. 81–95. Fordham University Press, New York.
1985 *The Perfume Industry of Mycenaean Pylos*. Studies in Mediterranean Archaeology Pocket Book 34. Paul Åströms, Göteborg.

1998 The Perfumed Oil Industry. In *Sandy Pylos: An Archaeological History from Nestor to Navarino,* edited by Jack L. Davis, pp. 101–109. University of Texas Press, Austin.

Shepard, Glenn H., Jr.

2004 A Sensory Ecology of Medicinal Plant Therapy in Two Amazonian Societies. *American Anthropologist* 106(2):252–266.

Sherratt, Andrew

1991a Palaeoethnobotany: From Crops to Cuisine. In *Paleoecologia e Arqueologia II. Trabalhos dedicados a A. R. Pinto da Silva,* edited by Francisco M. V. R. Queiroga and António P. Dinis, pp. 221–236. Centro de Estudos Arqueologicos Famalicenses, Vila Nova de Famalicão, Portugal.

1991b Sacred and Profane Substances: The Ritual of Narcotics in Later Neolithic Europe. In *Sacred and Profane: Proceedings of a Conference on Archaeology, Ritual, and Religion,* edited by Paul Garwood, David Jennings, Robin Skeates, and Judith Toms, pp. 50–64. Oxford University Committee for Archaeology Monographs 32. Oxbow, Oxford.

1995 Alcohol and Its Alternatives: Symbol and Substance in Pre-Industrial Cultures. In *Consuming Habits: Drugs in History and Anthropology,* edited by Jordan Goodman, Paul E. Lovejoy, and Andrew Sherratt, pp. 11–46. Routledge, London.

Sherratt, Andrew, and Susan Sherratt

1991 From Luxuries to Commodities: The Nature of Mediterranean Bronze Age Trading Systems. In *Bronze Age Trade in the Mediterranean. Papers Presented at the Conference Held at Rewley House, Oxford, in December 1989,* edited by Noël H. Gale, pp. 351–386. Studies in Mediterranean Archaeology 90. Paul Åströms, Jonsered.

Shoemaker, Anna C., Matthew I. J. Davies, and Henrietta L. Moore

2017 Back to the Grindstone? The Archaeological Potential of Grinding-Stone Studies in Africa with Reference to Contemporary Grinding Practices in Marakwet, Northwest Kenya. *African Archaeological Review* 34(3):415–435.

Shugart, Helene A.

2008 Sumptuous Texts: Consuming "Otherness" in the Food Film Genre. *Critical Studies in Media Communication* 25(1):68–90.

Sicard, Delphine, and Jean-Luc Legras

2011 Bread, Beer, and Wine: Yeast Domestication in the *Saccharomyces* sensu stricto Complex. *Comptes Rendus Biologies* 334(3):229–236.

Simoons, Frederick J.

1998 *Plants of Life, Plants of Death.* University of Wisconsin Press, Madison.

Simpson, St. John

2005 The Royal Table. In *Forgotten Empire: The World of Ancient Persia,* edited by John Curtis and Nigel Tallis, pp. 104–131. University of California Press, Berkeley.

Singh, Umaid

1995 Methods for Dehulling of Pulses: A Critical Appraisal. *Journal of Food Science and Technology* 32(2):81–93.

Singh Umaid, S. Sehgal, and Y. S. Tomer

2000 Influence of Dehulling, Soaking Solution, and Enzyme Treatment on the Cooking Quality of Improved Varieties of Pulses. *Journal of Food Science and Technology* 37(6):627–630.

Singleton, Vernon L.
1995 An Enologist's Commentary on Ancient Wines. In *The Origins and Ancient History of Wine*, edited by Patrick E. McGovern, Stuart J. Fleming, and Solomon H. Katz, pp. 67–77. Food and Nutrition in History and Anthropology 11. Gordon and Breach, Philadelphia.

Skibo, James M.
2013 *Understanding Pottery Function*. Manuals in Archaeological Method, Theory, and Technique. Springer, New York.

Skouteri-Didaskalou, Eleonora, and Evi Voutsina
2005 The Ultimate Simplicity of Greek Cuisine: A Well-Tempered Culinary Experience. In *Culinary Cultures of Europe: Identity, Diversity, and Dialogue*, edited by Darra Goldstein and Kathrin Merkle, pp. 195–208. Council of Europe, Strasbourg.

Smith, B. Holly
1984 Patterns of Molar Wear in Hunter-Gatherers and Agriculturalists. *American Journal of Physical Anthropology* 63(1):39–56.

Smith, Thyrza R.
1987 *Mycenaean Trade and Interaction in the West Central Mediterranean, 1600–1000 B.C.* British Archaeological Reports International Series 371. BAR, Oxford.

Smith, Bruce D.
1995 *The Emergence of Agriculture*. Scientific American Library Series 54. W. H. Freeman, New York.

Smith, Helen, and Glynis Jones
1991 Experiments on the Effects of Charring on Cultivated Grape Seeds. *Journal of Archaeological Science* 17(3):317–327.

Smith, Holly B.
1984 Patterns of Molar Wear in Hunter-Gatherers and Agriculturalists. *American Journal of Physical Anthropology* 63(1): 39–56.

Smith, Rachel K., Rebecca J. Stacey, Ed Bergström, and Jane Thomas-Oates
2018 Detection of Opium Alkaloids in a Cypriot Base-Ring Juglet. *Analyst* 143(21) 5127–5136.

Smith, Thyrza R.
1987 *Mycenaean Trade and Interaction in the West Central Mediterranean 1600–1000 B.C.* British Archaeological Reports International Series 371. B.A.R., Oxford.

Soares, Justin
2019 *The Extent of Eastern Influence: The Case for the Pre-Colonization of Iberia*. Master's thesis, Brandeis University, Waltham, Massachusetts.

Souvatzi, Stella, Agathe Reingruber, and Giorgos Toufexis
2021 Socialising the Landscape in the Early Neolithic of Thessaly, Greece. *Open Archaeology* 7(1):1176–1191.

Spagnoli, Federica
2016 Two Mycenaean Stirrup Jars from the Levant. *Mediterranean Archaeology and Archaeometry* 16(1):185–192.

Spengler, Robert N., III, Ilaria De Nigris, Barbara Cerasetti, Marialetizia Carra, and Lynne M. Rouse
2018 The Breadth of Dietary Economy in the Central Asian Bronze Age: Case Study from Adji Kui in the Murghab Region of Turkmenistan. *Journal of Archaeological Science: Reports* 22:372–381.

Staal, Frits
2001 How a Psychoactive Substance Becomes a Ritual: The Case of Soma. *Social Research* 68(3):745–778.

Stahl, Ann B.
1989 Plant-Food Processing: Implications for Dietary Quality. In *Foraging and Farming: The Evolution of Plant Exploitation,* edited by David R. Harris and Gordon C. Hillman, pp. 171–196. Unwin Hyman, London.

Stanley, Autumn
1981 Daughters of Isis, Daughters of Demeter: When Women Sowed and Reaped. *Women's Studies International Quarterly* 4(3): 289–304.

Stara, Kalliopi, Rigas Tsiakiris, Vasilis Nitsiakos, and John M. Halley
2016 Religion and the Management of the Commons: The Sacred Forests of Epirus. In *Biocultural Diversity in Europe,* edited by Mauro Agnoletti and Francesca Emanueli, pp. 283–302. Environmental History 5. Springer, Cham, Switzerland.

Stefani, Liana, Soultana Maria Valamoti, and Afroditi Konstandinidou
2006 Το ερευνητικό πρόγραμμα του προϊστορικού οικισμού Αγγελοχωρίου: Οι γεωργοκτηνοτροφικές δραστηριότητες και οι διατροφικές συνήθειες στο Αγγελοχώρι της ΥΕΧ [The Research Programme in the Prehistoric Settlement at Angelochori: Farming, Stockbreeding, and Diet at Angelochori in the Late Bronze Age]. *Proceedings of the Archaeological Work in Macedonia and Thrace (AEMTH)* 18, 2004:431–438.

Stern, Ben, Carl Heron, Lindsey Corr, Margaret Serpico, and Janine Bourriau
2003 Compositional Variations in Aged and Heated Pistacia Resin Found in Late Bronze Age Canaanite Amphorae and Bowls from Amarna, Egypt. *Archaeometry* 45(3):457–469.

Stika, Hans-Peter
2011 Early Iron Age and Late Medieval Malt Finds from Germany: Attempts at Reconstruction of Early Celtic Brewing and the Taste of Celtic Beer. *Archaeological and Anthropological Sciences* 3(1):41–48.

Stika, Hans-Peter, Andreas G. Heiss, and Hanni Rützler
2013 Nearly Forgotten Cereals Re-Activated for Today's Slow Food Production in South-Western Germany and Eastern Austria. In *Book of Abstracts of 16th Conference of the International Work Group for Palaeoethnobotany, Thessaloniki 17–21 June 2013,* edited by Soultana Maria Valamoti, pp. 219–220. Aristotle University of Thessaloniki. Stoddart, Simon, and Caroline Malone. 2001. Editorial. *Antiquity* 75(289):459–482.

Stocker, Sharon R., and Jack L. Davis
2004 Animal Sacrifice, Archives, and Feasting at the Palace of Nestor. *Hesperia* 73(2):179–195

Stockman, Ralph
1917 Lathyrism. *Edinburgh Medical Journal* 19(5): 277–296.

Stoddart, Simon, and Caroline Malone
2001 Editorial. *Antiquity* 75(289): 459–482.

Strašil, Zdeněk, and Martin Káš
2005 The Effect of Nitrogen Fertilisation, Sowing Rates, and Weather Conditions on Yields and Yield Components of *Lallemantia iberica* (L.) Fisch. et Mey. *Scientia Agriculturae Bohemica* 36(1):15–20.

Strasser, Thomas F., Eleni Panagopoulou, Curtis N. Runnels, Priscilla M. Murray, Nicholas Thompson, Panayiotis Karkanas, Floyd W McCoy, and Karl W. Wegmann

2010 Stone Age Seafaring in the Mediterranean: Evidence from the Plakias Region for Lower Palaeolithic and Mesolithic Habitation of Crete. *Hesperia* 79(2):145–190.

Stratouli, Georgia, Niels H. Andreasen, Evita Kalogiropoulou, Nikos Katsikaridis, Dimitris Kloukinas, Georgia Koromila, Evi Margaritis, and Tasos Bekiaris

2015 "Σπίτια και αυλές" στον νεολιθικό οικισμό Αυγής Καστοριάς: Το Κτήριο 5 και οι γειτονικοί ανοιχτοί χώροι ["Houses and Yards" at the Neolithic Settlement of Avgi, Kastoria: Building 5 and the Adjacent Open Areas]. *Proceedings of the Archaeological Work in Macedonia and Thrace (AEMTH)* 25, 2011:7–18.

Stroulia, Anna

2010 *Flexible Stones: Ground Stone Tools from Franchthi Cave*. Excavations at Franchthi Cave, Greece, Vol. 14. Indiana University Press, Bloomington.

2018 Macrolithics: Ordinary Things in an Extraordinary Place. In *Neolithic Alepotrypa Cave in the Mani, Greece. In Honor of George Papathanassopoulos,* edited by Anastasia Papathanasiou, William A. Parkinson, Daniel J. Pullen, Michael L. Galaty, and Panagiotis Karkanas, pp. 200–241. Oxbow, Oxford.

Stroulia, Anna, and Danai Chondrou

2013 Destroying the Means of Production: The Case of Ground Stone Tools from Kremasti-Kilada, Greece. In *Destruction: Archaeological, Philological, and Historical Perspectives,* edited by Jan Driessen, pp. 109–131. Presses Universitaires de Louvain.

Stummer, Albert

1911 Zur Urgeschichte der Rebe und des Weinbans. *Mitteilungen der Anthropologischen Gesellschaft in Wien* 41:283–296.

Stylianakou, Charoula

2013 *Αρχαιοβοτανικά δεδομένα από το Νεολιθικό οικισμό του Κλείτου Κοζάνης* [Archaeobotanical Data from the Neolithic Settlement of Kleitos in Kozani]. Master's thesis, Aristotle University of Thessaloniki.

Stylianakou, Charoula, and Soultana Maria Valamoti

2013 Plant Remains from the Final Neolithic and Early Bronze Age Settlement of Vassilara Rahi Velvento, Northern Greece. Poster presented at the 16th Conference of the International Work Group for Palaeoethnobotany, 17–22 June 2013, Thessaloniki. Program and Abstracts:39, 225.

Svenaeus, Fredrik

2001 *The Hermeneutics of Medicine and the Phenomenology of Health: Steps Towards a Philosophy of Medical Practice*. Kluwer Academic, Dordrecht.

Tanno, Ken-ichi, and George Willcox

2006a How Fast Was Wild Wheat Domesticated? *Science* 311 (5769):1886.

2006b The Origins of Cultivation of *Cicer arietinum* L. and *Vicia faba* L.: Early Finds from Tell el-Kerkh, North-West Syria, Late 10th Millennium B.P. *Vegetation History and Archaeobotany* 15(3):197–204.

Tartaron, Thomas F.

2013 *Maritime Networks in the Mycenaean World*. Cambridge University Press, Cambridge.

Teall, Emily K.
2014 Medicine and Doctoring in Ancient Mesopotamia. *Grand Valley Journal of History* 3(1):Article 2.
Tekle-Haimanot, Redda, Berhanu M. Abegaz, Elizabeth Wuhib, Angelina Kassina, Yemane Kidane, Naod Kebede, Tadesse Alemu, and Peter S. Spencer
1993 Pattern of *Lathyrus sativus* (Grass Pea) Consumption and beta-N-Oxalyl-α-β-Diaminoproprionic Acid (β-ODAP) Content of Food Samples in the Lathyrism Endemic Region of Northwest Ethiopia. *Nutrition Research* 13(10):1113–1126.
Tekle-Haimanot, Redda, Yemane Kidane, Elizabeth Wuhib, Angelina Kalissa, Tadesse Alemu, Zein Ahmed Zein, and Peter S. Spencer
1990 Lathyrism in Rural Northwestern Ethiopia: A Highly Prevalent Neurotoxic Disorder. *International Journal of Epidemiology* 19(3):664–672.
Telioridou, Eleni
2014 *Αρχαιοβοτανική έρευνα της φάσης 3 στο Αρχοντικό Γιαννιτσών* [Archaeobotanical Study from Phase III of Archondiko Giannitswn]. Master's thesis, Aristotle University of Thessaloniki.
Terral, Jean Fédérick, Elidie Tabar, Laurent Bouby, Sarah Ivorra, Thierry Pastor, Isabel Figueiral, Sandrine Picq, Jean-Baptiste Chevance, Cécile Jung, Laurent Fabre, Christophe Tardy, Michel Compan, Roberto Bacilieri, Thierry Lacombe, and Patrice This
2010 Evolution and History of Grapevine (*Vitis vinifera*) under Domestication: New Morphometric Perspectives to Understand Seed Domestication Syndrome and Reveal Origins of Ancient European Cultivars. *Annals of Botany* 105(3):443–455.
Tharanathan, Rudrapatnam Narayanaswamy, and Siddaiah Mahadevamma
2003 Grain Legumes—A Boon to Human Nutrition. *Trends in Food Science and Technology* 14(12):507–518.
Theocharis, Dimitrios
1967 *Η Αυγή της Θεσσαλικής Προϊστορίας* [The Dawn of Thessalian Prehistory]. Philarchaeos Society of Volos.
1973 *Νεολιθική Ελλάς* [Neolithic Greece]. National Bank of Greece, Athens.
Theodosaki, Pelagia-Maria
2021 *Αρχαιοβοτανικές έρευνες στη Μακεδονία κατά την Εποχή του Χαλκού: προσεγγίζοντας τη σχέση ανθρώπου και βλάστησης* [Archaeobotanical Research in Macedonia during the Bronze Age: Approaching the Relationship between Humans and Vegetation]. Master's thesis, Aristotle University of Thessaloniki.
Thomas, Julian
2003 Thoughts on the "Repacked" Neolithic Revolution. *Antiquity* 77(295):67–74.
Thorpe, I. J.
2003 *The Origins of Agriculture in Europe*. Routledge, London.
Tiverios, Michael, Eleni Manakidou, Despoina Tsiafaki, Soultana Maria Valamoti, Tatiana Theodoropoulou, and Eugenia Gatzogia
2013 Cooking in an Iron Age Pit in Northern Greece: An Interdisciplinary Approach. In *Diet, Economy, and Society in the Ancient Greek World: Towards a Better Integration of Archaeology and Science. Proceedings of the International Conference Held at the Netherlands Institute at Athens on 22–24 March 2010*, edited by Sofia

Voutsaki and Soultana Maria Valamoti, pp. 205–214. Pharos Supplement 1. Peeters, Leuven.

Tiwari, Brijesh K., and Narpinder Singh
2012 *Pulse Chemistry and Technology*. RSC Publishing, Cambridge.

Tomkins, Peter
2007 Communality and Competition: The Social Life of Food and Containers at Aceramic and Early Neolithic Knossos, Crete. In *Cooking Up the Past: Food and Culinary Practices in the Neolithic and Bronze Age Aegean*, edited by Christopher Mee and Josette Renard, pp. 174–199. Oxbow, Oxford.

Toulemonde, Françoise, Frédérique Durand, Laura Berrio, Emmanuelle Bonnaire, Geneviève Daoulas, and Julian Wiethold
2015 Records of "New" Glume Wheat in France: A Review. *Vegetation History and Archaeobotany* 24(1):197–206.

Tourloukis, Evangelos, and Ourania Palli
2009 The First Mesolithic Site of Thesprotia. In *Thesprotia Expedition I: Towards a Regional History*, edited by Björn Forsén, pp. 25–38. Papers and Monographs of the Finnish Institute at Athens XV. Foundation of the Finnish Institute at Athens, Helsinki.

Tournavitou, Iphigenia
1999 Hearths in Non-palatial Settlement Contexts. The LBA Period in the Peloponnese. In *MELETEMATA. Studies in Aegean Archaeology Presented to Malcolm H. Wiener as he Enters his 65th Year*, edited by Philip P. Betancourt, Vassos Karageorghis, Robert Laffineur and Wolf-Dietrich Niemeier, pp. 833–840. Aegaeum 20. Université de Liège, Liège/University of Texas at Austin. 840. Aegaeum 20. Université de Liège, Liège/University of Texas at Austin.

Townsend, J. B.
1997 Shamanism. In *Anthropology of Religion: A Handbook*, edited by Stephen D. Glazier, pp. 429–470. Praeger, Westport, Connecticut.

Trantalidou, Katerina, Georgios Lazaridis, Konstantinos Trimmis, Katarina Gerometta, Yiannis Maniatis, Vassiliki Milidaki, Antigone Papadea, Haricleia-Aida Zikidi, Georgia Kotzamani, Katerina Papayianni, and Filippos Stephanou
2019 Consumed by the Darkness: The Archaeological Assemblages Uncovered during the 2011 Excavation Season at the Kataphygadi Cave, on Kythera. *Aegean Archaeology* 12:65–100.

Triantaphyllou, Sevi
2001 A Bioarchaeological Approach to Prehistoric Cemetery Populations from Western and Central Greek Macedonia. British Archaeological Reports International Series 976. Archaeopress, Oxford.

Tsartsidou, Georgia, and Kostas Kotsakis
2020 Grinding in a Hollow? Phytolith Evidence for Pounding Cereals in Bedrock Mortars at Paliambela Kolindros, an Early Neolithic Site in Macedonia, North Greece. *Archaeological and Anthropological Sciences* 12(8):173.

Tsikritzi-Momtsiou, Matina, and Fani Ftaka-Tsikritzi
2006 *Γεύσεις από παλιά Κοζάνη, Τομ. 2* [Tastes from Old Kozani, Vol. 2]. Institute of Book and Reading, Kozani.

Tsirtsoni, Zoï, Pascal Darcque, Chaido Koukouli-Chryssanthaki, Dimitra Malamidou, and René Treuil
2018 The Chronological and Social Dimensions of the Late Neolithic I–II and the

Late Neolithic–Early Bronze Age Transitions in a Long-Lived Settlement in Northern Greece (Dikili Tash, Kavala District). In *Communities in Transition: The Circum-Aegean Area during the 5th and 4th Millennia BC*, edited by Søren Dietz, Fanis Mavridis, Žarko Tankosić and Turan Takaoğlu, pp. 197–210. Oxbow, Oxford.

Tsountas, Christos
1908 *Αι Προϊστορικαί Ακροπόλεις Διμηνίου και Σέσκλου* [The Prehistoric Akropoleis of Dimini and Sesklo]. Monographs 14. The Library of the Archaeological Society at Athens.

Turner C., and Maria Fernanda Sánchez-Goñi
1997 Late Glacial Landscape and Vegetation in Epirus. In *Klithi: Palaeolithic Settlement and Quaternary Landscapes in Northwest Greece, Vol. 2: Klithi in Its Local and Regional Setting*, edited by Geoff Bailey, pp. 559–585. McDonald Institute for Archaeological Research, Cambridge.

Turner, Judith, and James R. A. Greig
1975 Some Holocene Pollen Diagrams from Greece. *Review of Palaeobotany and Palynology* 20(3):171–204.

Tzahili, Iris
1997 *Υφαντική και υφάντρες στο Προϊστορικό Αιγαίο, 2000–1000 π.Χ.* [Weaving and Weavers in the Prehistoric Aegean Sea, 2000–1000 BC]. University of Crete Press, Herakleion.

Tzedakis, Yannis, and Holley Martlew (editors)
1999 *Minoans and Mycenaeans: Flavours of Their Time*. Exhibition Catalog, National Archaeological Museum of Athens, 12 July–27 November 1999. Kapon, Athens.

Ucchesu, Mariano, Martino Orrù, Oscar Grillo, Gianfranco Venora, Alessandro Usai, Pietro Francesco Serreli, and Gianluigi Bacchetta
2015 Earliest Evidence of a Primitive Cultivar of *Vitis vinifera* L. during the Bronze Age in Sardinia (Italy). *Vegetation History and Archaeobotany* 24(5):587–600.

Ulaş, Burhan, and Girolamo Fiorentino
2020 Recent Attestations of "New" Glume Wheat in Turkey: A Reassessment of Its Role in the Reconstruction of Neolithic Agriculture. *Vegetation History and Archaeobotany* 30(5):685–701.

Ungar, Peter S., and Matt Sponheimer
2011 The Diets of Early Hominins. *Science* 334(6053):190–193.

Urem-Kotsou, Dushka, Anastasia Dimoula, Gazmend Elezi, Trisevgeni Papadakou, Anna Papaioannou, Niki Saridaki, Ioanna Siamidou, Teresa Silva, Eirini Tzemopoulou, and Kostas Kotsakis
2018 Patterns in Contemporaneous Ceramic Traditions: Interregional Relations between Thessaly and Macedonia during the Early and Middle Neolithic. In *Communities, Landscapes, and Interaction in Neolithic Greece: Proceedings of the International Conference, Rethymno 29–30 May 2015*, edited by Apostolos Sarris, Evita Kalogiropoulou, Tuna Kalayci, and Lia Karimali, pp. 324–338. International Monographs in Prehistory: Archaeological Series 20. Berghahn, New York.

Urem-Kotsou, Dushka, and Kostas Kotsakis
2007 Pottery Cuisine and Community in the Neolithic of North Greece. In *Cooking up the Past: Food and Culinary Practices in the Neolithic and Bronze Age Aegean*, edited by Christopher Mee and Josette Renard, pp. 225–246. Oxbow, Oxford.

Urem-Kotsou, Dushka, Anna Papaioannou, Trisevgeni Papadakou, Niki Saridaki, and Zoe Intze

2014 Pottery and Stylistic Boundaries. Early and Middle Neolithic Pottery in Macedonia. In *1912–2012: Εκατό χρόνια έρευνας στην προϊστορική Μακεδονία, Πρακτικά διεθνούς συνεδρίου, Αρχαιολογικό Μουσείο Θεσσαλονίκης, 22–24 Νοεμβρίου 2012* [1912–2012: A Century of Research in Prehistoric Macedonia, International Conference Proceedings, Archaeological Museum of Thessaloniki, 22–24 November 2012], edited by Evaggelia Stefani, Nikos Merousis, and Anastasia Dimoula, pp. 505–517. Archaeological Museum of Thessaloniki.

Usai, Alessandro

2017 Sardinia and the Aegean World in the Bronze Age: Advances in Understanding. In *Έσπερος/Hesperos: The Aegean Seen from the West. Proceedings of the 16th International Aegean Conference, University of Ioannina, Department of History and Archaeology, Unit of Archaeology and Art History, 18–21 May 2016*, edited by Michael Fotiadis, Robert Laffineur, Yannos Lolos, and Andreas Vlachopoulos, pp. 73–80. Aegaeum 41. Peeters, Leuven/Liège.

Ustinova, Yulia

2017 *Divine Mania: Alteration of Consciousness in Ancient Greece*. Routledge, New York.

Vaiglova, Petra, Amy Bogaard, Matthew Collins, William Cavanagh, Christopher Mee, Josette Renard, Angela Lamb, Armelle Gardeisen, and Rebecca Fraser

2014 An Integrated Stable Isotope Study of Plants and Animals from Kouphovouno, Southern Greece: A New Look at Neolithic Farming. *Journal of Archaeological Science* 42:201–215.

Valamoti, Soultana Maria

1992 Απανθρακωμένα φυτικά κατάλοιπα από την προϊστορική Θέρμη: Μία προκαταρκτική μελέτη [Charred Plant Remains from Prehistoric Thermi: A Preliminary Examination]. *Makedonika* 18:443–455.

1995 Γεωργικά προϊόντα από το νεολιθικό οικισμό Γιαννιτσά Β: Μία προκαταρκτική προσέγγιση μέσω των αρχαιοβοτανικών δεδομένων [Agricultural Products from the Neolithic Settlement Giannitsa B: A Preliminary Approach through the Archaeobotanical Data]. *Proceedings of the Archaeological Work in Macedonia and Thrace (AEMTH)* 6, 1992:177–184.

1997a Αρχαιοβοτανικά κατάλοιπα από τον οικισμό του Αρχοντικού: Ανασκαφική περίοδος Σεπτεμβρίου 1993 [Archaeobotanical Remains from Archondiko: Excavation Season September 1993]. *Proceedings of the Archaeological Work in Macedonia and Thrace (AEMTH)* 7, 1993:155–158.

1997b Προϊστορική βοτανοθεραπεία: Οι πιθανές χρήσεις των φαρμακευτικών και αρωματικών φυτών στους προϊστορικούς χρόνους [Prehistoric Herbal Healing: The Potential Uses of Medicinal and Aromatic Plants in Prehistoric Times]. *Kathimerini Newspaper, Issue "Epta Imeres,"* 29 June:3–4. Athens.

1998 Η εκμετάλλευση της αμπέλου στο χώρο της Μακεδονίας και της Θράκης κατά τη Νεολιθική και την Πρώιμη Εποχή του Χαλκού [The Exploitation of the Grapevine in Macedonia and Thrace during the Neolithic and the Early Bronze Age]. In *Αμπελοοινική ιστορία στο χώρο της Μακεδονίας και της Θράκης* [The Viticultural History of Macedonia and Thrace]. *Proceedings of the Three-Day Working Meeting on the Viticultural History of Macedonia and Thrace, Naoussa, 17–19*

September 1993, pp. 137–149. ETVA Cultural and Technological Foundation, Athens.

1999 Αρχοντικό 1997: Προκαταρκτικά δεδομένα από την αρχαιοβοτανική έρευνα στον οικισμό, Παράρτημα [Archondiko 1997: Preliminary Evidence from the Archaeobotanical Research at the Settlement, Appendix]. *Proceedings of the Archaeological Work in Macedonia and Thrace (AEMTH)* 11, 1997:172–175.

2001a *An Archaeobotanical Investigation of Late Neolithic and Early Bronze Age Agriculture and Plant Exploitation in Northern Greece*. PhD dissertation, University of Sheffield.

2001b Φαρμακευτικά και αρωματικά φυτά στην προϊστορική βόρεια Ελλάδα: Τα αρχαιοβοτανικά δεδομένα [Medicinal and Aromatic Plants in Prehistoric Northern Greece: The Archaeobotanical Evidence]. In *Φαρμακευτικά και Αρωματικά Φυτά* [Pharmaceutical and Aromatic Plants]. *Proceedings of the 7th Three-Day Working Meeting on Pharmaceutical and Aromatic Plants—Traditional Uses and Their Potential for Utilisation, Paralimni, Cyprus, 21–25 March 1997*, pp. 23–38. ETVA Cultural and Technological Foundation, Athens.

2002a Food Remains from Bronze Age Archondiko and Mesimeriani Toumba in Northern Greece? *Vegetation History and Archaeobotany* 11(1–2):17–22.

2002b Απανθρακωμένα αρχαιοβοτανικά κατάλοιπα από τη Μεσημεριανή Τούμπα: Προκαταρκτικές παρατηρήσεις, παράρτημα Γ [Charred Plant Remains from Mesimeriani Toumba: Preliminary Observations, Appendix C]. In *Ανασκαφή στον Προϊστορικό Οικισμό "Μεσημεριανή Τούμπα" Τριλόφου, Ν. Θεσσαλονίκης (Ανασκαφικές Περίοδοι 1992, 1994–1996, 2000, 2001)* [Excavations in the Prehistoric Settlement of "Mesimeriani Tumulus" at Trilofos, Thessaloniki (Excavation Seasons of 1992, 1994–1996, 2000, 2001)], edited by Dimitrios Grammenos and Stavros Kotsos, pp. 315–319. Publications of Archaeological Institute of Northern Greece 1. Archaeological Institute of Northern Greece, Thessaloniki.

2003 Neolithic and Early Bronze Age "Food" from Northern Greece: The Archaeobotanical Evidence. In *Food, Culture, and Identity in the Neolithic and Early Bronze Age*, edited by Michael Parker-Pearson, pp. 97–112. British Archaeological Reports International Series 1117. Archaeopress, Oxford.

2004 *Plants and People in Late Neolithic and Early Bronze Age Northern Greece: An Archaeobotanical Investigation*. British Archaeological Reports International Series 1258. Archaeopress, Oxford.

2006a Η συμβολή των αρχαιοβοτανικών καταλοίπων στη διερεύνηση των εκτεταμένων οικισμών: Τα δεδομένα από το νεολιθικό οικισμό Άψαλος-Γραμμή [The Contribution of Archaeobotanical Evidence to the Investigation of Flat-Extended Settlements: The Archaeobotanical Assemblage from Neolithic Apsalos-Grammi]. *EGNATIA (Annals of the Department of History and Archaeology, Aristotle University of Thessaloniki)* 10:9–130.

2006b *Τρόφιμα φυτικής προέλευσης στην προϊστορική βόρεια Ελλάδα* [Plant Foods in Prehistoric Northern Greece]. Research Committee AUTh, Thessaloniki.

2007a Traditional Food and Culinary Novelties in Neolithic and Bronze Age Northern Greece: An Overview of the Archaeobotanical Evidence. In *Cooking Up the Past: Food and Culinary Practices in the Neolithic and Bronze Age Aegean*, edited by Christopher Mee and Josette Renard, pp. 89–107. Oxbow, Oxford.

2007b Food across Borders: A Consideration of the Neolithic and Bronze Age Archaeobotanical Evidence from Northern Greece. In *Between the Aegean and Baltic Seas: Prehistory across Borders. Proceedings of the International Conference Bronze and Early Iron Age Interconnections and Contemporary Developments between the Aegean and the Regions of the Balkan Peninsula, Central and Northern Europe, University of Zagreb, 11–14 April 2005*, edited by Ioanna Galanaki, Helena Tomas, Yannis Galanakis, and Robert Laffineur, pp. 281–293. Aegaeum 27. Université de Liège/University of Texas at Austin.

2007c Agriculture and Use of Space at Promachon/Topolnitsa: Preliminary Observations on the Archaeobotanical Material. In *The Struma/Strymon River Valley in Prehistory: Proceedings of the International Symposium Strymon Praehistorikus, Kjustendil, Blagoevgrad (Bulgaria) and Serres—Amphipolis (Greece), 27/9–1/10/2004*, edited by Henrieta Todorova, Mark Stefanovich, and Georgi Ivanov, pp. 523–529. In the Steps of James Harvey Gaul, Vol. 2. Gerda Henkel Stiftung, Sofia.

2009 *Η αρχαιοβοτανική έρευνα της διατροφής στην προϊστορική Ελλάδα* [An Archaeobotanical Investigation of Diet in Prehistoric Greece]. University Studio Press, Thessaloniki.

2010a Η ανθρώπινη δραστηριότητα στο Αγγελοχώρι μέσα από τα φυτά και τις χρήσεις τους: Η συμβολή των αρχαιοβοτανικών δεδομένων [Human Activity at Angelochori through Plants and their Uses: The Contribution of Archaeobotanical Remains]. In *Angelochori Emathias: A Late Bronze Age Settlement, Vol. 1*, edited by Evangelia Stefani, pp. 171–197, 269–288, 327–330. Kyriakidis, Thessaloniki.

2010b Γεύσεις και τοπία μέσα από τα προϊστορικά φυτά της Ελλάδας [Tastes and Landscapes through the Plants of Prehistoric Greece]. In *Κιβωτός: Παλαιοί Σπόροι για Νέες Καλλιέργειες* [The Arc: Old Seeds for New Cultures], *12th Biennale of Architecture in Venice, 29 August–21 November 2010, Greek participation*, edited by Phoebe Giannisi and Zisis Kotionis, pp. 276–281. University of Thessaly Press, Volos.

2011a Σπόροι για τους νεκρούς? Αρχαιοβοτανικά δεδομένα από τη Μαυροπηγή Κοζάνης, θέση Φυλλοτσαΐρι [Grain for the Dead? Archaeobotanical Remains from Mavropigi Near Kozani, Site Fillotsairi]. *AEAM (Proceedings of the Archaeological Work in Upper Macedonia)* 1, 2009:245–256.

2011b Ground Cereal Food Preparations from Greece: The Prehistory and Modern Survival of Traditional Mediterranean "Fast Foods." *Archaeological and Anthropological Sciences* 3(1):19–39.

2011c Flax in Neolithic and Bronze Age Greece: Archaeobotanical Evidence. *Vegetation History and Archaeobotany* 20(6):549–560.

2013a Towards a Distinction between Digested and Undigested Glume Bases in the Archaeobotanical Record from Neolithic Northern Greece: A Preliminary Experimental Investigation. *Environmental Archaeology* 18(1):31–42

2013b Healing with Plants in Prehistoric Northern Greece: A Contribution from Archaeobotany. In *Von Sylt bis Kastanas: Festschrift für Helmut Johannes Kroll zum 65. Geburtstag*, edited by Claus von Carnap-Bornheim, Walter Dörfler, Wiebke Kirleis, Johannes Müller, and Ulrich Müller, pp. 479–494. *Offa* 69/70 (2012/2013). Wachholtz, Neumünster.

2013c Greek Archaeobotany: Modern Culinary Traditions and Myths. In *Book of*

Abstracts of 16th Conference of the International Work Group for Palaeoethnobotany, Thessaloniki 17–21 June 2013, edited by Soultana Maria Valamoti, pp. 221. Aristotle University of Thessaloniki.

2015a Harvesting the "Wild"? Exploring the Context of Fruit and Nut Exploitation at Neolithic Dikili Tash, with Special Reference to Wine. *Vegetation History and Archaeobotany* 24(1):35–46.

2015b Τα αρχαιοβοτανικά δεδομένα από το νεολιθικό σπίτι της Σωσάνδρας [The Archaeobotnical Data from the Neolithic House at Sosandra]. In *Το νεολιθικό σπίτι της Σωσάνδρας* [The Neolithic House at Sosandra], edited by Anastasia Georgiadou, pp. 163–166. University Studio Press, Thessaloniki.

2016 Millet the Late Comer: On the Tracks of *Panicum miliaceum* in Prehistoric Greece. *Archaeological and Anthropological Sciences* 8(1):51–63.

2017 Culinary Landscapes and Identity in Prehistoric Greece: An Archaeobotanical Exploration. In *Balkan Dialogues: Negotiating Identity between Prehistory and the Present*, edited by Maja Gori and Maria Ivanova, pp. 169–194. Routledge Studies in Archaeology 25. Routledge, London.

2018 Brewing Beer in Wine Country? First Archaeobotanical Indications for Beer Making in Early and Middle Bronze Age Greece. *Vegetation History and Archaeobotany* 27(4):611–625.

Valamoti, Soultana Maria, and Stella Anastasaki

2007 A Daily Bread—Prepared but Once a Year. *Petits Propos Culinaires* 84:75–100.

Valamoti, Soultana Maria, and Felix Bittmann

2015 Proceedings of the 16th Conference of the International Work Group for Palaeoethnobotany, Thessaloniki 2013. *Vegetation History and Archaeobotany* 24(1):5–7.

Valamoti, Soultana Maria, and Michael Charles

2005 Distinguishing Food from Fodder through the Study of Charred Plant Remains: An Experimental Approach to Dung-Derived Chaff. *Vegetation History and Archaeobotany* 14(4):528–533.

Valamoti, Soultana Maria, Danai Chondrou, and Lamprini Papadopoulou

2013 Plant Food Processing and Ground Stone Equipment in Prehistoric Greece: An Experimental Investigation Using Seeds of Einkorn and Grass-Pea. In *Regards Croisés sur les Outils Liés au Travail des Végétaux/An Interdisciplinary Focus on Plant-Working Tools: Proceedings of the 33rd International Meeting of Archaeology and History, Antibes, 23–25 October 2012*, edited by Patricia C. Anderson, Carole Cheval, and Aline Durand, pp. 169–187. Association pour la promotion et la diffusion des connaissances archéologiques, Antibes.

Valamoti Soultana Maria, Pascal Darcque, Chaido Koukouli-Chrysanthaki, Dimitra Malamidou, and Zoï Tsirtsoni

2015 An Archaeobotanical Investigation of Prehistoric Grape Vine Exploitation and Wine Making in Northern Greece: Recent Finds from Dikili Tash. In *Olive Oil and Wine Production in Eastern Mediterranean during Antiquity. Proceedings of the International Symposium, 17–19 November 2011, Urla, Turkey*, edited by Adnan Diler, Ahmet Kaan Şenol, and Ümit Aydınoğlu, pp. 125–139. Ege Üniversitesi Yayınları, İzmir.

Valamoti, Soultana Maria, Vasilis Fyntikoglou, and Kostas Symponis

2022a *Food Crops in Ancient Greek Cuisine: An Archaeobotanical and Textual Study*. University Studio Press, Thessaloniki.

Valamoti, Soultana Maria, Stefanie Jacomet, Hans-Peter Stika, and Andreas G. Heiss

2017 The PlantCult Project: Identifying the Plant Food Cultures of Ancient Europe. *Antiquity* 91(358), Project Gallery:e9:1–7.

Valamoti, Soultana Maria, and Glynis Jones

2003 Plant Diversity and Storage at Mandalo, Macedonia, Greece: Archaeobotanical Evidence from the Final Neolithic and Early Bronze Age. *Annual of the British School at Athens* 98:1–35.

2010 Bronze and Oil: A Possible Link between the Introduction of Tin and *Lallemantia* to Northern Greece. *Annual of the British School at Athens* 105:83–96.

Valamoti, Soultana Maria, and Kostas Kotsakis

2007 Transitions to Agriculture in the Aegean: The Archaeobotanical Evidence. In *The Origins and Spread of Domestic Plants in Southwest Asia and Europe*, edited by Sue Colledge and James Conolly, pp. 75–91. Left Coast Press, Walnut Creek.

Valamoti, Soultana Maria, Angela Kreuz, Chryssa Petridou, Eugenia Gkatzogia, Angeliki Karathanou, Martha Kokkidou, Pavlos Lathiras, Stavroula Michou, Pelagia Paraskevopoulou, and Charoula Stylianakou

2022b Plant Ingredients Archived with ArboDat—Evaluating Regional Preferences of Crop Spectra and Culinary Changes with the New, Greek Archaeobotanical Database. In *Cooking with Plants in Ancient Europe and Beyond*, edited by Soultana Maria Valamoti, Anastasia Dimoula, and Maria Ntinou. Sidestone Press, in press.

Valamoti, Soultana Maria, and Maria Mangafa

1993 Σαράντα αιώνες κρασί: Αμπελοκαλλιέργεια και οινοποιία στην προϊστορική Μακεδονία και Θράκη [Forty Centuries of Wine: Vine Growing and Wine Making in Prehistoric Macedonia and Thrace]. *Kathimerini Newspaper, Issue "Epta Imeres"* 11 October:5. Athens, Greece.

Valamoti, Soultana Maria, Maria Mangafa, Chaido Koukouli-Chrysanthaki, and Dimitra Malamidou

2007 Grape-Pressings from Northern Greece: The Earliest Wine in the Aegean? *Antiquity* 81(311):54–61.

Valamoti, Soultana Maria, Elena Marinova, Andreas G. Heiss, Ivanka Hristova, Chryssa Petridou, Tzvetana Popova, Stavroula Michou, Lambrini Papadopoulou, Panagiotis Chrysostomou, Pascal Darcque, Dimitrios Grammenos, Stanislav Iliev, Stavros Kotsos, Chaido Koukouli-Chrysanthaki, Krassimir Leshtakov, Dimitra Malamidou, Nikos Merousis, Vassil Nikolov, Krassimir Nikov, Krastina Panayotova, Aikaterini Papanthimou, Hristo Popov, Liana Stefani, Zoï Tsirtsoni, and Tatjana Kanceva Ruseva

2019 Prehistoric Cereal Foods of Southeastern Europe: An Archaeobotanical Exploration. *Journal of Archaeological Science* 104:97–113.

Valamoti, Soultana Maria, and Ioanna Mimi

2016 Preserving *Pyrus amygdaliformis* for Later Consumption: Archaeobotanical and Ethnographic Observations. Poster presented at the 17th Conference of the International Work Group for Palaeoethnobotany, 4–9 July 2016, Paris. Program and Abstracts: 73.

Valamoti, Soultana Maria, Aikaterini Moniaki, and Angeliki Karathanou

2011 An Investigation of Processing and Consumption of Pulses among Prehistoric Societies: Archaeobotanical, Experimental, and Ethnographic Evidence from Greece. *Vegetation History and Archaeobotany* 20(5):381–396.

Valamoti, Soultana Maria, Clémence Pagnoux, Maria Ntinou, Laurent Bouby, Vincent Bonhomme, and Jean-Frédéric Terral
2020 More Than Meets the Eye: New Archaeobotanical Evidence on Bronze Age Viticulture and Wine Making in the Peloponnese, Greece. *Vegetation History and Archaeobotany* 29(1):35–50.

Valamoti, Soultana Maria, Aikaterini Papanthimou, and Angeliki Pilali
2008a Cooking Ingredients from EBA Archondiko: The Archaeobotanical Evidence. In *Proceedings of the 4th Symposium of the Hellenic Society for Archaeometry, National Hellenic Research Foundation, Athens 28–31 May 2003*, edited by Yorgos Fakorellis, Nikos Zacharias, and Kiki Polykreti, pp. 187–194. British Archaeological Reports International Series 1746. Archaeopress, Oxford.

Valamoti Soultana Maria, and Chryssa Petridou
2022 Μαγειρεύοντας με δημητριακά στις κουζίνες του Αρχοντικού Γιαννιτσών: Μια αρχαιοβοτανική διερεύνηση της φάσης IV [Cooking with Cereals at the Kitchens of Archondiko Giannitson: An Archaeobotanical Exploration of Phase IV]. *Myrinni*, in press.

Valamoti, Soultana Maria, Chryssa Petridou, Marian Berihuete-Azorín, Hans-Peter Stika, Lambrini Papadopoulou, Ioanna Mimi
2021 Deciphering Ancient "Recipes" from Charred Cereal Fragments: An Integrated Methodological Approach Using Experimental, Ethnographic, and Archaeological Evidence. *Journal of Archaeological Science* 128:105347.

Valamoti, Soultana Maria, Delwen Samuel, Mustafa Bayram, and Elena Marinova
2008b Prehistoric Cereal Foods from Greece and Bulgaria: Investigation of Starch Microstructure in Experimental and Archaeological Charred Remains. *Vegetation History and Archaeobotany* 17, Supplement 1:S265–S276.

Valamoti, Soultana Maria, and Hans-Peter Stika
2019 Alcoholic Drinks of Prehistoric Europe: Exploring the Archaeobotanical Evidence from the Aegean to Central Europe in the Context of ERC Project PlantCult. In *Was Tranken die Frühen Kelten? Bedeutungen und Funktionen mediterraner Importe im früheisenzeitlichen Mitteleuropa. Internationale Konferenz Kloster Weltenburg 28.04.–01.05.2017*, edited by Philipp W. Stockhammer and Janine Fries-Knoblach, pp. 113–134. Sidestone Press, Leiden.

Valamoti, Soultana Maria, and Charoula Stylianakou
2015 Προϊστορικός οικισμός Κλείτου Κοζάνης: Προκαταρτικά αποτελέσματα από την αρχαιοβοτανική έρευνα στον οικισμό [Prehistoric Site of Kleitos Near Kozani: Preliminary Results from the Archaeobotanical Investigation of the Settlement]. *Proceedings of the Archaeological Work in Macedonia and Thrace (AEMTH)* 25, 2011:61–71.

Valamoti, Soultana Maria, Charoula Stylianakou, and Angeliki Karathanou
2018 Προϊστορικές καλλιέργειες στη Δυτική Μακεδονία: Τα αρχαιοβοτανικά δεδομένα [Prehistoric Cultivation in Western Macedonia: The Archaeobotanical Data]. *AEAM (Proceedings of the Archaeological Work in Upper Macedonia)* 3, 2013:231–239.

Vallega, Victor
1992 Agronomical Performance and Breeding Value of Selected Strains of Diploid Wheat, *Triticum monococcum*. *Euphytica* 61(1):13–23.

Van Andel, Tjeerd H., and Judith C. Shackleton
1982 Late Paleolithic and Mesolithic Coastlines of Greece and the Aegean. *Journal of Field Archaeology* 9(4):445–454.
Van der Wiel, A. M., and T. A. Wijmstra
1987 Palynology of the Lower Part (78–120 M) of the Core Tenaghi Philippon II, Middle Pleistocene of Macedonia, Greece. *Review of Palaeobotany and Palynology* 52(2–3):73–88.
Van der Veen, Marijke
1992 Crop Husbandry Regimes: An Archaeobotanical Study of Farming in Northern England 1000 BC–AD 500. Sheffield Archaeological Monographs 3. J. R. Collis, Sheffield.
Van der Veen, Marijke, and Jacob Morales
2015 The Roman and Islamic Spice Trade: New Archaeological Evidence. *Journal of Ethnopharmacology* 167:54–63.
Velitzelos, Evangelos
1990 New Palaeobotanical Data for the Evolutionary History of Plants in the Aegean Area, with Special Reference to the Palaeoflora of Thera. In *Thera and the Aegean World III: Proceedings of the Third International Congress, Santorini, Greece, 3–9 September 1989, Vol. 1: Archaeology*, edited by David A. Hardy, Christos G. Doumas, John A. Sakellarakis, and Peter M. Warren, pp. 406–409. Thera Foundation, London.
Vencl, S.
1996 Acorns as Food: Again. *Památky archeologické* 87:95–111.
Ventris, Michael, and John Chadwick
1973 *Documents in Mycenaean Greek*. 2nd ed. Cambridge University Press, Cambridge.
Vetters, Melissa, Ann Brysbaert, Maria Ntinou, Georgia Tsartsidou, and Evi Margaritis
2016 People and Plants: Piecing Together Archaeological and Archaeobotanical Data to Reconstruct Plant Use and Craft Activities in Mycenaean Tiryns. *Opuscula* 9:93–132.
Vickery, Kenton Frank
1936 *Food in Early Greece*. Illinois Studies in the Social Sciences 20. University of Illinois Press, Urbana.
Vidal-Naquet, Pierre
1983 *Ο Μαύρος Κυνηγός* [Le Chasseur Noir]. Translated by Giangos Andreades and Pepi Pigopoulou. Livanis, Athens.
Vitelli, Karen D.
1989 Were Pots First Made for Foods? Doubts from Franchthi. *World Archaeology* 21(1):17–29.
1993 *Franchthi Neolithic Pottery, Vol. 1: Classification and Ceramic Phases 1 and 2*. Excavations at Franchthi Cave, Greece, Vol. 8. Indiana University Press, Bloomington.
1995 Pots, Potters, and the Shaping of Greek Neolithic Society. In *The Emergence of Pottery: Technology and Innovation in Ancient Societies*, edited by William K. Barnett and John W. Hoopes, pp. 55–63. Smithsonian Institution Press, Washington, DC.
2000 Observations on Palaeodietary Research in the Aegean. In *Palaeodiet in the Aegean. Papers from a Colloquium Held at the 1993 Meeting of the Archaeological*

Institute of America in Washington, DC, edited by Sarah J. Vaughan and William D. E. Coulson, pp. 10–12. Wiener Laboratory Monograph 1. Oxbow, Oxford.

Voutiras, Emmanouil

2016 Poppies for Demeter. In *ἠχάδιν: Τιμητικός τόμος για τη Στέλλα Δρούγου* [Volume in Honour of Stella Drougou], edited by Mimika Giannopoulou and Chrysanthi Kallini, pp. 278–284. Fund of Archaeological Proceeds (TAPA), Athens.

Voutsina, Evi

2000 *To Ψωμί* [Bread]. Kastaniotis, Athens.

2008 *The Cookery of Lefkada*. Fagotto, Athens.

Wace, Alan J. B.

1921 Excavations at Mycenae VII, 2: The Granary. *Annual of the British School at Athens* 25:38–61.

Währen, M.

2002 Pain, pâtisserie et religion en Europe pré- et protohistorique. In *Pains, Fours et Foyers Domestiques des Temps Passés—Breads, Ovens, and Hearths of the Past*, edited by Marianne Mesnil and Kaï Fechner, pp. 381–400. *Civilisation* 49(1–2). Institut de Sociologie de l' Université de Bruxelles.

Wallace, Michael, Glynis Jones, Michael Charles, Emily Forster, Eleanor Stillman, Vincent Bonhomme, Alexandra Livarda, Colin P. Osborne, Mark Rees, Georg Frenck, and Catherine Preece

2018 Re-Analysis of Archaeobotanical Remains from Pre- and Early Agricultural Sites Provides no Evidence for a Narrowing of the Wild Plant Food Spectrum during the Origins of Agriculture in Southwest Asia. *Vegetation History and Archaeobotany* 28(4):449–463.

Walmsley, Emily

2005 Race, Place, and Taste: Making Identities through Sensory Experience in Ecuador. *Etnofoor* 18(1):43–60.

Wandsnider, LuAnn

1997 The Roasted and the Boiled: Food Composition and Heat Treatment with Special Emphasis on Pit-Hearth Cooking. *Journal of Anthropological Archaeology* 16(1):1–48.

Wang, Jiajing, Li Liu, Terry Ball, Linjie Yu, Yuanqing Li, and Fulai Xing

2016 Revealing a 5,000-Year-Old Beer Recipe in China. *Proceedings of the National Academy of Sciences of the United States of America (PNAS)* 113(23):6444–6448.

Warren, Meredith J. C.

2017 Tastes from Beyond: Persephone's Pomegranate and Otherworldly Consumption in Antiquity. In *Taste and the Ancient Senses*, edited by Kelli C. Rudolph, pp. 104–119. The Senses in Antiquity. Routledge, London.

Watrous, L. Vance, D. Matthew Buell, John C. McEnroe, John G. Younger, Lee Ann Turner, Brian S. Kunkel, Kevin Glowacki, Scott Gallimore, Angus Smith, Panagiota A. Pantou, Anne Chapin, and Evi Margaritis

2015 Excavations at Gournia, 2010–2012. *Hesperia* 84(3):397–465.

Watson, Patty Jo

2003 Investigating Agricultural Transitions: A Comparative Study. In *The Widening Harvest: The Neolithic Transition in Europe—Looking Back, Looking Forward*, edited by Albert J. Ammerman and Paolo Biagi, pp. 27–41. AIA Colloquia and Conference Papers 6. Archaeological Institute of America, Boston.

Weckerle, Caroline S., Franz K. Huber, Yang Yongping, and Sun Weibang
2005 The Role of Barley among the Shuhi in the Tibetan Cultural Area of the Eastern Himalayas. *Economic Botany* 59(4):386–390.
Weiberg, Erika, Ingmar Unkel, Katerina Kouli, Karin Holmgren, Pavlos Avramidis, Anton Bonnier, Flint Dibble, Martin Finné, Adam Izdebski, Christos Katrantsiotis, Sharon R. Stocker, Maria Andwinge, Kalliopi Baika, Meighan Boyd, and Christian Heymann
2016 The Socio-Environmental History of the Peloponnese during the Holocene: Towards an Integrated Understanding of the Past. *Quaternary Science Reviews* 136:40–65.
Weiner, Michael A.
1980 *Earth Medicine, Earth Food: Plant Remedies, Drugs, and Natural Foods of the North American Indians.* Macmillan, New York.
Wells, Throth
1990 *The New Internationalist Food Book: Recipes from Africa, Asia, and Latin America for Western Kitchens.* New Internationalist, Oxford.
Wiessner, Polly W.
2001 Of Feasting and Value: Enga Feasts in a Historical Perspective (Papua New Guinea). In *Feasts: Archaeological and Ethnographic Perspectives on Food, Politics, and Power,* edited by Michael Dietler and Brian Hayden, pp. 115–143. Smithsonian Institution Press, Washington, DC.
Wijmstra, T. A.
1969 Palynology of the First 30 Metres of a 120 M Deep Section in Northern Greece. *Acta Botanica Neerlandica* 18(4):511–527.
Willcox, George
1999 Archaeobotanical Evidence for the Beginnings of Agriculture in South-West Asia. In *The Origins of Agriculture and Crop Domestication Proceedings of the Harlan Symposium, 10–14 May 1997, Aleppo, Syria,* edited by Ardeshir B. Damania, Jan Valkoun, George Willcox, and Calvin O. Qualset, pp. 25–38. ICARDA, Aleppo.
2000 Nouvelles données sur l'origine de la domestication des plantes au Proche Orient. In *Premiers paysans du monde: Naissances des agricultures,* edited by Jean Guilaine, pp. 121–140. Errance, Paris.
2005 The Distribution, Natural Habitats, and Availability of Wild Cereals in Relation to Their Domestication in the Near East: Multiple Events, Multiple Centres. *Vegetation History and Archaeobotany* 14(4):534–541.
2013 The Roots of Cultivation in Southwestern Asia. *Science* 341(6141):39–40
Willcox, George, Ramon Buxó, and Linda Herveux
2009 Late Pleistocene and Early Holocene Climate and the Beginnings of Cultivation in Northern Syria. *The Holocene* 19(1):151–158.
Willcox, George, Sandra Fornite, and Linda Herveux
2008 Early Holocene Cultivation before Domestication in Northern Syria. *Vegetation History and Archaeobotany* 17(3):313–325.
Willcox, George, Mark Nesbitt, and Felix Bittmann
2012 From Collecting to Cultivation: Transitions to a Production Economy in the Near East. *Vegetation History and Archaeobotany* 21(2):81–83.
Willerding, Ulrich
1973 Bronzezeitliche Pflanzenreste aus Iria und Synoro. In *Tiryns: Forschungen und Berichte VI,* edited by Ulf Jantzen, pp. 221–240. Philip von Zabern, Mainz.

Willis, Katherine J.
1992 The Late Quaternary Vegetational History of Northwest Greece. III. A Comparative Study of Two Contrasting Sites. *The New Phytologist* 121(1):139–155.
1994 The Vegetational History of the Balkans. *Quaternary Science Reviews* 13(8):769–788.
Wilson, D. G.
1984 The Carbonisation of Weed Seeds and Their Representation in Macrofossil Assemblages. In *Plants and Ancient Man: Studies in Palaeoethnobotany: Proceedings of the Sixth Symposium of the International Work Group for Palaeoethnobotany, Groningen, 30 May–3 June 1983*, edited by Willem Van Zeist and Will A. Casparie, pp. 201–206. Balkema, Groningen.
Wilson, Kax
1979 *A History of Textiles*. Westview Press, Boulder, Colorado.
Wiltshire, Patricia E. J.
1995 The Effect of Food Processing on the Palatability of Wild Fruits with High Tannin Content. In *Res Archaeobotanicae: Proceedings of the 9th Symposium of the International Work Group for Palaeoethnobotany, Kiel 1992*, edited by Helmut Kroll and Rainer Pasternak, pp. 385–397. Oetker-Voges, Kiel.
Winkelman, Michael
1997 Altered States of Consciousness and Religious Behavior. In *Anthropology of Religion: A Handbook*, edited by Stephen D. Glazier, pp. 393–428. Praeger, Westport, Connecticut.
Wollstonecroft, Michèle M.
2011 Investigating the Role of Food Processing in Human Evolution: A Niche Construction Approach. *Archaeological and Anthropological Sciences* 3(1):141–150.
Woods, Demetrius L., J. J. Capcara, and Keith R. Downey
1991 The Potential of Mustard (*Brassica juncea* [L.] Coss) as an Edible Oil Crop on the Canadian Prairies. *Canadian Journal of Plant Science* 71(1):195–198.
Woudhuizen, Fred C.
2009 Minoan and Mycenaean Overseas Contacts: The Epigraphic Evidence. *Dacia, Revue d'Archéologie et d'Histoire Ancienne* 53:5–11.
Wrangham, Richard
2009 *Catching Fire: How Cooking Made Us Human*. Basic Books, New York.
2017 Control of Fire in the Paleolithic: Evaluating the Cooking Hypothesis. *Current Anthropology* 58, Supplement 16:S303–S313.
Wright, James C.
2004a A Survey of Evidence for Feasting in Mycenaean Society. *Hesperia* 73(2):133–178.
Wright, James C. (editor)
2004b *The Mycenaean Feast*. The American School of Classical Studies at Athens, Princeton.
Wright, James C., Evangelia Pappi, Sevasti Triantaphyllou, Mary K. Dabney, Panagiotis Karkanas, Georgia Kotzamani, and Alexandra Livarda
2008 Nemea Valley Archaeological Project, Excavations at Barnavos: Final Report. *Hesperia* 77(4):607–654.
Wright, Katherine I.
1991 The Origins and Development of Ground Stone Assemblages in Late Pleistocene Southwest Asia. *Paléorient* 17(1):19–45.

Wright, Herbert Edgar, Jr.
1972 Vegetation History. In *The Minnesota Messenia Expedition: Reconstructing a Bronze Age Regional Environment*, edited by William A. McDonald and George Rapp Jr., pp. 188–199. University of Minnesota Press, Minneapolis.

Yasur-Landau, Assaf
2012 The Role of the Canaanite Population in the Aegean Migration to the Southern Levant in the Late 2nd Millennium BC. In *Materiality and Social Practice: Transformative Capacities of Intercultural Encounters*, edited by Joseph Maran and Philipp Stockhammer, pp. 191–197. Oxbow, Oxford.

Yiouni, Paraskevi
2004 Counting Pots in Early Neolithic Greece. *Annual of the British School at Athens* 99:1–22.

Zanicovic, Vladimir
2001 Similarities between Tropical Spastic Paraparesis (TSP) and Neurolathyrism. *Lathyrus Lathyrism Newsletter* 2:11–14.

Zapata-Peña, Lydia
2007 First Farmers along the Coast of the Bay of Biscay. In *The Origins and Spread of Domestic Plants in Southwest Asia and Europe*, edited by Sue Colledge and James Conolly, pp. 189–208. Left Coast Press, Walnut Creek.

Zarnkow, Martin, Adelheid Otto, and Berthold Einwag
2011 Interdisciplinary Investigations into the Brewing Technology of the Ancient Near East and the Potential of the Cold Mashing Process. In *Liquid Bread: Beer and Brewing in Cross-Cultural Perspective*, edited by Wulf Schiefenhövel and Helen Macbeth, pp. 47–54. Berghahn, New York.

Zeder, Melinda A.
2011 The Origins of Agriculture in the Near East. *Current Anthropology* 52, Supplement 4:S221–S235.
2015 Core Questions in Domestication Research. *Proceedings of the National Academy of Sciences of the United States of America (PNAS)* 112(11):3191–3198.

Zeist, Willem van
1988 Some Aspects of Early Neolithic Plant Husbandry in the Near East. *Anatolica* 15:49–67.

Zeist, Willem van, and Sytze Bottema
1971 Plant Husbandry in Early Neolithic Nea Nikomedeia, Greece. *Acta Botanica Neerlandica* 20(5):524–538.

Ziota, Christina
2014 The Settlement of Kleitos Kozanis in Its Wider Natural and Anthropogenic Environment during the Late and Final Neolithic Periods. In *1912–2012: Εκατό χρόνια έρευνας στην προϊστορική Μακεδονία, Πρακτικά διεθνούς συνεδρίου, Αρχαιολογικό Μουσείο Θεσσαλονίκης, 22–24 Νοεμβρίου 2012* [1912–2012: A Century of Research in Prehistoric Macedonia, International Conference Proceedings, Archaeological Museum of Thessaloniki, 22–24 November 2012], edited by Evaggelia Stefani, Nikos Merousis, and Anastasia Dimoula, pp. 323–336. Archaeological Museum of Thessaloniki.

Zohary, Daniel
1995 The Domestication of the Grapevine *Vitis vinifera* L. in the Near East. In *The Origins and Ancient History of Wine*, edited by Patrick E. McGovern, Stuart J.

Fleming, and Solomon H. Katz, pp. 23–30. Food and Nutrition in History and Anthropology 11. Gordon and Breach, Philadelphia.

1996 The Mode of Domestication of the Founder Crops of Southwest Asian Agriculture. In *The Origins and Spread of Agriculture and Pastoralism in Eurasia*, edited by David R. Harris, pp. 142–158. University College London Press, London.

1989 Domestication of the Southwest Asian Neolithic Crop Assemblage of Cereals, Pulses, and Flax: The Evidence from Living Plants. In *Foraging and Farming: The Evolution of Plant Exploitation*, edited by David R. Harris and Gordon C. Hillman, pp. 358–373. Unwin Hyman, London.

Zohary, Daniel, and Maria Hopf

2000 *Domestication of Plants in the Old World: The Origin and Spread of Cultivated Plants in West Asia, Europe, and the Nile Valley*. 3rd ed. Oxford University Press, Oxford.

Zohary, Daniel, Maria Hopf, and Ehud Weiss

2012 *Domestication of Plants in the Old World: The Origin and Spread of Domesticated Plants in Southwest Asia, Europe, and the Mediterranean Basin*. 4th ed. Oxford University Press, Oxford.

Zohary, Daniel, and Pinhas Spiegel-Roy

1975 Beginnings of Fruit Growing in the Old World. *Science* 187(4174):319–27.

Zubaida, Sami

1994 National, Communal, and Global Dimensions in Middle Eastern Food Cultures. In *A Taste of Thyme: Culinary Cultures of the Middle East*, edited by Sami Zubaida and Richard Tapper, pp. 33–45. I. B. Tauris, London.

Zubaida, Sami, and Richard Tapper (editors)

1994 *A Taste of Thyme: Culinary Cultures of the Middle East*. I. B. Tauris, London.

Index